Batteries

Materials principles and characterization methods

IOP Series in Renewable and Sustainable Power

The IOP Series in Renewable and Sustainable Power aims to bring together topics relating to renewable energy, from generation to transmission, storage, integration, and use patterns, with a particular focus on systems-level and interdisciplinary discussions. It is intended to provide a state-of-the-art resource for all researchers involved in the power conversation.

Series Editor
Professor David Elliott
Open University, UK

About the Editor
David Elliott is emeritus Professor of Technology Policy at the Open University, where he developed courses and research on technological innovation, focusing on renewable energy policy. Since retirement, he has continued to write extensively on that topic, including a series of books for IOP Publishing and a weekly blog post for *Physics World* (physicsworld.com/author/david-elliott)

About the Series
Renewable and sustainable energy systems offer the potential for long-term solutions to the world's growing energy needs, operating at a broad array of scales and technology levels. The IOP Series in Renewable and Sustainable Power aims to bring together topics relating to renewable energy, from generation to transmission, storage, integration, and use patterns, with a particular focus on systems-level and interdisciplinary discussions. It is intended to provide a state-of-the-art resource for all researchers involved in the power conversation.

We welcome proposals in all areas of renewable energy including (but not limited to): wind power, wave power, tidal power, hydroelectric power, PV/solar power, geothermal power, bioenergy, heating, grid balancing and integration, energy storage, energy efficiency, carbon capture, fuel cells, power to gas, electric/green transport, and energy saving and efficiency.

Authors are encouraged to take advantage of electronic publication through the use of colour, animations, video, data files, and interactive elements, all of which provide opportunities to enhance the reader experience.

Do you have an idea for a book you'd like to explore?
We are currently commissioning for the series; if you are interested in writing or editing a book please contact Caroline Mitchell at caroline.mitchell@ioppublishing.org.

Batteries

Materials principles and characterization methods

Edited by
Chen Liao
Chemical Sciences and Engineering Division, Argonne National Laboratory, Lemont, USA

IOP Publishing, Bristol, UK

ISBN 978-0-7503-2682-7 (ebook)
ISBN 978-0-7503-2680-3 (print)
ISBN 978-0-7503-2683-4 (myPrint)
ISBN 978-0-7503-2681-0 (mobi)

DOI 10.1088/978-0-7503-2682-7

Version: 20211201

IOP ebooks

British Library Cataloguing-in-Publication Data: A catalogue record for this book is available from the British Library.

Published by IOP Publishing, wholly owned by The Institute of Physics, London

IOP Publishing, Temple Circus, Temple Way, Bristol, BS1 6HG, UK

US Office: IOP Publishing, Inc., 190 North Independence Mall West, Suite 601, Philadelphia, PA 19106, USA

Contents

Preface

This book is intended for the readers with an interest in learning about some of the most recent experimental analysis techniques.

Here is the flow of the book: the first chapter is focused on the basic understanding and techniques, while rest of the chapters each focus on one advanced characterization technique. We also limit the book to cover only experimental aspects and not modeling, such as the database and large throughput screening. If readers are interested in those topics, they can explore other available tools such as the Materials Project (https://materialsproject.org/).

To differentiate this book from other characterization books, the emphasis will not be on the materials but on the principles, to enable the readers to explore references and literature on their own with the necessary education. The other distinctive feature of our book is the *in situ* feature of the many techniques listed here. Depending on the cycling conditions, batteries can exhibit very complexed behavior and having *in situ* tools allow us to decipher the fading mechanisms and design better materials and engineering.

A brief glimpse of the book chapters is provided here.

1 Introduction

A very brief introduction to batteries is provided, with a focus on lithium-ion batteries and the common materials.

2 Electrochemical characterization and modeling for batteries

Dr Jae Jin Kim and co-authors provide a concise account of both electrochemical modeling approaches (empirical and physics-based models) and experimental characterization (DC-and AC-based techniques), widely employed to characterize materials' fundamental properties used in batteries and their change/interaction with adjacent components during battery operation. The performance of batteries is determined by redox reaction thermodynamics and transport kinetics, which cover a wide range of time and length scales. In this context, synergistic integration of modeling and characterization can provide more holistic understanding of complex electrochemical reaction mechanisms in batteries, by validating each other, bridging their fundamental limits to time and space, and uncovering unknown region with more valid prediction.

3 Synchrotron-based battery imaging with multi-modal x-ray signals

Dr Yijing Liu and co-authors provide a detailed picture of how the synchrotron can be used for batteries characterizations, covering areas such as multi-modal synchrotron-based probes to conduct multi-scale and high-dimensional imaging of batteries with structural, compositional, chemical, and morphological sensitivities. Synchrotron-based battery research is focused on the structural and chemical complexities in batteries. Spectroscopic and scattering signals are probed using a synchrotron beam in a spatially resolved manner for imaging batteries with different contrast mechanisms. Specific case studies are also shown where both cutting-edge synchrotron techniques and advanced computational developments benefit the characterization technique.

4 FTIR

Dr Sang-Don Han and Dr Bertrand Tremolet de Villers describe infrared (IR) and Raman spectroscopy studies. Not only are the fundamental principles and measurement methods or FTIR and Raman presented, but also *in situ* spectroscopic analysis during battery cycling. The advantage of the *in situ* method is the additional insight into the chemical, thermal, and electrical mechanisms underlying the battery performance that are largely inaccessible with traditional *ex situ* measurements. The future perspectives on the continuing evolution of IR spectroscopy in electrochemical energy storage research are also presented.

5 OEMS

We provide an updated description of the state-of-art online electrochemical mass spectroscopy (OEMS), with the focus on systems that can be applied to batteries.

6 Electron microscopy

Dr Layla Mehdi and co-authors provide an overview of electron microscopies (EMs), including scanning electron microscopy (SEM) and (scanning) transmission electron microscopy ((S)TEM). The electron microscopy is important to characterize the micro- to atomic-scale structural and compositional properties for battery materials, some specific *in situ* TEM examples are also given.

7 NMR

Dr Greenbaum and co-authors begin with a review of background and implementation of standard NMR methods including 1-D and 2-D spectroscopy, relaxometry, and diffusometry, and some specialized techniques including imaging and *operando* methods. Most of the investigations reviewed here, some from our own laboratory, concern structural and dynamical aspects of Li ion battery materials, in particular carbonaceous anodes, transition metal oxide cathodes, and both liquid and solid electrolytes.

8 X-ray photoelectron spectroscopy for batteries

Dr Justin Connell outlines the important operating principles, applications, and limitations of XPS applied to battery materials. Rather than a comprehensive overview, this chapter is intended as an operator's guide for how to practically (and skeptically) approach the XPS technique when investigating battery systems and ensure that the analysis of XPS data is presented fairly and with appropriate qualifications and/or complementary measurements that support the proposed interpretation. There are multiple examples of the types of battery-relevant questions XPS is well-suited to answering, and discussion is also provided on emerging *in situ* and *operando* lab- and synchrotron-based techniques that are pushing the frontier of what can be done with XPS. Finally, some practical considerations for measuring and interpreting x-ray photoelectron spectra are provided as a roadmap to aid with the appropriate interpretation, and perhaps more importantly, qualification of the conclusions drawn from XPS measurements.

9 Scanning electrochemical microscopy: a versatile tool for inspecting the reactivity of battery electrodes

Dr Rodgiguez Joaquin *et al* describe an emerging technique of scanning electrochemical microscopy (SECM) for understanding the complex processes and performance of materials within batteries. SECM utilizes a small electrode probe to map and quantify localized electrochemical phenomena over a sample surface and provide *in situ* information on each major component in a typical battery—anode, cathode, electrolyte, and the separator. The power of SECM lies in its ability to study ion and electron transfer—both of which are the fundamental processes driving all batteries. Further, SECM provides insight on the complex interfacial electrochemistry that occurs within battery systems at length scales ranging from μm to nm. For instance, this dynamic range is particularly useful in exploring the formation and progression of the solid–electrolyte interphase (SEI), which is critical to the safe operation of Li-ion batteries.

10 Small-angle x-ray scattering for battery research

Dr Tao Li *et al* describe the technique of SAXS for understanding of nanostructure information, especially the presence of phase-separated polymer membranes, changes in nanomaterials' structure, and sub-nano structures in liquid systems. SAXS uses the scattering signal generated in a small angle range close to the beam by x-ray after passing through the sample to obtain nanostructure information with different electron cloud densities. Therefore, SAXS has low requirements for test samples but can obtain abundant material structure information. They review the principles of the SAXS test and the corresponding testing sample requirements. The representative applications of SAXS in separator, liquid electrolyte, and redox flow battery and *in situ* or *operando* applications are then introduced and prospected.

Editor biography

Chen Liao

Dr Chen Liao (PI) is a Chemist at the Electrochemical Storage System Department in Chemical Sciences and Engineering Division at ANL. Dr Liao serves as the subthrust leader of Joint Center of Energy Storage Research (JCESR) and a co-PI (principle investigator) for Department Of Energy projects for advanced material research for lithium-ion and multivalent ion batteries. She also led a LDRD (laboratory directed research and development) project from 2018–2020 on the design and syntheses of single-ion conducting solid state electrolytes (SIEs) for high voltage materials and investigated their structure–electrochemical property relationship.

She has over 15 years of research experience in both materials' development and electrochemical energy storage materials and has published over 70 peer reviewed papers with 10+ issued patents.

List of contributors

Nigel D Browning
Department of Mechanical, Materials and Aerospace Engineering, University of Liverpool, Liverpool, L69 3GH, UK

David Clarkson
Department of Physics and Astronomy, Hunter College of the City University of New York, New York, NY 10065 USA

Justin G Connell
Joint Center for Energy Storage Research, Argonne National Laboratory, 9700 South Cass Avenue, Lemont, Illinois 60439, USA

and

Materials Science Division, Argonne National Laboratory, 9700 South Cass Avenue, Lemont, Illinois 60439, USA

Mounesha N Garaga
Department of Physics and Astronomy, Hunter College of the City University of New York, New York, NY 10065 USA

Jihyeon Gim
Chemical Sciences and Engineering Division, Argonne National Laboratory, 9700 South Cass Avenue, Lemont, Illinois 60439, USA

Zachary T Gossage
Department of Chemistry, University of Illinois at Urbana Champaign, USA

Steve Greenbaum
Department of Physics and Astronomy, Hunter College of the City University of New York, New York, NY 10065 USA

Sang-Don Han
Materials, Chemical, and Computational Science Directorate, National Renewable Energy Laboratory, Golden, Colorado 80401, USA

Kendrich O Hatfield
Department of Chemistry, University of Illinois at Urbana Champaign, USA

Hong Keun Kim
Department of Mechanical Engineering, Inha University, 100 Inha-ro, Michuhol-Gu, Incheon 22212, South Korea

Jae Jin Kim
Chemical Sciences and Engineering Division, Argonne National Laboratory, 9700 South Cass Avenue, Lemont, Illinois 60439, USA

Eungje Lee
Chemical Sciences and Engineering Division, Argonne National Laboratory, 9700 South Cass Avenue, Lemont, Illinois 60439, USA

Tao Li
Department of Chemistry and Biochemistry, Northern Illinois University, DeKalb, Illinois 60115, USA
and
X-ray Science Division, Argonne National Laboratory, Lemont, Illinois 60439, USA

Weiqun Li
Department of Mechanical, Materials and Aerospace Engineering, University of Liverpool, Liverpool, L69 3GH, UK

Chen Liao
Chemical Sciences and Engineering Division, Argonne National Laboratory, 9700 South Cass Avenue, Lemont, Illinois 60439, USA
and
Joint Center of Energy Storage Research, Argonne National Laboratory, 9700 South Cass Avenue, Lemont, Illinois 60439, USA

Kewei Liu
Chemical Sciences and Engineering Division, Argonne National Laboratory, 9700 South Cass Avenue, Lemont, Illinois 60439, USA

Yijin Liu
Stanford Synchrotron Radiation Lightsource, SLAC National Accelerator Laboratory, Menlo Park, CA 94025, USA

B Layla Mehdi
Department of Mechanical, Materials and Aerospace Engineering, University of Liverpool, Liverpool, L69 3GH, UK

Abhiroop Mishra
Department of Materials Science and Engineering, University of Illinois at Urbana Champaign, USA

Piero Pianetta
Stanford Synchrotron Radiation Lightsource, SLAC National Accelerator Laboratory, Menlo Park, CA 94025, USA

Kun Qian
Department of Chemistry and Biochemistry, Northern Illinois University, DeKalb, Illinois 60115, USA

Joaquín Rodríguez-López
Department of Chemistry, University of Illinois at Urbana Champaign, USA

Dipobrato Sarbapalli
Department of Materials Science and Engineering, University of Illinois at Urbana Champaign, USA

Boyu Shi
Chemical Sciences and Engineering Division, Argonne National Laboratory, 9700 South Cass Avenue, Lemont, Illinois 60439, USA

Seoung-Bum Son
Chemical Sciences and Engineering Division, Argonne National Laboratory, 9700 South Cass Avenue, Lemont, Illinois 60439, USA

Juhyun Song
Battery Algorithm Team at Apple, Inc., USA

Bertrand J Tremolet de Villers
Materials, Chemical, and Computational Science Directorate, National Renewable Energy Laboratory, Golden, Colorado 80401, USA

Guibin Zan
Stanford Synchrotron Radiation Lightsource, SLAC National Accelerator Laboratory, Menlo Park, CA 94025, USA

Yun Zhao
Department of Chemistry and Biochemistry, Northern Illinois University, DeKalb, Illinois 60115, USA

IOP Publishing

Batteries
Materials principles and characterization methods
Chen Liao

Chapter 1

Fundamentals of rechargeable lithium ion and beyond lithium ion batteries

Chen Liao

In the modern era, transportation is becoming electrified, and portable electronics are widely utilized at work and in daily life. Accompanying this electrification is a quest for batteries with high energy density and long lifetime, and lithium ion batteries are favored because of their combination of high voltage and high specific capacity. However, the inherent chemical complexity and structural hierarchy in battery materials necessitate a deep understanding of both the bulk materials composing electrolytes and electrodes and the interfaces between them. A brief introduction to the fundamentals of batteries is presented here.

Glossary

All-solid-state battery (ASSB): Battery consisting of cathode, anode, and solid electrolyte.

Area Specific Impedance (ASI): Impedance measured for charge and discharge pulses. See equation (1.9) for definition.

Coulombic efficiency: The ratio of discharge capacity to the charge capacity of the preceding cycle.

Crystal field stabilization energy theory (CFSE): Depicts the stability that results from placing a transition metal ion in the crystal field generated by a set of ligands.

Density of states: In solid state physics and condensed matter physics, the proportion of states that are to be occupied by the system at each energy.

***dQ/dV*:** Differential capacity versus differential voltage; depicts the plateau information associated with phase equilibria information regarding the intercalation and deintercalation of Li^+ (or other active ions).

***dV/dQ*:** Differential voltage versus differential capacity; depicts the phase transition (the voltage drop/increase in the voltage profile).

EV: Electric vehicle.

Faraday's Law: A statement of the relationship between the current that runs through an external circuit and the quantities of active materials that are consumed/produced in the ensuing electrochemical reaction.

Gr: Graphite anode.

Gr-Si: A composite anode consisting of graphite (Gr) and silicon (Si).

Highest Occupied Molecular Orbital (HOMO): The highest occupied molecular orbital in frontier molecular orbital theory, usually manifested as the oxidation potential.

Hybrid Pulse Power Characterization (HPPC): A test profile specifically designed to elicit the dynamic power capability of a Li-ion battery for electric vehicles (EVs) at a given state of charge and aging status.

Li^+ inventory: Total amount of cyclable Li^+, usually referring to a full cell setup using a lithiated transition metal oxide as the cathode and graphite as the anode, with all cyclable Li^+ inventory coming from the cathode.

Lithium cobalt oxide ($LiCoO_2$ or LCO): A layered cathode, consisting of lithium, cobalt and oxygen elements, that was widely used in volumetrically dense lithium ion batteries.

LMR: Lithium manganese rich

Lowest Unoccupied Molecular Orbital (LUMO): The lowest unoccupied molecular orbital in frontier molecular orbital theory, usually manifested as the reduction potential.

NMC: Lithium nickel–manganese cobalt oxide cathode materials.

OTMS: Trimethylsilyl oxyl group, with a molecular formula of $-O-Si(CH_3)_3$.

$P(OTMS)_2F$: Bis(trimethylsilyl) fluorophosphate, with a molecular formula of $P(-O-Si(CH_3)_3)_2F$.

$P(OTMS)_3$: Tris(trimethylsilyl) phosphite, with a molecular formula of $P(-O-Si(CH_3)_3)_3$.

$P(OTMS)F_2$: Trimethylsilyl bisfluorophosphite, with a molecular formula of $P(-O-Si(CH_3)_3)F_2$.

RE: Reference electrode.

Solid–electrolyte interphase (SEI): A layer, consisting of both organic and inorganic materials, that exists as the medium between two surfaces; historically, the SEI in lithium ion batteries consists of a mosaic-like array of 3-D materials, with components including Li_2CO_3, LiF, RCOOLi, R′OLi (the R′ is used to represent a different alkyl group than the previous R group) and many others [1].

State of charge (SOC): Level of charge of an electric battery relative to its capacity.

Transition metal oxides (TMOs): Prevalent cathode materials that consist of transition metals and their oxides.

List of symbols

c_i	concentration of species I, mol m^{-3}
D_i	molecular diffusivity of species I, $m^2 \cdot s^{-1}$
E_F	Fermi level
F	Faraday's constant, 96 485 C per equivalent
I	current density, A m^{-2}
N_{AV}	molar flux of species I relative to stationary coordinates, mol m^{-2} s^{-1}
N_{AVO}	Avogadro's number
N_i	molar flux of species I relative to stationary coordinates, mol m^{-2} s^{-1}
q	fundamental unit of charge
R	universal gas constant, 8.314 J $mol^{-1}K^{-1}$
T	temperature, K or °C
u_i	mobility of ion, m^2 mol $J^{-1}s^{-1}$
V_{oc}	open-circuit voltage
z_i	charge number
α	Debye constant for solvent, $(kg\ mol^{-1})^{1/2}$
β	symmetry factor, unitless
$\gamma_{\pm}$	activity coefficient, unitless
η_s	surface overpotential, V
μ_c	potential of cathode, V
μ_a	potential of anode, V
v	velocity, m s^{-1}
ϕ	potential, V

Since the emergence of Sony's commercial lithium ion battery in 1991, society has embraced new technologies ranging from battery-operated power tools and portable consumer electronics such as smart phones to electric vehicles (EVs). It was no surprise that the 2019 Nobel Prize in Chemistry was jointly awarded to John B Goodenough, M Stanley Whittingham, and Akira Yoshino for 'the development of lithium ion batteries (LIBs).' [2]

There are still hurdles to overcome before EVs capable of a long driving range (200 miles and above) on a single charge, and other more far-reaching possibilities, such as Unmanned Aerial Vehicles, become mass-market technologies. A few of the dominant hurdles are achieving higher energy density (of cathodes, anodes and electrolytes) for: longer mileage and usage time; a long lifetime; low cost; pathways to reuse, recycling, and reduction of carbon footprint; and the ability to withstand extreme temperature changes.

The following text is divided into three sections. The first section is an introduction to batteries for researchers from all scientific disciplines; the second section describes current study areas in more detail, centering on battery components; and the last is a brief section of conclusion.

1.1 Introduction to batteries

Rechargeable lithium ion batteries and beyond lithium ion batteries are attracting more attention recently. Conventional lithium ion batteries (LIBs) are rechargeable electrochemical energy storage devices based on the intercalation mechanism

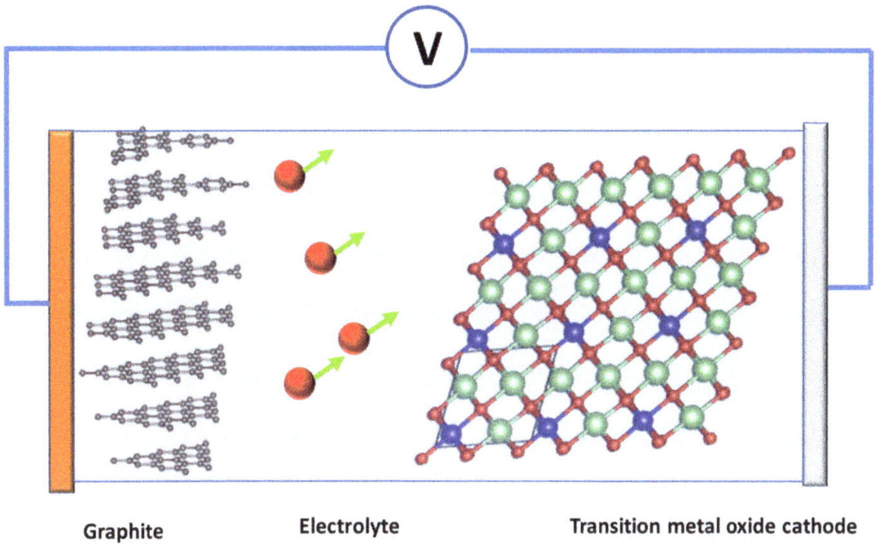

Figure 1.1. Schematic representation of basic components of a classic intercalation lithium ion battery with a transition metal oxide cathode, a graphite anode, and a nonaqueous electrolyte.

(figure 1.1): upon discharging, through the movement and intercalationof Li^+ ions into the cathode, a concomitant electron moves into the cathode and generates current to satisfy the charge balance. Since the battery is inside a closed circuit, the electron will go through the circuit to the anode side, and one electron will come out together with deintercalation of a Li^+. The process is reversed for the charging process. Overall, batteries are relatively low in power generation (pulse) and high in energy storage. For processes such as braking or accelerating in EVs, assistance in power generation is required, as a large current is passed through the system during these processes.

Other than materials that use the intercalation mechanism, cathodes for batteries can also be of conversion materials, such as in the cases of the low-cost/high-energy-density Li–S and Li–air batteries. Conversion reactions can be expressed in the following equation for a conversion type material of M:

$$M + xLi^+ + xe^{-1} \rightarrow Li_xM \quad (1.1)$$

1.1.1 Lithium ion batteries

1.1.1.1 Fundamentals

As shown in figure 1.1, a conventional intercalation battery setup includes three major components: cathode, anode (a graphite anode, or Gr, is used as an example), and electrolyte. An example of an intercalation cathode material is lithium cobalt oxide ($LiCoO_2$), or LCO, which is widely adopted for portable electronics. Because cobalt resources are limited and geopolitical tensions arise from cobalt mining,

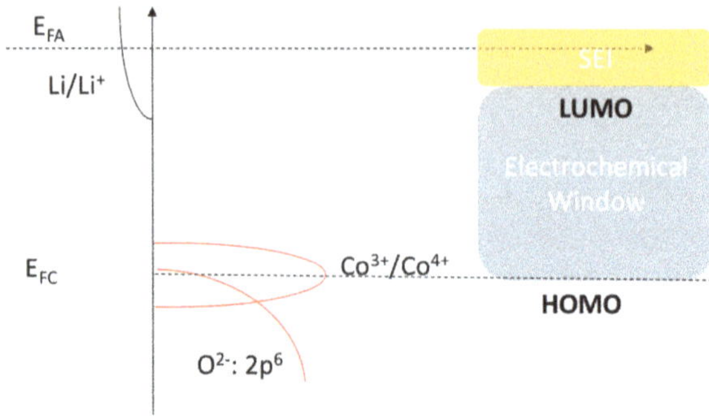

Figure 1.2. Schematic of the density of states (DOS) of $LiCoO_2$, in which the Fermi energy levels of anode and cathode (E_{FA} and E_{FC}) are represented by horizontal dotted lines.

current research is devoted to development of transition metal oxides (TMOs) that are low in cobalt but high in nickel and manganese. However, because of the high volumetric energy density of LCO, it is still favored over other TMOs for portable electronics. EVs, on the other hand, are more forgiving with respect to volumetric energy density, and the emphasis there is on high gravimetric energy density and cost effectiveness.

The anode for lithium ion batteries is usually graphite due to its low potential, safety, and high specific capacity. More specific details are given in section 1.2.

An easy way to characterize a nonaqueous electrolyte is through its electrochemical window, a practical value that can be associated with the theoretical quantity E_g. The traditional view is that the Lowest Unoccupied Molecular Orbital (LUMO) is associated with cathodic stability, the Highest Occupied Molecular Orbital (HOMO) is associated with anodic stability, and the difference between LUMO and HOMO characterizes the electrolyte's electrochemical window. However, it is better to explain the electrochemical window in parallel with the transition-metal energy band, as many of the physical and chemical phenomena in batteries are related. The potentials of the cathode (μ_c) and anode (μ_a) are determined by their Fermi energies (E_F) in the itinerant-electron band, and the μ_c and μ_a consequently determine the cell's open-circuit voltages $V_{oc} = (\mu_a - \mu_c)/e$. As shown in figure 1.2, the μ_c and μ_a should ideally be within the bracket of LUMO and HOMO. However, in most LIBs, because of the low reduction potential of the Gr, 0.1 V versus Li/Li^+ (all voltages subsequently cited in this paper will use Li/Li^+ as the reference), almost no electrolytes will have a higher LUMO than lithiated graphite.

The practical way to expand HOMO and LUMO is to have a solid–electrolyte interface (SEI) that successfully blocks the reduction of electrolyte in contact with the anode, and yet effectively transports the Li^+ ions.

In a typical LIB with a TMO cathode and a Gr, different electrochemical reactions take place at the cathode and the anode simultaneously. The general Faraday's Law can be used to calculate the theoretical capacity of electrode materials; it depicts the relationship between the current that runs through the external circuit and the amount of the corresponding active materials that are consumed/produced in the electrochemical reaction. For example, in a typical graphite reduction,

$$C_6 + Li^{+} + e^{-} \rightarrow LiC_6 \tag{1.2}$$

Using Faraday's Law, $F = qN_{AV}$: where q is the fundamental unit of charge, and N_{AVO} is Avogadro's number, Faraday's constant F has a value of 96 485 C mol^{-1}. The molecular weight of C_6 (graphite) is 72 g mol^{-1}, and the theoretical capacity of graphite can be calculated as

$$1\ \mathrm{mAhg^{-1}} = 10^{-3} \cdot 1\ \mathrm{C} \cdot \mathrm{s^{-1}} \cdot 1\ h\ \cdot 3600 \cdot \frac{s}{h} \cdot \mathrm{g^{-1}} \tag{1.3}$$

$$1\ \mathrm{C} \cdot \mathrm{g^{-1}} = \frac{1}{3.6}\ \mathrm{mAhg^{-1}} \tag{1.4}$$

$$\text{Capacity} = \frac{96485\ \text{C/mole}}{72\text{g/mole}} = 984\ \mathrm{C} \cdot \mathrm{g^{-1}} = \frac{984}{3.6}\ \mathrm{mAhg^{-1}} = 273\ \mathrm{mAhg^{-1}} \tag{1.5}$$

From this example, a general formula can be deduced for the calculation of a theoretical capacity:

$$\text{Capacity} = \frac{\frac{96485}{3.6} \cdot z}{\mathrm{Mw}} = \frac{26800 \cdot z}{\mathrm{Mw}} \mathrm{mAhg^{-1}}, \tag{1.6}$$

where z is the number of electrons generated per molecule.

1.1.1.2 Electrochemistry

The fundamentals of electrochemistry are covered well in the classical electrochemistry books. A recent book by Fuller and Harb also offers both the basics and in-depth study materials for topics such as thermodynamics, electrochemical kinetics, and transport, with examples of how each concept is used [3]. We encourage readers to refer to those books for an understanding of the basics. Three fundamental equations, the Debye–Hückel equation (activity coefficient), Butler–Volmer equation (electrochemical kinetics), and Nernst–Plant equation (transport properties), are given here for the readers' reference:

$$\text{Debye-Hückel equation } \ln \gamma_{\pm} = z_{+}z_{-}\alpha\sqrt{I} \tag{1.7}$$

$$\text{Butler-Volmer equation } i = i_a + i_c = i_0\left[\exp\left(\frac{(1-\beta)F\eta_s}{RT}\right) - \exp\left(-\frac{\beta F\eta_s}{RT}\right)\right] \tag{1.8}$$

$$\text{Nernst-Plant equation } N_j = \underbrace{-z_i u_i F c_i \nabla\phi}_{\text{migration}} \underbrace{- D_i \nabla c_i}_{\text{diffusion}} + \underbrace{c_i v}_{\text{convection}} \tag{1.9}$$

Three parameters are identified as the important criteria for measuring the performance of a battery: coulombic efficiency, specific capacity, and capacity retention. Recently, more detailed studies have been implemented to get a deeper understanding of the operation of a battery, mainly addressing: (1) the individual voltages of cathodes and anodes using a reference cell; (2) detection of lithium slippage; (3) application of dV/dQ and dQ/dV; and (4) impedance studies on both the anodes and cathodes.

Figure 1.3 illustrates the overall voltages of the batteries, as well as the individual voltages experienced on the cathode and anode. The method commonly used to measure the individual voltages is to use a third Li metal reference electrode that can either be made in-house or purchased. The same concept can be adapted for measurement of individual impedance on the cathode and anode. Interestingly, in an NMC//Gr cell, the initial cell voltage after assembly is ~0 V; however, through the initial charge pulse (for example, a current of 1.5 C rate), the cathode stays at a relatively stable voltage of 3.7 V [4].

A typical coin-cell anatomy is shown in figure 1.3(A). A more advanced version, a so-called reference cell, is shown in figure 1.3(B). This type of cell consists of two 20.3-cm^2 electrode disks separated by two layers of a 25-μm-thick microporous separator (Celgard 2325). The unique feature of this reference cell is the Cu–Li alloy wire used as a reference. Lithium metal was plated *in situ* onto the tip of this

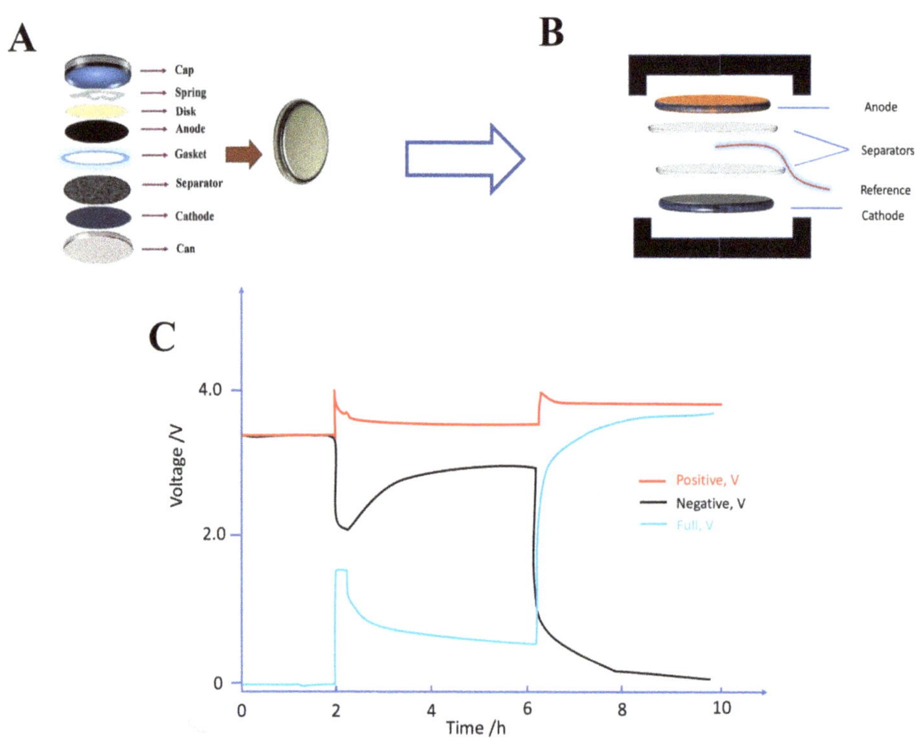

Figure 1.3. (a) Schematic representation of a basic coin-cell setup. (b) A custom reference cell for measurement of both *in situ* potential and impedance versus a reference. (c) A representative voltage profile of a cathode and anode that is measured from a reference cell.

thin (25-μm-dia.) Cu wire to form a microprobe reference electrode (RE). Only a 2- to 3-mm portion of this wire was exposed to the electrolyte and coated with Li; the remainder was insulated by a polyurethane coating. The Cu wire was lithiated at 5 μA for 6 h, with the cathode as the source of lithium. The advantage of such a Li–Cu alloy wire is that it provides an accurate reference for both the potential and the impedance. A typical compiled voltage profile consisting of the anode potential, the cathode potential, and the full cell potential is shown in figure 1.3(C).

Cycling protocols (determining which electrochemical condition the cells will experience) are also crucial for battery performance evaluation and mechanism diagnosis. For example, a comprehensive cycling protocol was implemented at Argonne National Laboratory to impose harsher conditions than normal cycling, and triplicate cells were used to provide for repeatable examination of the cells [5]. In general, the cycling procedure can be divided into four electrochemical steps: formation, preparation, HPPC (Hybrid Pulse Power Characterization), and aging. The formation step consists of five slow C/20 cycles to ensure the formation of a robust SEI layer. Preparation consists of a slow C/10 cycle. HPPC cycles are specifically designed to mimic the EV operating conditions occurring at the 'accelerating' or 'braking' stages, with large pulses going through the cell and the impedance recorded at different states of charge (SOCs). Over the course of 119 cycles, periods of slow C/20 cycles were inserted before each of the 20 fast C/3 cycles (classified as the 'aging' step) so that capacity could be indicated with less influence from the impedance. The HPPC segments were purposely reinstituted before each of the 20 cycles of the C/3 aging. A rather aggressive cycling protocol was used: the galvanostatic charge (constant current) was followed by a constant voltage hold of 3 h. Through a standard cycle package of 119 cycles, information is obtained with respect to (1) the Area Specific Impedance (ASI) change during the cycling [6]; (2) aging, reflecting the capacity retention (through both the major C/3 cycles and periodic C/20 cycles); and (3) dQ/dV, from the C/20 cycles [5]. ASI can be measured through the HPPC cycles using the following equation, with the parameters V_t, V_0, t_0, and t_1 shown in figure 1.4(b) [6]:

$$\mathrm{ASI} = \frac{V_{t_0} - V_{t_1}}{I_{t_1} - I_{t_0}} \tag{1.10}$$

Another key concept in battery research is the slippage of voltage. The original voltage profiles for a representative NMR//Gr cell is shown in figure 1.5(a), the V_{c0} at 4.22 V and V_{a0} at 0.12 V, once the cycled electrodes reach the charge voltage and in the absence of the reference cell, the voltage reflected through the cell potential is the difference between the positive and negative electrode ($V_c = V_c - V_a$). Using an NMC//Gr-Si cell as an example, the slippage of capacity results in a large increase in the electrode potential of the anode. In one scenario where no Li^+ inventory loss and no cathode degradation occurs, the endpoint voltages and discharge/charge profiles of the cathode and the anode will remain the same. However, using the NMC cathode as an example, when a charge cutoff of 4.1 V is maintained, as the cathode capacity decreases (anode capacity slips to the right), the anode voltage will slightly

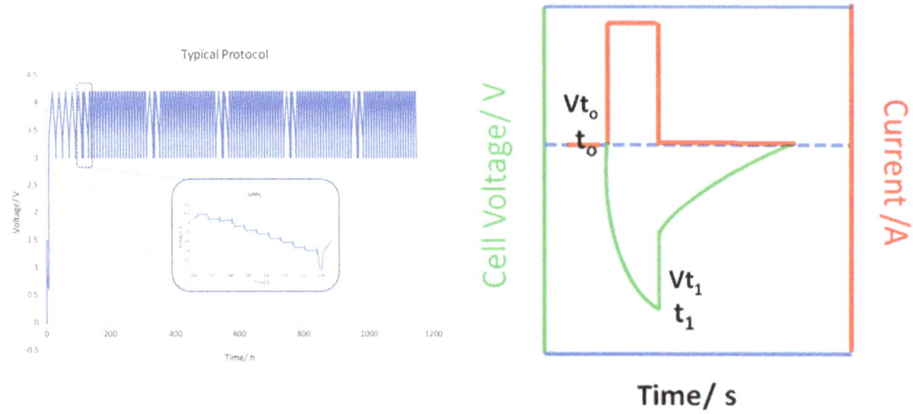

Figure 1.4. (a) Typical protocol (voltage versus time) for testing a full cell consisting of a graphite anode and TMO cathode; the cutoff voltages are 3 and 4.2 V, and the inset shows the voltage profile of the HPPC cycle. (b) Schematic representation of the cell voltage and current for the ASI measurement.

increase (V_{a1} = 0.15 V). In the next cycle, the cathode needs to reach a higher individual voltage to maintain the $V_{c1}-V_{a1}$, resulting in higher polarization of the cathode ($V_{c1} > V_{c0}$). The voltage profile of graphite is featured with plateaus at ~ 120 mV and 90 mV, and loss of the capacity when cells are cycled at the same upper cutoff voltages will push the voltage in the anode to the right with respect to the cathode. For example, in figure 1.5(b), as the cell ages, the capacity decreases, as denoted by the gray area of Li^+ inventory loss, resulting in the increase of anode voltage at 0.15 V. As shown in figure 1.5(c), when a Li^+ gain occurs in the anode side, slippage towards the right of the anode is observed (V_{a2} = 0.09 V).

1.1.2 Beyond lithium ion batteries

'Beyond lithium ion batteries' refers to technologies that do not revolve around the traditional rocking-chair Li intercalation–deintercalation mechanism; instead, they comprise frontier technologies such as Na and K ion batteries, lithium–air batteries, batteries with multivalent ions (where the mobile charge carrier has a valence number of 2 or more), lithium metal batteries, and all-solid-state batteries (ASSBs) [8].

The electrochemistry and fundamentals of beyond lithium ion batteries, including SEI formation and intercalation mechanisms for cathodes, resemble those of the LIBs and will not be described here. The important features and characteristics of beyond lithium ion batteries can be found in several recent reviews and perspectives [8, 9].

1.2 Basic components of batteries

1.2.1 Cathodes

As shown in table 1.1, intercalation cathodes can be loosely categorized into layered materials (LCO, NMC, LMR-NMC), polyanion materials ($LiCoPO_4$, $LiFePO_4$), and spinel materials ($LiMn_2O_4$, $LiNi_{0.5}Mn_{1.5}O_2$), and conversion materials on the

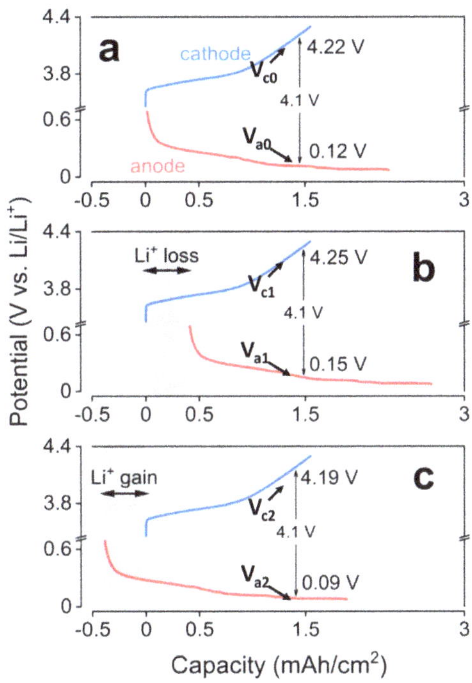

Figure 1.5. Schematic representation of the slippages in the electrodes. (a) the absence of capacity fade, electrode potentials remain stationary (a), while net capacity loss or gain causes the anode to shift to the right (b) or to the left (c), respectively, relative to the cathode. As in figure 1.1, capacity loss leads to higher electrode potentials at the same full-cell voltage cutoff of 4.1 V. Reproduced from [7].

basis of their framework structures. The structures of the former three materials are shown in figure 1.6. The emphasis here will be on the layered materials because of their commercialization as cathode materials in the EV market [10, 11].

1.2.1.1 Layered materials

Nowadays, LCO is still the cathode of choice for portable electronics because of its high volumetric energy density. However, the dominant usage of LCO raises concerns over the geopolitical and environmental aspects of cobalt mining. The desire for a reduction in the cobalt component, and for other characteristics including higher voltage, high rate capability, long cycle life, and high energy density, has motivated research on cathodes for LIBs. Concerted efforts to improve the performance of LCO have included structural modifications to improve both its stability and resistivity, and the modification of the electrolyte–electrode interface to reduce cathode impedance rise [15]. In contrast to the high compact density of LCO (4.05–4.15 g cm^3) [16], NMC materials only have a compact density in the range of 3.1–3.6 g cm^3. Despite the theoretical capacity of 273 $mAhg^{-1}$, only half of the Li^+ inventory in LCO is practically usable with a cutoff voltage of less than 4.2 V versus Li/Li^+, which is necessary for maintaining structural stability over prolonged cycling [15, 17–19]. Recognized as one of the first layered materials through the work of Goodenough [20], LCO has a layered α-$NaFeO_2$ structure, featuring a rhombohedral

Table 1.1. Specific capacity and voltages of some common cathodes for lithium ion batteries.

Structure	Cathode	Specific gravimetric capacity (mAh g^{-1})	Volumetric capacity (theoretical/ practical, Wh l^{-1})	Average potential (V)	References
Layered	LCO	273/145	1363/550	3.8	[11]
	$LiNi_xMn_yCo_zO_2$	273/	1333/600	3.7	[11]
	LMR-NMC $Li_{1.2}Ni_{0.4}Mn_{0.4}O_2$ (=$0.5Li_2MnO_3$–$0.5LiNi_{0.5}Mn_{0.5}O_2$)	375/250–280 (1.2 li extraction)	>1000	3.7	[12]
Spinel	$LiMn_2O_4$	148/120	596	4.1	[11]
	$LiNi_{0.5}Mn_{1.5}O_2$	147		4.9	[11]
Polyanion (olivine)	$LiCoPO_4$	167/125	510	3.7	[11]
	$LiFePO_4$	170/165	589	3.4	[11]
Conversion	S	1675		2.8	[13]
	O_2	3840		3.1	[14]

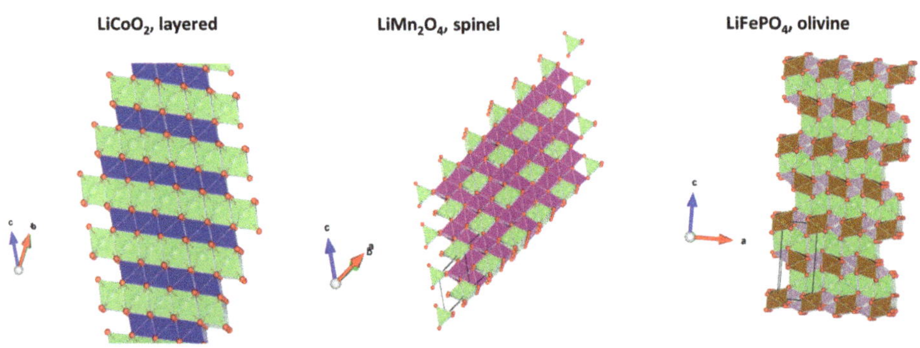

Figure 1.6. The crystal structures of layered $LiCoO_2$ (one-dimensional Li^+ transport), spinel $LiMn_2O_4$ (two-dimensional Li^+ transport), and olivine $LiFePO_4$ (three-dimensional Li^+ transport).

structure with an *R*3*m* space group, with the oxygens arranged in a cubic close-packed network and lattice constants of $a = 2.816$ Å and $c = 14.052$. Li and Co^{3+} alternate in the (111) planes to form CoO_6 and LiO_6 octahedra. The CoO_6 octahedra share edges and form CoO_2 sheets, separated by sheets of octahedral Li [21]. During delithiation, the structure first undergoes expansion of the *c* axis due to the loss of Li and weak Van der Waals expulsion between the CoO_2 sheets; however, further delithiation (to Li_xCoO_2 with $x < 0.5$) would be detrimental, resulting in structure collapse.

Other layered TMOs with a lower cobalt content include NMC materials with a general formula of $LiNi_xMn_yCo_zO_2$ (where $x + y + z = 1$), including the high-nickel

Table 1.2. Common anodes with specific gravimetric capacity and volumetric capacity.

Anode	Specific gravimetric capacity ($mAh\ g^{-1}$)	Volumetric capacity ($mAh\ c.c.^{-1}$)	Average potential (V)	References
Graphite (Gr)	360	570	0.1 V	[27]
Silicon	4200/3154[a]	800	0.06–0.12 V	[27, 28]
Lithium	3828	1700	0 V	[29]
Sodium	1165		0.1 V	[30]
Magnesium	2233	3833	0.6 V	
Calcium	1340	2073	0.1 V	

[a] Theoretical specific capacity/actual specific capacity

cathodes that are under industrial consideration for the realization of higher-energy-density batteries of 800 Wh kg^{-1} at the cell level.

Three other severe problems have been noted with cycling LCO at high voltages [22]: (1) the interface between the delithiated LCO and the electrolyte is unstable. As shown in figure 1.2, the thermodynamic stability window of the electrolyte should surpass the electrode electrochemical potential of both the cathode and the anode. As LCO is cycled to a higher voltage (> 4.45 V), the deterioration of electrolyte is exacerbated, with the formation of a resistive passivation layer and a large impedance rise. Coating or doping with oxides such as Al_2O_3 [23] or ZrO_2 [24] reduces the contact area between LCO and electrolyte and improves the performance under high voltages. (2) The Co dissolution increases with a higher cutoff voltage (> 4.2 V), together with the loss of O (peroxo species of O_2, O_2^-). The delithiation process lowers the Fermi level of Co^{3+} until the Co 3d orbital overlaps with the O 2p band (figure 1.4) and facilitates peroxo evolution [25, 26]. Evolution of these peroxo species causes irreversible surface decomposition and phase transformation of the cathodes, which eventually leads to capacity fade. (3) Degradation of other components such as the binders and the separators occurs, and the consequent detachment of cathode particles can lead to capacity loss.

1.2.2 Anodes

Anodes are important because of the possible high volumetric energy density associated with utilization of metal anodes such as Li, Mg and Ca. The reduction potentials of these metal anodes are compared with those of the traditional Gr in table 1.2. This chapter does not intend to give a complete discussion of all possible anodes, but will focus on the important characteristics of certain anodes, which are categorized loosely based on application: (1) graphites, (2) lithium metal anodes, and (3) other metal anodes, including alloys.

1.2.2.1 Graphite

In the context of battery materials, graphite refers to 'graphitic carbon,' with a structure consisting of a hexagonal network of carbon atoms lacking crystalline ordering in the *c* direction. A typical crystalline structure of graphite is illustrated in figure 1.7, with alternating A and B layers consisting of sp^2 hexagonal array of carbon atoms. Upon lithium intercalation, each C_6 can accommodate one Li^+ atom.

The basal planes, referring to the hexagonal C_6 planes, have a spacing (d_{002}) of 3.35 Å, about twice that of the Van der Waals distance for carbon atoms. Graphite displays a non-perfectness, and the basal planes demonstrate less reactivity than the edge planes. Upon lithiation, the Li^+ intercalates through the edge plane and imperfect basal plane, with three distinctive plateaus at 208, 116 and 83 mV (figure 1.8(c)). To circumvent the reactivity of the edge plane, SEI formation enables further intercalation at the expense of consumption of Li^+ inventory.

1.2.2.2 Li metal anodes

With the development of Li–S batteries and ASSBs, lithium metal anode research has demonstrated the potential for reaching the high theoretical gravimetric capacity of 3860 mAh g^{-1}. However, the promise of Li metal anodes is accompanied by serious issues, such as the extreme electropositivity of Li metal and its resulting reactivity with almost all organic components, as well as the large interfacial resistance of SEIs consisting of LiF, Li_2CO_3, ROLi, and R′COOLi [1].

1.2.2.3 Other metal anodes

The prevalent challenges associated with the application of metal anodes include particle pulverization, unstable SEI, and impedance rise from electrolyte decomposition on the interfaces. The large ionic radii of the relevant metal ions (Na^+, 1.02 Å; K^+, 1.38 Å; Mg^{2+}, 0.72 Å; Ca^{2+}, 1.00 Å [32]) and the charge density can also cause difficulty with intercalation into the cathode lattices.

Among the potential metal anodes for batteries, silicon has been researched extensively in both academia and industry. Co-utilization of graphite and silicon is a

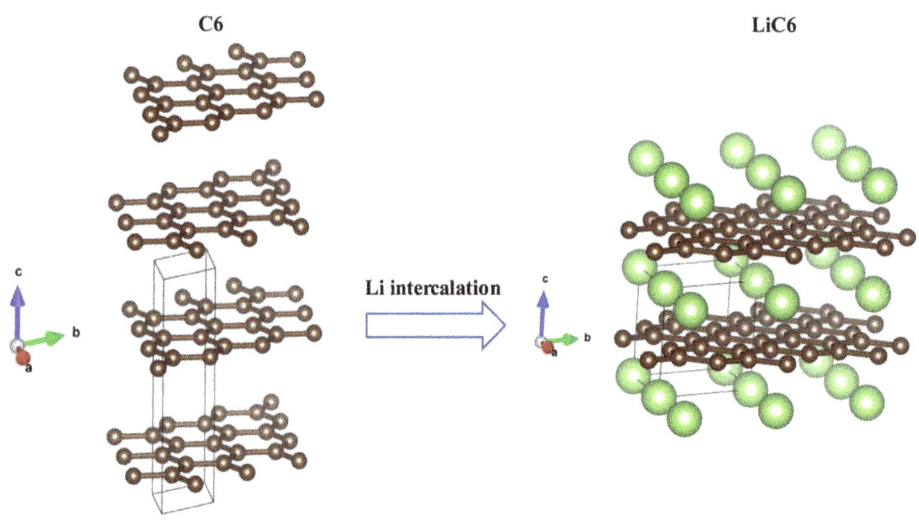

Figure 1.7. Schematic representation of the layered structure of graphite and how Li^+ intercalates between the graphite layers.

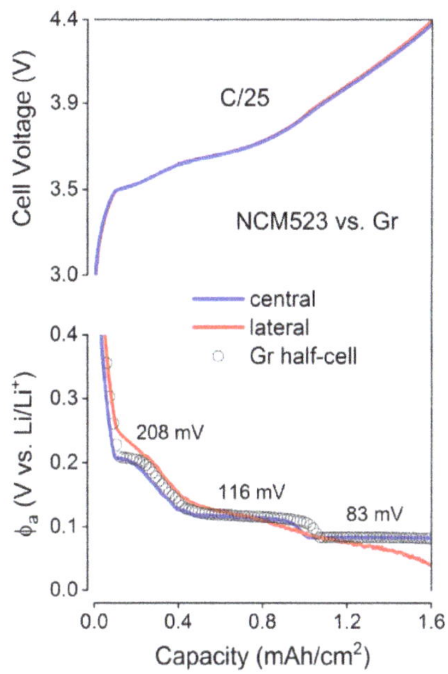

Figure 1.8. Schematic representation of the three distinctive plateaus observed upon Li^+ intercalation: the voltage profile of a full cell consisting of an NCM 523 cathode and a Gr was used as an example. Cell voltages and anode potentials (φ_a) were measured at C/25 using the RE arrangements. Graphite staging potentials determined using the centrally placed RE are indicated in the plot. The potential of a Li//Gr half-cell charged at a C/110 rate (gray open circles) is overlaid for comparison (from [31]).

common practice to alleviate the issues associated with using Si particles alone, and can increase the specific capacity [27].

The concerns about cost and long-term sustainability of Li resources also demand development of alternative low-cost and sustainable energy storage systems. Novel research, targeting energy based on low-cost and naturally abundant elements, is focusing on using Na and K. The abundance and low price of these elements enable the possibility of sustainably produced, low-cost batteries based on Na^+ and K^+ ions [8, 33].

Cathode materials utilizing Na^+ and K^+ intercalation that have been explored include layered TMOs, polyanionic compounds [33], and Prussian-blue analogs ($A_xM_A[M_B(CN)_6]_y$, where A = a combination of Li and Na, and M is a transition metal: Fe, Mn, Co, Ni, Cu, etc). Compared with their Li analog, the large size of Na^+ and K^+ can be of benefit for slowing the mobility and disordering of the transition metal. Cathodes with the layered structure of O_3–$NaNi_{0.5}Mn_{0.5}O_2$ can show good cycling stability, with the major issues arising mostly from the delithiated state with low Na/K content [8].

1.2.3 Electrolytes

For the sake of convenience, electrolytes used in LIBs and in beyond lithium ion batteries are discussed separately.

1.2.3.1 Lithium ion batteries

Without Yoshino Akira's introduction of the combination of Gr and carbonate (a unique compound with an –O–C(=O)–O moiety that is prone to react with underpotential Li to generate a SEI and impart extra thermodynamic stability), it would be hard to imagine developing the compositions of today's electrolytes. In particular, ethylene carbonate is responsible for formation of a stable SEI on the surface of a Gr. The success of the Gr anode in LIBs has also revealed another truth: the electrolyte has unique capabilities to modify the interlayer surface.

Despite the enormous progress researchers have made regarding cathodes and anodes, the fraction of LIB research devoted to electrolytes seems relatively small. The discussion of electrolytes below will center on the development of (1) lithium salts; (2) additives; (3) solvents; and (4) special applications involving extreme temperature, fast-charging conditions, and non-flammability.

1.2.3.1.1 Lithium salts

Lithium salts are essentially used for the purpose of transporting Li^+ under chemically and electrochemically active conditions. The electrolyte environment needs to be compatible with current collectors, cathodes—especially when in a delithiated state, with many energetic active sites to cause decomposition—and anodes in a lithiated state with a potential close to the Li redox potential (~ 100 mV).

The most common lithium salt used in commercial LIBs is $LiPF_6$, which seems to provide the best combination of ionic conductivity, electrochemical stability, stability with respect to the current collector (a safety criterion), and overall LIB performance. The basic properties of $LiPF_6$ and several other common lithium salts, such as LiTFSI ($Li(CF_3SO_2)_2N$), $LiAsF_6$, $LiClO_4$, $LiBF_4$, and $LiCF_3SO_3$, can be found in the literature [34]; these include such properties as ionic conductivity, ionic radii, temperature, and stability towards both Cu (the anode current collector) and Al (the cathode current collector). Relatively new salts—$LiN(SO_2C_2F_5)_2$ (LiBETI) [35], lithium bis (oxalate) borate (LiBOB) [36], $LiPF_3(C_2F_5)_3$ (LiFAP) [37], lithium 4,5-dicyano-2-(trifluoromethyl) imidazolide (LiTDI) [38], lithium 4,5-dicyano-2-(pentafluoroethyl) imidazolide (LiPDI), and lithium bis(fluorosulfonyl)imide (LiFSI)—are also available through exploratory design and syntheses, although their compatibility with coin cells and their electrochemical stability are still a matter of debate.

Below, the chemistry of $LiPF_6$ will be discussed in detail with respect to its reactivity towards H_2O [39] (an unavoidable impurity), solvent [40], and additives such as tris(trimethylsilyl) phosphite [41].

The cascading hydrolysis of $LiPF_6$ is illustrated in scheme 1.1. Overall, the reactivity of $LiPF_6$ with water is detrimental, leading to formation of both HF and oxyfluorophosphoric acid, which corrodes the cathodes and causes crossover of transition metal species (Mn^{2+}); these species cause the poisoning of the anode, leading to an electrocatalytic reduction of Li^+ inventory and capacity loss. Recent studies of hydrolysis of $LiPF_6$ under high voltage suggest that the detrimental catalytic effects of high voltage are much worse, with the majority of the hydrolysis products coming from the reaction between PF_6^- anion and H_2O under voltages above 4.5 V.

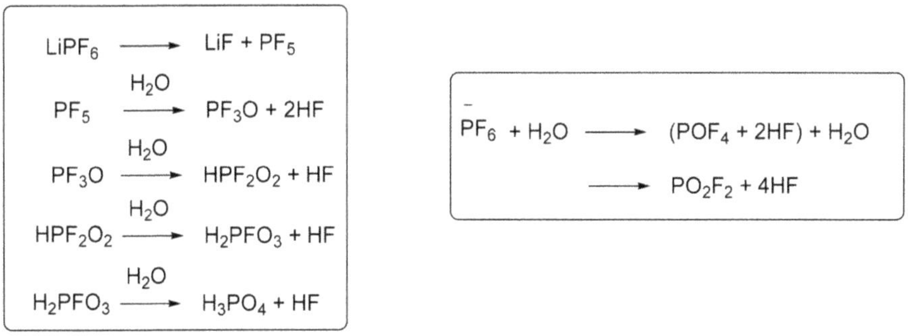

Scheme 1.1. (Left) the hydrolysis cascade reaction between $LiPF_6$ and H_2O. (Right) the proposed reaction of PF_6^- anion and H_2O under high voltages [42].

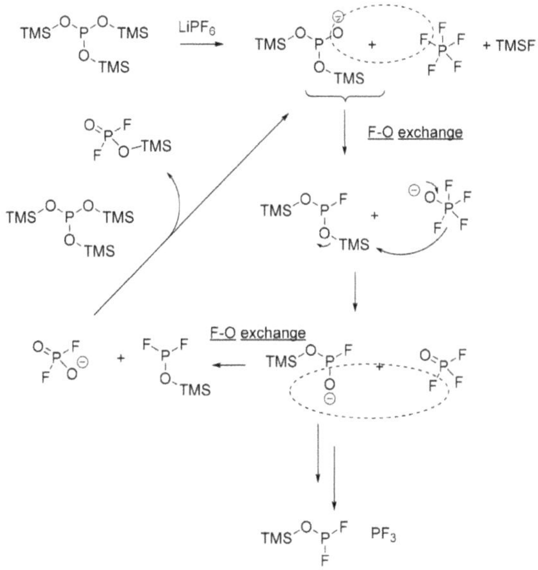

Scheme 1.2. Spontaneous reactivity of TMSPi with $LiPF_6$. The F–O exchange is emphasized with an underscore.

1.2.3.1.2 Additives

Additives are designed for task-specific applications such as anode SEI formation, cathode CEI formation, flame retardation, and other unique applications including fast-charge and extreme-temperature applications. The Dahn group, as well as Argonne National Laboratory researchers, have long been developing both singular and combinatorial additives [43, 44].

The reactivity of $LiPF_6$ with tris(trimethylsilyl) phosphite (TMSPi) was discovered by the Winter group [41]. As shown in scheme 1.2, the reaction involves an astounding F–O exchange between the P–F bond and the P–O–TMS bond,

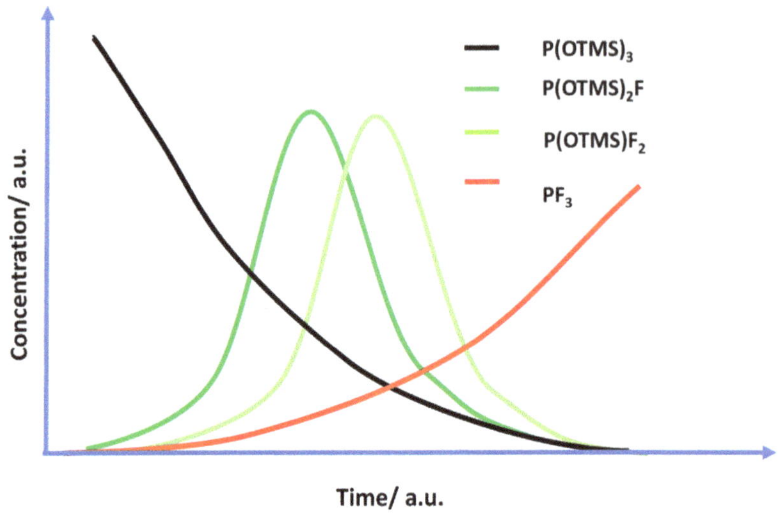

Figure 1.9. Schematic representation of species evolution with time for the reaction of the additive TMSPi with a solution of $LiPF_6$ in carbonate solvent.

resulting in formation of three compounds: $P(OTMS)_2F$, $P(OTMS)F_2$ and PF_3. The intermediate compound $O{=}P(-F)_2-O^-$ can also react with $P(OTMS)_3$ to generate $P(OTMS)_2F$ and $O{=}P(-F)_2-OTMS$; note the presence of an extra O=P in the latter compound. Overall, a spontaneous cascading reaction is observed with aging, with little control of either the speciation or concentration (figure 1.9). In general, it is predicted that $P(OTMS)_3$ will show a consistent decrease with time, while both the $P(OTMS)_2F$ and $P(OTMS)F_2$ intermediates will show an initial increase before reaching a maximum concentration, followed by a decrease. The final product PF_3 appears last in the reaction and is expected to show a steady increase until the original $P(OTMS)_3$ is totally consumed.

The control of the spontaneity of the $LiPF_6$ reactivity, however, has been showcased in our group recently by a unique *in situ* method for formation of cathode-protective additives with structures identified as lithium oxyfluorophosphates (**2a** and **2b**), as illustrated in scheme 1.3. These *in situ* reactions take advantage of the reactivity of $LiPF_6$ with the bis(trimethylsilyl) esters **1a** and **1b**, and a series of extremely moisture-sensitive tetrafluorophosphates can be easily obtained at a nearly 100% yield in the solution (scheme 1.4) [45]. In contrast to the reaction in scheme 1.2, there is only one designated product from the reaction between $LiPF_6$ and bis(trimethylsilyl)ester. As discussed by Yang *et al*, the strain of the carboxylate linker in the tetrafluorophosphate is important: strong preference goes to the 5-membered and 6-membered rings because of the lesser ring strain and torsional strain. A similar protective effect on the cathode surface was also discovered: a 6-membered ring was found to have a superior effect on the reduction of impedance rise in an NMC//Gr full cell [45].

LIBs for industrial applications require a combination of excellent thermal stability, wetting capability, and SEI formation capability, resulting in the presence

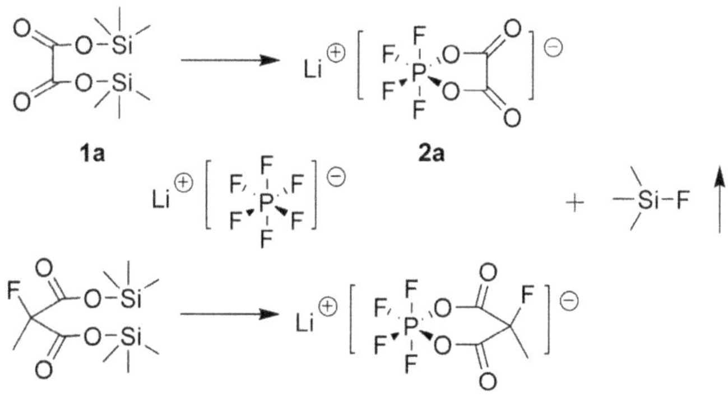

Scheme 1.3. *In situ* synthesis of anions **2a** and **2b** from molecular precursors **1a** and **1b**.

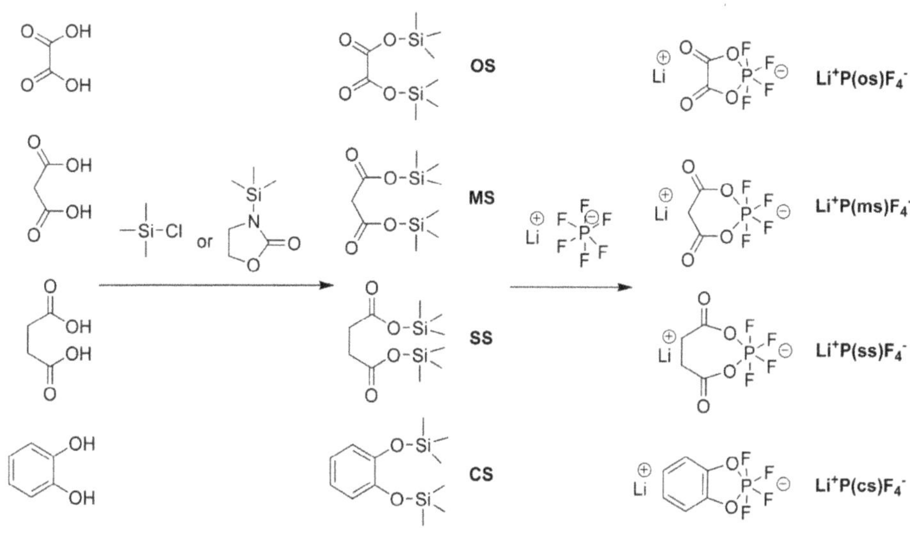

Scheme 1.4. Synthesis and structural formulas of hypothesized monoanion products.

of many additives and co-solvents of propionate compounds [46]. The additives in commercial LIBs are targeted for different applications and can be loosely classified as anode additives, cathode (interfacial) additives, overcharge protection, and alloy additives [43, 47]. Some common additives are shown in figure 1.10.

Despite the large number of additives that are already available, new additives and uses for those additives are continuing to be invented [48].

1.2.3.1.3 Solvents

The solvent component for LIBs has not been changed significantly since its invention in 1991 by Sony. Owing to the bulk amount of solvent compared with other components in electrolytes (salts and additives), the following properties of solvents are usually considered most important: high electrochemical stability

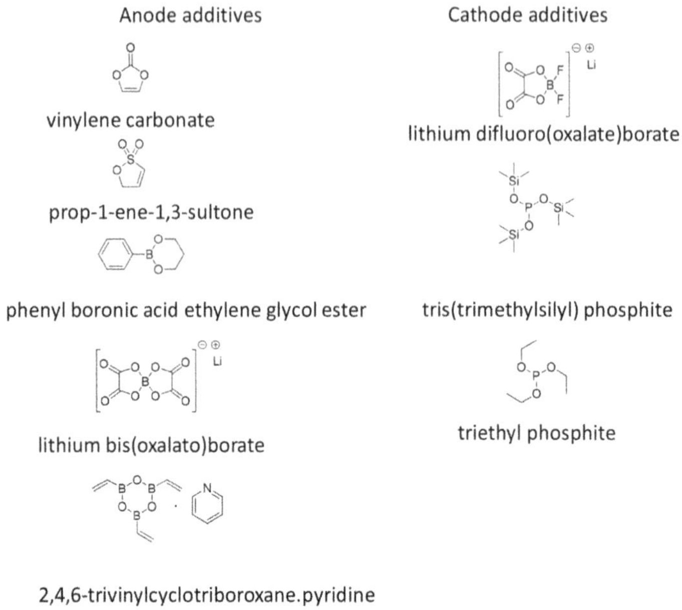

Figure 1.10. Exemplary structures of additives used in lithium ion batteries.

towards cathodes and anodes (graphite), high transport rate, good wetting ability, and good thermal properties (ability to function reasonably well in a wide temperature range). Despite the critical role that ethylene carbonate plays in formation of the SEI, this component has problems such as low electrochemical stability and large impedance rise [49]. The use of co-solvent is a strategy that can significantly change the Li^+ ion transport speed, wetting capability, and, potentially, electrochemical properties. Issues with fast charging and temperature intolerance usually relate to the extremely slow transport of Li^+ relative to the fast electron movement; Li plating occurs as a result of overpotential buildup and slow solid-state Li^+ transport. To achieve fast charging (usually with EV applications in mind) and the lofty goal of 80% capacity retention within 15 min, solvents have been sifted through a myriad of criteria to better support Li^+ ion transport [46]. Esters, albeit possessing low electrochemical stability, have found applications in industrial-scale electrolyte formulation for LIBs because of several favorable properties, including high ionic conductivity, low viscosity, and good wetting properties. Using esters as an example, one can compare the performance of co-solvents with respect to the following criteria: electrochemical window, relative permittivity, viscosity, self-discharge voltages, discharge capacity at 40 °C from 3.0 to 4.2 V, coulombic efficiency, gas generation during formation, storage, and stability at 60 °C 500 h^{-1}. Several new categories of co-solvents, including formates and nitriles, have proven to be beneficial for fast-charging applications. This discussion is not meant to provide an inclusive list of all the chemicals that have been studied; however, it is representative enough to encourage more exploration, especially centering on the categories of nitriles and esters (propionates and formates).

1.2.3.1.4 Special applications: extreme-temperature, fast-charging, and flame-retardant electrolytes

Task-specific electrolytes can be designed for special applications: solvents, salts, and additives are all important for these non-conventional or extreme applications. Additives such as high-boiling-point nitrile (HTCN) are important in an industrial setting to allow electrolytes to tolerate the high-temperature formation and storage processes that industrial cells usually go through. Nitrile compounds (propionitrile or butyronitrile) can also enable low-temperature operation of energy storage systems in the range of −20 °C to −110 °C [50]. High-temperature operation of LIBs requires appropriate choices of both electrolytes (solvents, salts and additives) and cathodes, using advanced thermal analysis methods such as differential scanning calorimetry and accelerating-rate calorimetry, and *in situ* gas analysis [51] via differential electrochemical mass spectroscopy.

1.2.3.2 Beyond lithium ion batteries

1.2.3.2.1 Lithium metal batteries

The electrolytes used for lithium metal-based batteries, including Li–air batteries and Li–S batteries, are usually LiTFSI dissolved in an ethereal solvent. The advantages of Li–S batteries include the high theoretical specific energy of sulfur (2600 W h kg^{-1}), its natural abundance (3.96%–4.77%), and low cost (\$50 ton^{-1}). However, several well-known problems associated with Li–S batteries prevent their commercial application: the shuttling of soluble polysulfide species, the poisoning of polysulfide at the lithium metal anode, and low loading of the active materials on the cathode part. Since a coin-cell setup can lead to misleadingly 'good' performance of Li–S batteries, with flooded cells leading to an unlimited Li^+ inventory, Li–S batteries have been vigorously tested in a pouch cell, with controlled Li^+ inventory and electrolyte amount (electrolyte/sulfur ratio) [13].

Li–air batteries and Li–oxygen batteries, despite their promise, have practical issues, including kinetic issues manifested as high overpotential and low discharge voltages, and the detrimental interference of persistent impurity components in the air (CO_2, H_2O, N_2, etc). A materials-to-system analysis of mass, volume, and cost values reveals that Li–air batteries are unlikely to be useful in EVs because of the large penalty resulting from utilization of a gaseous reactant [29]. Such a systems-level engineering simulation also identifies Si/NMC as a moderate-risk pathway to achieve commercial application in EV markets. The electrolytes for Li–air batteries consist of a lithium salt in a variety of solvents (carbonates, dimethylsulfoxide, ethers, etc) [52].

1.2.3.2.2 Multivalent ion batteries

Multivalent ion batteries, especially Mg^{2+} and Ca^{2+} ion batteries, hold the promise of ensuring high energy density, but so far have fallen short on the energy density that can be reached. A holistic approach is being taken, designing a low-barrier, high-mobility cathode as well as a highly electrochemically stable electrolyte with a high coulombic efficiency of plating/stripping. Electrolyte development has taken a big leap during the last five years, with, for example, Mg^{2+} ion electrolytes, where an evolution has occurred from organohalogenated aluminate ($MgAl(EtCl_2)$, with electrochemical stability ~2.0 V versus Mg/Mg^{2+}) [53] to a weakly coordinating ion ($Mg[Al\{OC(CF_3)_3\}_4]_2$) with an electrochemical window of ~ 5.0 V versus Mg/Mg^{2+} [54]. The coulombic efficiency of plating/stripping of these electrolytes at the Mg anode is also high for multiple-cycling applications. In recent years, cathode development has focused on spinel TMOs [55–57].

1.3 Conclusion

The complexity of lithium ion battery and beyond lithium ion materials (cathodes, anodes, and electrolytes) demands advanced techniques for detailed diagnosis and understanding of the changes and mechanisms associated with battery cycling. Further understanding of the battery degradation and mechanism is beyond the scope of this chapter, and readers can refer to other reviews for in-depth research. Overall, the future decarbonization relies on the parallel development of both the renewable energy generation (wind, tide, solar, etc) and storage. New forms of energy storage are also emerging, with a focus on either electric vehicles or grid storage. New forms such as structural batteries [58] and thermal batteries are finding application to niche areas. Another hotspot for the batteries is the implementation of machine learning/artificial intelligence: with automated algorithms to discover materials of interest for energy storage application [59]. We hope that with efforts from all scientific disciplines, the energy storage systems can reach unprecedented high energy, and ultimately be used in a mission such as aviation.

Acknowledgements

This work was partially supported by the Joint Center for Energy Storage Research, an Energy Innovation Hub funded by the U.S. Department of Energy and also partially supported the Vehicle Technologies Office (VTO), David Howell (Manager), Battery R&D, Peter Faguy (Technology Manager), at the U.S. Department of Energy, Office of Energy Efficiency and Renewable Energy. The submitted manuscript has been created by UChicago Argonne, LLC, Operator of Argonne National Laboratory ('Argonne'). Argonne, a U.S. Department of Energy Office of Science laboratory, is operated under Contract No. DE-AC02–06CH11357.

References

[1] Peled E and Menkin S 2017 Review—SEI: past, present and future *J. Electrochem. Soc.* **164** A1703–19

[2] https://nobelprize.org/prizes/chemistry/2019/summary/

[3] Fuller T F and Harb J N 2018 *Electrochemical Engineering* (New York: Wiley)

[4] Bloom I *et al* 2005 Differential voltage analyses of high-power, lithium-ion cells: 1. Technique and application *J. Power Sources.* **139** 295–303

[5] Long B R *et al* 2016 Enabling high-energy, high-voltage lithium-ion cells: standardization of coin-cell assembly, electrochemical testing, and evaluation of full cells *J. Electrochem. Soc.* **163** A2999–3009
[6] Gallagher K G, Nelson P A and Dees D W 2011 Simplified calculation of the area specific impedance for battery design *J. Power Sources.* **196** 2289–97
[7] Rodrigues M-T F *et al* 2020 Insights on the cycling behavior of a highly-prelithiated silicon–graphite electrode in lithium-ion cells *J. Phys.: Energy.* **2** 024002
[8] Tian Y *et al* 2020 Promises and challenges of next-generation 'beyond Li-ion' batteries for electric vehicles and grid decarbonization *Chem. Rev.*
[9] Park M J, Yaghoobnejad Asl H and Manthiram A 2020 Multivalent-ion versus proton insertion into battery electrodes *ACS Energy Lett.* **5** 2367–75
[10] Schmuch R *et al* 2018 Performance and cost of materials for lithium-based rechargeable automotive batteries *Nat. Energy.* **3** 267–78
[11] Nitta N *et al* 2015 Li-ion battery materials: present and future *Mater. Today.* **18** 252–64
[12] Thackeray M M *et al* 2018 The quest for manganese-rich electrodes for lithium batteries: strategic design and electrochemical behavior *Sustain. Energy Fuels.* **2** 1375–97
[13] Zhao C *et al* 2020 A high-energy and long-cycling lithium–sulfur pouch cell via a macroporous catalytic cathode with double-end binding sites *Nat. Nanotechnol.* **16** 166–73
[14] Beattie S D, Manolescu D M and Blair S L 2009 High-capacity lithium–air cathodes *J. Electrochem. Soc.* **156** A44
[15] Wang L *et al* 2018 Reviving lithium cobalt oxide-based lithium secondary batteries-toward a higher energy density *Chem. Soc. Rev.* **47** 6505–602
[16] https://novarials-store.com/products/lithium-cobalt-oxide
[17] Ellis B L, Lee K T and Nazar L F 2010 Positive electrode materials for Li-ion and Li-batteries *Chem. Mater.* **22** 691–714
[18] Kraytsberg A and Ein-Eli Y 2012 Higher, stronger, better… a review of 5 volt cathode materials for advanced lithium-ion batteries *Adv. Energ. Mater.* **2** 922–39
[19] Kalluri S *et al* 2017 Surface engineering strategies of layered $LiCoO_2$ cathode material to realize high-energy and high-voltage li-ion cells *Adv. Energ. Mater.* **7** 1601507
[20] Manthiram A and Goodenough J B 2021 Layered lithium cobalt oxide cathodes *Nat. Energy.* **6** 323–3
[21] Amatucci G G, Tarascon J M and Klein L C 1996 CoO_2, the end member of the Li_xCoO_2 solid solution *J. Electrochem. Soc.* **143** 1114–23
[22] Li H *et al* 2009 Research on advanced materials for Li-ion batteries *Adv. Mater.* **21** 4593–607
[23] Liu L *et al* 2004 Electrochemical and in situ synchrotron XRD studies on Al_2O_3-coated $LiCoO_2$ cathode material *J. Electrochem. Soc.* **151** A1344–51
[24] Chen Z and Dahn J R 2002 Effect of a ZrO_2 coating on the structure and electrochemistry of Li_xCoO_2 when cycled to 4.5 V *Electrochem. Solid State Lett.* **5** A213–16
[25] Ensling D *et al* 2014 Nonrigid band behavior of the electronic structure of $LiCoO_2$ thin film during electrochemical Li deintercalation *Chem. Mater.* **26** 3948–56
[26] Renfrew S E and McCloskey B D 2019 Quantification of surface oxygen depletion and solid carbonate evolution on the first cycle of $LiNi_{0.6}Mn_{0.2}Co_{0.2}O_2$ electrodes *ACS Appl. Energ. Mater.* **2** 3762–72
[27] Chae S *et al* 2020 Integration of graphite and silicon anodes for the commercialization of high-energy lithium-ion batteries *Angew. Chem. Int. Ed.* **59** 110–35

[28] Tornheim A, Trask S E and Zhang Z 2019 Communication—effect of lower cutoff voltage on the 1st cycle performance of silicon electrodes *J. Electrochem. Soc.* **166** A132–34
[29] Gallagher K G *et al* 2014 Quantifying the promise of lithium–air batteries for electric vehicles *Energy Environ. Sci.* **7** 1555–63
[30] Slater M D *et al* 2013 Sodium-ion batteries *Adv. Funct. Mater.* **23** 947–58
[31] Rodrigues M-T F *et al* 2019 Fast charging of Li-ion cells: part I. using Li/Cu reference electrodes to probe individual electrode potentials *J. Electrochem. Soc.* **166** A996–A1003
[32] Liu Q *et al* 2019 Multi-ion strategies towards emerging rechargeable batteries with high performance *Energy Storage Mater.* **23** 566–86
[33] Jin T *et al* 2020 Polyanion-type cathode materials for sodium-ion batteries *Chem. Soc. Rev.* **49** 2342–77
[34] Park J-K 2012 *Principles and Applications of Lithium Secondary Batteries* (New York: Wiley)
[35] Webber A 1991 Conductivity and viscosity of solutions of $LiCF_3SO$, $Li(CF_3SO_2)_2N$, and their mixtures *J. Electrochem. Soc.* **138** 2586–90
[36] Xu W and Angell C A 2001 Weakly coordinating anions, and the exceptional conductivity of their nonaqueous solutions *Electrochem. Solid-State Lett.* **4** E1
[37] Gnanaraj J S *et al* 2003 On the use of $LiPF_3(CF_2CF_3)_3$ (LiFAP) solutions for Li-ion batteries. Electrochemical and thermal studies *Electrochem. Commun.* **5** 946–51
[38] Lau K-C *et al* 2019 Widening electrochemical window of Mg salt by weakly coordinating perfluoroalkoxyaluminate anion for Mg battery electrolyte *J. Electrochem. Soc.* **166** A1510–19
[39] Terborg L *et al* 2012 Ion chromatographic determination of hydrolysis products of hexafluorophosphate salts in aqueous solution *Anal. Chim. Acta.* **714** 121–26
[40] Wiemers-Meyer S, Winter M and Nowak S 2016 Mechanistic insights into lithium ion battery electrolyte degradation—a quantitative NMR study *Phys. Chem. Chem. Phys.* **18** 26595–601
[41] Qi X *et al* 2016 Lifetime limit of tris(trimethylsilyl) phosphite as electrolyte additive for high voltage lithium ion batteries *RSC Adv.* **6** 38342–49
[42] Liu M *et al* 2021 Hydrolysis of $LiPF_6$-containing electrolyte at high voltage *ACS Energy Lett.* 2096–102
[43] Tornheim A *et al* 2017 Evaluating electrolyte additives for lithium-ion cells: a new figure of merit approach *J. Power Sources.* **365** 201–9
[44] Logan E R *et al* 2019 Electrolyte development for high-performance li-ion cells: additives, solvents, and agreement with a generalized molecular model *Electrochem. Soc. Interface.* **28** 49–53
[45] Yang J *et al* 2019 Structural underpinnings of cathode protection by in situ generated lithium oxyfluorophosphates *J. Power Sources.* **438** 227039
[46] Hall D S *et al* 2018 Exploring classes of co-solvents for fast-charging lithium-ion cells *J. Electrochem. Soc.* **165** A2365–73
[47] Xu K 2014 Electrolytes and interphases in Li-ion batteries and beyond *Chem. Rev.* **114** 11503–618
[48] Shkrob I A *et al* 2019 Facile in situ syntheses of cathode protective electrolyte additives for high energy density Li-ion cells *Chem. Mater.* **31** 2459–68
[49] Li W *et al* 2019 Ethylene carbonate-free electrolytes for high-nickel layered oxide cathodes in lithium-ion batteries *Adv. Energy Mater.* **9** 1901152
[50] Cho Y-G *et al* 2014 Nitrile-assistant eutectic electrolytes for cryogenic operation of lithium ion batteries at fast charges and discharges *Energy Environ. Sci.* **7** 1737–43

[51] Liu X *et al* 2020 Probing the thermal-driven structural and chemical degradation of Ni-rich layered cathodes by Co/Mn exchange *J. Am. Chem. Soc.* **142** 19745–53
[52] Balaish M, Kraytsberg A and Ein-Eli Y 2014 A critical review on lithium–air battery electrolytes *Phys. Chem. Chem. Phys.* **16** 2801–22
[53] Aurbach D *et al* 2000 Prototype systems for rechargeable magnesium batteries *Nature.* **407** 724–27
[54] Lau K-C *et al* 2019 Widening electrochemical window of Mg salt by weakly coordinating perfluoroalkoxyaluminate anion for Mg battery electrolyte *J. Electrochem. Soc.* **166** A1510–19
[55] Kwon B J *et al* 2020 Intercalation of Mg into a few-layer phyllomanganate in nonaqueous electrolytes at room temperature *Chem. Mater.* **32** 6014–25
[56] Kwon B J *et al* 2019 Probing electrochemical Mg-ion activity in $MgCr_{2\text{-}x}V_xO_4$ spinel oxides *Chem. Mater.*
[57] Kwon B J *et al* 2020 High voltage Mg-ion battery cathode via a solid solution Cr–Mn spinel oxide *Chem. Mater.* **32** 6577–87
[58] Wang H *et al* 2020 Energy storing bricks for stationary PEDOT supercapacitors *Nat. Commun.* **11** 3882
[59] McDermott M J, Dwaraknath S S and Persson K A 2021 A graph-based network for predicting chemical reaction pathways in solid-state materials synthesis *Nat. Commun.* **12** 3097

Chapter 2

Electrochemical characterization and modeling for batteries

Jae Jin Kim, Hong Keun Kim, Jihyeon Gim, Seoung-Bum Son and Juhyun Song

2.1 Introduction

2.1.1 Electrochemistry in batteries

Electrochemical cells, including batteries, fuel cells and electrolyzers, involve the conversion of chemical energy into electrical energy, and the other way around. Figure 2.1 illustrates the electrochemical reaction in an electrochemical cell, comprising an electrolyte between two reactants (A, B) in each electrode for the chemical reaction of interest (e.g., $A + B \rightarrow C$). In the cell, the electrolyte is designed to allow passage of only ionic species, being an electronic insulator. Therefore, in order for the reaction in this example to proceed, the reactant A in the left electrode is oxidized to A^+ ions ($A \rightarrow A^+ + e^-$) at the electrode/electrolyte interface and travels through the electrolyte to the other side of the cell. At the same time, liberated electrons must be collected by a current collector and be delivered through an external circuit to the right electrode/electrolyte interface where the reduction of A^+ ions ($A^+ + e^- \rightarrow A$) and the subsequent chemical reaction ($A + B \rightarrow C$) finally occurs. This separate transport of charge species (ions and electrons) by redox reactions at the electrodes/electrolyte interfaces is distinguished from direct chemical reactions in a vessel. With this mechanism, an electrochemical cell can provide useful electrical work, i.e., direct current of electricity. The reversed flows of both ionic and electronic charges can be electrically driven, which is how rechargeable batteries operate during the charging process for repetitive use, as converting electrical energy into chemical energy. As the convention for batteries, electrodes are labeled by considering reactions in the discharging mode when chemical energy is transduced into electrical energy; an electrode where oxidation occurs is called an anode, while the other electrode where reduction occurs is called a cathode.

Electrodes in commercial batteries are often made in a porous structure (i.e., porous electrode) that consists of an active material, conducting agent, and binder,

doi:10.1088/978-0-7503-2682-7ch2

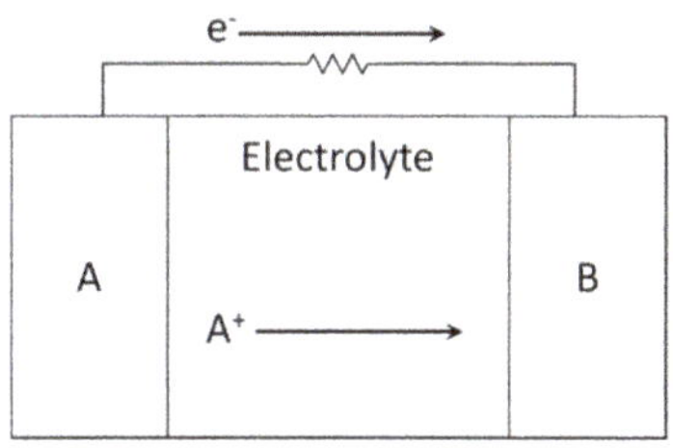

Figure 2.1. Simple schematic model of an electrochemical cell with A^+ ions as a major ionic species in an electrolyte.

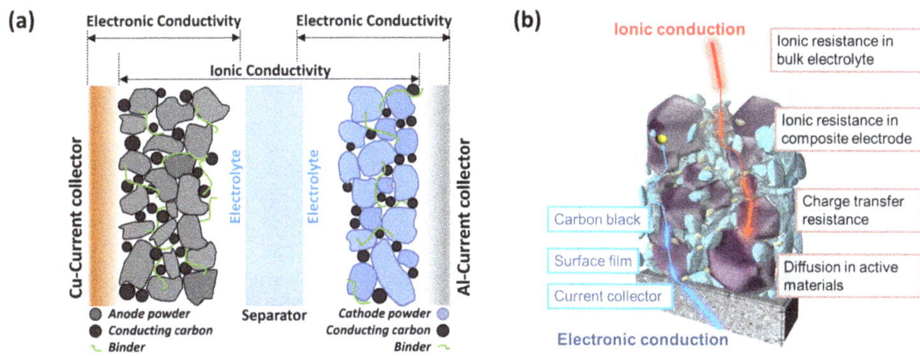

Figure 2.2. (a) Schematic illustration of the composite electrode indicating the predominant ionic/electronic conducting components in lithium-ion batteries and (b) the 3D cross-sectional illustration of the pathways for the ionic/electronic conduction. Reproduced with permission from [8]. Copyright (2016) with permission of Springer.

as shown in figure 2.2(a). The active material is one that actually takes lithium ions in and out in battery operation. Typical cathode materials for Li ion batteries include $LiCoO_2$, $LiFePO_4$, $LiNi_xMn_yCo_{(1-x-y)}O_2$, and $LiNi_xAl_yCo_{(1-x-y)}O_2$, where x and y vary depending on applications and cell manufacturers [1, 2]. For anode, graphite is a dominant active material, while silicon and $Li_4Ti_5O_{12}$ are used for special applications [3, 4]. Multiple active materials are occasionally employed, aiming to enhance performance [5, 6]. The conducting agent provides a pathway for electric conduction. Typically, carbon black is used for conducting agent, but it can be omitted when an active material has sufficient electric conductivity. Binder is a sticky polymeric material that provides mechanical stability of porous structure. Styrene butadiene (SBR) and polyvinylidene fluoride (PVDF) are typical binders for anode and cathode, respectively [6]. Lastly, the pore space is filled by electrolyte which provides an ionic conduction path. Electrolyte is usually a mixture of ethylene, dimethyl, diethyl, and ethyl-methyl carbonates (EC, DMC, DEC, and EMC, respectively) with $LiPF_6$ as a salt [7]. The porous electrodes are advantageous in power capability by providing: (1) faster paths for both ionic and electronic charge transport throughout the electrode; and (2) larger surface area for electrochemical reaction, as shown in figure 2.2(b) [8]. The pore space also accommodates expansion and contraction of the active material during battery operation, which helps a battery to have longer lifetime. Both cathode and anode porous electrodes are coated on a current collector (in general, Cu for anode and Al for cathode) where the electric current is collected from the porous electrode and directed to the battery terminal. The anode and cathode are physically separated by a separator which also has a porous structure where the electrolyte is filling the pore space.

The performance of electrochemical energy conversion cells is determined by the thermodynamic driving force of the redox reactions, as well as their reaction kinetics. Here, the driving force is the difference between the standard Gibbs free energy of products and reactants in the overall electrochemical reactions including both anode and cathode (i.e., full-cell reaction), that is:

$$\Delta G_r^\circ = \Sigma \Delta G_f^\circ(\text{products}) - \Sigma \Delta G_f^\circ(\text{reactants}) \tag{2.1}$$

And, ideally, this decrease in chemical energy due to the chemical reaction occurring in the cell is transduced to electrical energy conveyed to the external circuit, which can be expressed as

$$\Delta G_r^\circ = -nFE^\circ \tag{2.2}$$

Here, n is the charge number of the mobile ionic species, F the Faraday constant (96 485 C mol^{-1}, the elementary charge per mole of electrons), and E° the difference in standard electrical potential of electrodes, in other words, the equilibrium voltage or open circuit voltage (OCV), measured in volts (V). This relation highlights how internal transport of ionic species and corresponding chemical reaction, driven by chemical driving force, is connected to external flow of electrons and resultant electrical work, driven by electrostatic driving force.

The cell voltage under operation differs from the equilibrium voltage by overpotential, which is the voltage drop required to sustain the current through the various kinetic mechanisms in batteries that span multiple scales in time and length [9]. In battery systems, at the shortest time scale, under microseconds, the electronic conduction puts an Ohmic resistance to the cell. Then between microseconds to milliseconds, via the ionic migration, the double layer throughout the porous electrode is charged progressively starting near the separator. In a few seconds, the charge transfer reaction at the porous electrode–electrolyte interface gradually takes over, which could be affected by any passivated layers on the interface (e.g., solid–electrolyte interphase, SEI). Later, in tens of seconds, the concentration polarization develops in the electrolyte along the thickness of the porous electrode. Lastly, the solid-state diffusion in the active material particles plays a role with a time constant between tens of minutes and hours depending on particle size and diffusivity. In addition to a wide range of time-scales for reaction kinetics, electrochemistry in battery systems involves diverse phenomena that occur across various length scales. For example, Li ion batteries require controlled migration of ions, electrons, and molecules through crystal lattice (~angstroms), particles (~nanometers), porous/composite electrode (~micrometers), and unit cell structure (~millimeters). As battery-powered applications become more complex and bigger in size, the significance of the multiscale and comprehensive framework for better understanding and control of battery behavior is drawing more attention [10].

In this context, synergistic integration of modeling and characterization is essential because it can provide more holistic understanding of complex electrochemical reaction mechanisms in batteries by validating each other, bridging their fundamental limits to time and space, and uncovering unknown region with more valid prediction. With this reason, both electrochemical modeling and experimental characterization techniques for batteries are introduced and discussed in this chapter.

2.1.2 Frequently used parameters

Practical parameters that are widely used to describe and/or evaluate the behavior of batteries are listed here.

- Theoretical capacity—the theoretical capacity of the material can be calculated using an equation $Q_{\text{theoretical}} = nF/M_w$, where n is number of charge carrier, F is the Faraday constant, and M_w is molecular weight of the material.
- Specific capacity—the gravimetric (mAh g^{-1}) and volumetric (mAh cm^{-3}) capacity are generally used.
- Depth of discharge (DoD)—the percentage of the capacity removed from the fully charged battery.
- Energy density—the amount of energy that can be stored per unit volume.
- Power density—the amount of power that can be stored per unit volume.
- State of charge (SoC)—the current value of the fraction of the maximum capacity that is still available to be supplied.
- State of health (SoH)—the ratio of the real capacity to its initial capacity.
- Operating voltage—the electrical potential difference between two electrodes during battery charging and discharging; specifically, the output voltage for the latter.
- Voltage hysteresis—the difference in voltage values between the positive-going and the negative-going input in voltage profile.
- Cycle life—the number of complete discharge/charge cycles of a secondary battery before its capacity degrades under certain limit (e.g. 80% of the original capacity).
- Shelf life—the ability of a battery to maintain its cycle life capability during storage. It can be influenced by cell design, ambient temperature, etc.

2.2 Electrochemical models for batteries

2.2.1 Classification of electrochemical models

Modeling efforts in the battery community can be broadly classified into two groups: the first-principle calculations and the electrochemical models. The first-principle calculations predict material-level properties based on the density functional theory (DFT) [11, 12], classical molecular dynamic (MD) simulations [13, 14], and kinetic Monte Carlo (kMC) [15]. They have been widely used to estimate atomic/electronic structural properties, reaction kinetics, diffusivity and interface behaviors. On the other hand, the electrochemical models describe the electric response of battery systems based on continuum thermodynamic and kinetic theories [16–18]. In this chapter, we focus on the electrochemical models as they can be more directly implemented to analyze experimental characterizations. There are various electrochemical modeling approaches for Li-ion batteries, which can be again grouped into two: empirical models and physics-based models.

Typically, empirical models extract a function with model parameters, or a trained model based on experimental calibration. These models are widely used in the system-level development due to their simplicity and low computational cost. The equivalent circuit model (ECM) [19–21], one of the representative empirical models, assumes a battery as a simple circuit using resistors and capacitors. To implement the dynamic behavior of the battery, first-order resistance-capacitance (1-RC) and second-order resistance-capacitance (2-RC) are widely used as ECMs. Since the ECM is sufficiently simple and reliable, this model has been applied to battery management system (BMS) for real-time control applications. However, empirical models are not able to provide insight on underlying physical phenomena inside battery. In addition, when formulated with limited experimental data, they may lead inaccurate prediction of battery dynamics.

Physics-based models are, on the other hand, based on mathematical description of electrochemical kinetics and transport phenomena. Thus, these models can capture the non-linear characteristics of Li-ion batteries and provide detailed information that is difficult to be measured by an experimental method during operation. The pseudo-two-dimensional (P2D) model [22–24] and the single particle model (SPM) [25, 26] are the most popular physics-based models. They have been extensively used in the performance and life predictions due to their relatively good accuracy. However, since these models solve the governing equation numerically, they are more complex and require longer simulation times than empirical models. Therefore, it is unsuitable for application to on-board systems that require real-time control.

Nowadays, there are efforts to improve battery models by compensating for the limitations of each model. For example, some researchers have proposed modified ECMs which can improve accuracy under various operating conditions by linking electrical circuit dynamics with electrochemical processes in batteries [27, 28]. To reduce the computational time, the physics-based models have been proposed to simplify their equations to achieve robust BMS which can estimate the performance and lifetime prediction [29–31]. As such, many researchers are moving toward developing simple and reliable models to get a better understanding of the battery characteristics as well as predicting battery responses.

2.2.2 Equivalent circuit models

The equivalent circuit models (ECMs) approximate dynamic response of batteries using electric circuit elements such as resistor, capacitor, and inductor. The ECMs are typically simpler than other modeling approaches by lumping electrochemical processes and other physical mechanisms, which are usually nonlinear, into linear circuit elements. Nevertheless, the ECMs are able to approximate dominant processes in battery response so that they are widely adopted in both analyzing experimental data in lab-scale evaluation and simulating battery performance with limited computational resources, for instance, in BMS for electronic devices and electric vehicles.

The ECMs could either be purely empirical or be based on electrochemical processes taking place in batteries. Empirical ECMs are typically a combination of multiple RC circuits, each of which have a resistor (R) and a capacitor (C) in parallel [19, 32]. The number and arrangement of RC circuits are determined by looking at the relaxation behavior or the impedance response of the battery to reflect the time constants observed from experiments. Rather than taking on a predefined circuit model, it is also possible to obtain the distribution of relaxation times (DRT) [33–37] or the distribution of diffusion times (DDT) [38] for an infinite array of RC circuits or diffusion elements, respectively, from experimental data.

On the other hand, there are ECMs derived from theoretical models describing the kinetic mechanisms in batteries, as discussed in the Introduction. These types of ECMs typically have separate parts assigned to battery cell components such as anode, cathode, electrolytes, and current collectors [20, 21, 39]. The electrode parts then consist of circuit elements representing the charge transfer reaction, double layer, as well as diffusion in active particles [40, 41]. To model the dynamics of porous electrodes, transmission-line type ECMs are also used [34, 42–44]. By using physics-based ECMs, it is possible to assign the electric potentials and currents to physically meaningful variables, which allows more informed decisions in battery design and control. For example, the potential drop across the anode could be an indicator of lithium-plating during battery charging, and one can optimize charging protocol to keep the potential drop under a marginal value.

The ECMs and their circuit elements will be further discussed later in section 2.4.2, with graphical representation in electrochemical impedance spectroscopy (EIS).

2.2.3 Physics-based models

2.2.3.1 Pseudo-two-dimensional (P2D) model

The pseudo-two-dimensional (P2D) model developed by Newman and coworkers [18, 22–24, 45] is one of the most well-known physics-based models for lithium-ion batteries. The P2D model describes dynamic behavior of lithium-ion battery at micro and meso-scales based on the porous-electrode theory, the concentrated-solution theory, the modified Ohm's law and charge transfer kinetics (refer to [18, 22–24] and [45] for detailed information).

A schematic of a lithium-ion battery in the P2D model is shown in figure 2.3. This model includes (1) a 1D electrode domain across anode, separator, and cathode, and (2) a spherical particle domain representing the active material in the electrodes. As explained in the Introduction, both composite electrodes are coated on thin metal current collectors and liquid electrolyte is filled in void pores of the solid composite matrix. First, to explain mass transport in liquid electrolyte and charge balance across solid composite electrodes and liquid electrolyte, a porous electrode model is

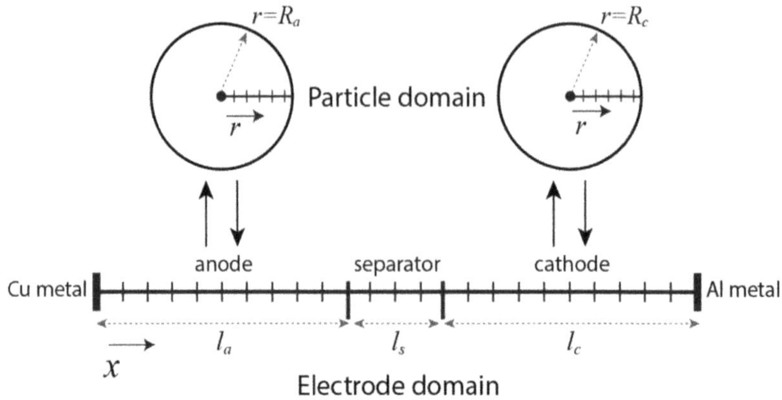

Figure 2.3. Schematic of a pseudo two-dimensional (P2D) model: spherical particle model for the particle domain and porous electrode sub-model for the electrode domain.

adopted as a sub-model for the 1D electrode domain. It is assumed that each composite electrode has homogeneous porous media across the x-direction and there is no gas generation or volume expansion/contraction during operation. The solution variables in this 1D electrode domain are solid composite electrodes potentials ($\phi_{s,a}$, $\phi_{s,c}$), liquid electrolyte potential (ϕ_e), and lithium concentration in the liquid electrolyte (c_e). Note that subscripts a and c mean anode and cathode electrodes, respectively. On the other hand, a spherical particle model is employed as a sub-model for the electrode particle domain. In this sub-model, the active material particles for both electrodes are assumed to be a perfect sphere, and the computational mesh are generated from center to surface as a spherical coordinate with r-direction. The electrochemical phenomena at the interfaces between particle surface and electrolyte are solved by using Butler–Volmer kinetics equation. The mass transport of lithium in the active material by intercalation or de-intercalation is described by Fick's diffusion law in this sub-model. The main solution variables are the local current density at the particle surface (i'') and the lithium concentration within a solid particle phase ($c_{s,a}$, $c_{s,c}$) in the particle domain. Note that the solution variables in particle and electrode domain models are solved simultaneously in a coupled manner, and the information is exchanged between domains. All governing equations and boundary conditions for P2D model are summarized in table 2.1.

2.2.3.2 P2D model coupled with lumped thermal model

The P2D model discussed in the previous section is an isothermal electrochemical model assuming constant cell temperature during operation. In other words, its approach does not address changes in electrochemistry due to temperature variation in a battery cell. Therefore, it shows relatively good agreement with experimental values under low C-rate conditions, but it could no longer explain experimentally observed phenomena, for example, when C-rate and consequently cell temperature increase. To address this shortcoming, Wang and coworkers have focused on an electrochemical–thermal coupled model for lithium-ion batteries and show good

Table 2.1. Summary of governing equations and boundary conditions for pseudo-two-dimensional (P2D) model [18, 45]. Symbols used in the equations are listed in appendix A.

Governing equations	Boundary conditions
– Spherical particle model	
• Mass transport in solid phase (active material)	
$\frac{\partial c_{s,i}}{\partial t} = \frac{D_s}{r^2}\frac{\partial}{\partial r}\left(r^2\frac{\partial c_{s,i}}{\partial r}\right),\ (i = a, c)$	$\frac{\partial c_{s,i}}{\partial r}\Big\vert_{r=0} = 0,\ \frac{\partial c_{s,i}}{\partial r}\Big\vert_{r=R} = -\frac{i''}{D_s F}$
• Charge transfer kinetic (Bulter–Volmer equation)	
$i'' = i_0''\left\{\exp\left[\frac{\alpha_a F}{RT}\eta\right] - \exp\left[\frac{\alpha_c F}{RT}\eta\right]\right\}$ where $i'' = Fk_i(c_e)^{\alpha_a}(c_{s,\max} - c_{s/e})^{\alpha_a}(c_{s/e})^{\alpha_c}$ and $\eta = \phi_s - \phi_e - U$	
– Porous electrode model	
• Mass transport in liquid phase (in liquid electrolytes)	
$\frac{\partial(\varepsilon_e c_e)}{\partial t} = \frac{\partial}{\partial x}\left(D_e^{\mathrm{eff}}\frac{\partial}{\partial x}c_e\right) + \frac{\left(1 - t_+^0\right)}{F}a_s i'' - \frac{\mathbf{i}_e \cdot \nabla t_+^0}{F}$	$\frac{\partial c_e}{\partial x}\Big\vert_{x=0} = 0,\ \frac{\partial c_e}{\partial x}\Big\vert_{x=l_a+l_s+l_c} = 0$
• Charge conservation in liquid phase (in liquid electrolytes)	
$\frac{\partial}{\partial x}\left(\kappa^{\mathrm{eff}}\frac{\partial}{\partial x}\phi_e\right) + \frac{\partial}{\partial x}\left(\kappa_D^{\mathrm{eff}}\frac{\partial}{\partial x}\ln c_e\right) + a_s i'' = 0$	$\frac{\partial \phi_e}{\partial x}\Big\vert_{x=0} = 0,\ \frac{\partial \phi_e}{\partial x}\Big\vert_{x=l_a+l_s+l_c} = 0$
• Charge conservation in solid phase (in solid composite matrix)	
$\frac{\partial}{\partial x}\left(\sigma^{\mathrm{eff}}\frac{\partial}{\partial x}\phi_{s,i}\right) - a_s i'' = 0,\ (i = a, c)$	$\phi_{s,a}\Big\vert_{x=0} = 0,\ \frac{\partial \phi_{s,a}}{\partial x}\Big\vert_{x=l_a} = 0,\ \frac{\partial \phi_{s,c}}{\partial x}\Big\vert_{x=l_a+l_s} = 0,\ \phi_{s,c}\Big\vert_{x=l_a+l_s+l_c} = V_{\mathrm{cell}}$

agreement between modeling and experimental data under various conditions [46, 47]. In their approaches, the P2D type electrochemical model is coupled with lumped thermal model which assumes uniform temperature over the cell volume across the unit-cell.

$$\frac{d(\rho c_p T)}{dt} = hA_s(T - T_{\text{air}}) + q_{\text{tot}},\ q_{\text{tot}} = (q_{\text{rxn}} + q_{\text{ohm}} + q_{\text{rev}}) \tag{2.3}$$

where h is the convection coefficient, A_s is the cooling surface area, T_{air} is the ambient temperature. The total heat generation q_{tot} is determined by integration of the reaction heat (q_{rxn}), the ohmic heat across the electrode (q_{ohm}), and the reversible entropic heat (q_{rev}) as given below.

$$q_{\text{rxn}} = a_s i_r^{''}(\phi_s - \phi_e - U) \tag{2.4}$$

$$q_{\text{ohm}} = \sigma^{\text{eff}}\nabla\phi_s \cdot \nabla\phi_s + \left(\kappa^{\text{eff}}\nabla\phi_e \cdot \nabla\phi_e + \kappa_D^{\text{eff}}\nabla\ln c_e \cdot \nabla\phi_e\right) \tag{2.5}$$

$$q_{\text{rev}} = a_s i_r^{''}\left(T\frac{\partial U}{\partial T}\right) \tag{2.6}$$

To connect the lumped thermal model with P2D model, temperature-dependence of some of the physicochemical properties, such as the diffusion coefficient, conductivity of electrolyte, kinetic reaction rate, are adopted as the Arrhenius-type equation.

$$\Phi = \Phi_{\text{ref}}\exp\left[\frac{E_a}{R}\left(\frac{1}{T_{\text{ref}}} - \frac{1}{T}\right)\right] \tag{2.7}$$

where Φ is a general physiochemical property, Φ_{ref} is the property value at the reference temperature (T_{ref}), and E_a is activation energy for Φ.

2.2.3.3 Single particle model (SP model)

The P2D model has the disadvantage of long simulation time due to large number of non-linear equations. To reduce its complexity of the full electrochemical model, the reduced-order models (ROMs), which use simplified equations or reduce the number of equations by applying several assumptions, have been suggested. One of the most commonly used ROMs is the single particle (SP) model [25, 26, 30], which describes each electrode as a single spherical particle, as depicted in figure 2.4. The main solution variables for the SP model are lithium concentration in solid particles and molar ionic fluxes generated by electrochemical reaction, which can be determined using the same equations of the P2D model in table 2.1. This model is based on assumptions that electrolyte concentration is constant and there are no spatial variations of ionic current in the electrolyte. It means that this model accounts for only diffusion transport in solid phase and charge transfer kinetics at the interface between electrode and electrolyte, ignoring mass transport in the electrolyte. Therefore, the SP model has the advantage of reduced calculation time but is only

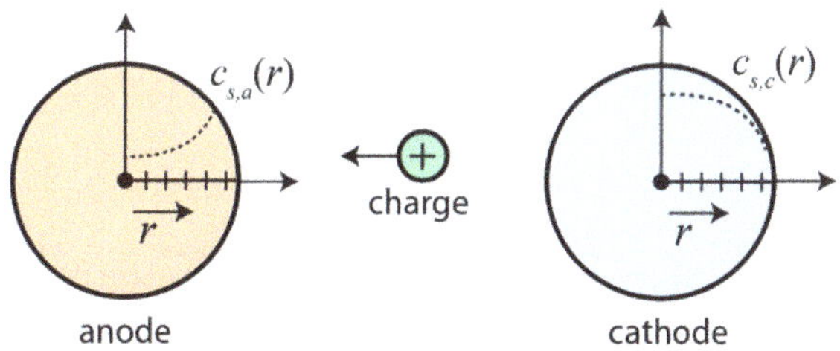

Figure 2.4. Schematic of the single particle (SP) model where each electrode is represented by a sphere. Similar to the pseudo-two-dimensional (P2D) model, the SP model also describes the lithium concentration inside the active material having a gradient from the surface to the center as shown by the dotted lines. However, the SP model neglects the non-uniform behavior of the concentration of lithium ions and potentials in an electrolyte.

valid for low-to-medium current scenarios when spatial imbalance in the electrolyte is negligible.

2.2.3.4 Multi-dimensional (2-D or 3-D) models

The physics-based models discussed above have relatively good accuracy in the case of small-format cells where spatial imbalance of potential, lithium-ion concentration, and temperature can be negligible. However, as a cell size increases, the non-uniformity within a cell also increases during operation. For example, current flow from the current collectors or concentrated current near the tabs can lead to potential distribution and temperature variation, which, in turn, can result in localized changes in the physical and chemical properties, and eventually cell performance. Thus, there are limitations for predicting the large-format cells with the P2D model or the SP model.

Various model approaches, which take multi-dimensional method to overcome these limitations, have been developed to describe the electrical-electrochemical-thermal coupled behaviors of large lithium-ion batteries. In figure 2.5(a), Kwon *et al* [48] and Kim *et al* [49] proposed the Newman, Tiedemann, Gu, and Kim (NTGK) model, which is a two-dimensional semi-empirical model. The NTGK model uses the relationship between the current density (j) and the two potential values at positive or negative current collectors (ϕ_+ and ϕ_-) as below.

$$\nabla^2\phi_+ = -R_+ j \tag{2.8}$$

$$\nabla^2\phi_- = +R_- j \tag{2.9}$$

$$j = a\Upsilon\left[U - (\phi_+ - \phi_-)\right] \tag{2.10}$$

where R_+ and R_- are electrical resistance of current collectors, a is the specific area of the electrodes and two parameters, Υ and U, are functions of DoD and temperature. Note that Υ and U can be determined by the curve-fitting on the voltage–current response of experimental data. This model is useful for solving the non-uniform electrical potential, SoC, and heat distribution in the large-format batteries (refer to [48] and [49] for modeling examples). However, this model has

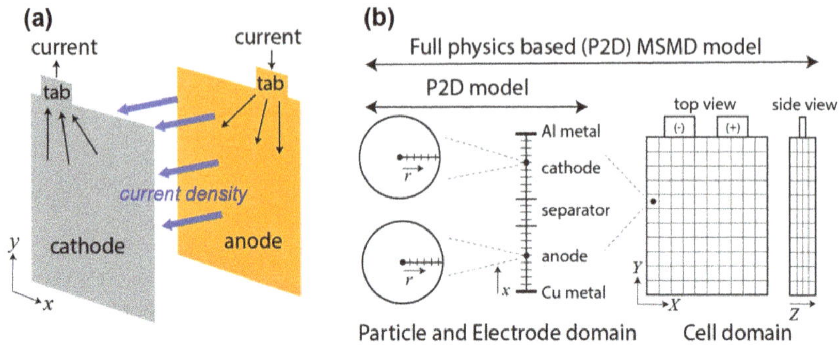

Figure 2.5. Schematics of (a) the Newman, Tiedemann, Gu, and Kim (NTGK) model and (b) the multi-scale multi-dimensional (MSMD) model based on P2D model.

limitations in predicting the dynamic response of a battery when Υ and U are obtained from insufficient experimental data.

The multi-scale multi-dimensional (MSMD) model was proposed by Kim *et al* [50], and other researchers have employed this model to perform large-format battery simulations [51–54]. As shown in figure 2.5(b), the MSMD model has a hierarchical structure and independent sub-models to describe the phenomena in the particle, electrode and cell domains. At the particle and electrode domains as lower levels, the simplified ROMs or P2D model can be adopted to solve the Li^+ transport, electrochemical reaction, and charge conservation at the unit-cell level. Then, the solution variables calculated in these domains are transferred to the upper level, i.e. the cell domain, to solve the electric potentials in the current collectors and temperature distribution over the cell volume. This model enables one not only to handle the cell shapes such as a pouch, cylindrical and prismatic cells but also to investigate the effect of tab location. However, the number of computational nodes of the MSMD model is large due to full resolution from particle to cell level resulting in long calculation time. Therefore, this model has limitation for simulations that require long operation times, such as life-time prediction. The NTGK model and the MSMD model are fully implemented in ANSYS Fluent in a battery module. There are also other commercial software tools such as COMSOL, Star-CCM+ and Battery Design Studio, all of which include various electrochemical battery models.

2.2.4 Models in battery characterization

Understanding electrochemical models helps us identify which parameters need to be evaluated in characterizing batteries, and the model sensitivity informs us how to estimate them. For example, the traditional P2D model tells us some of the critical parameters determining the battery behavior, including the exchange current density (i_0'') and the lithium diffusivity (D_s) in active materials. Also, the diffusivity determines the battery response in long-time scales, inversely implying that the diffusivity can be estimated by looking at long-time scale response. Furthermore, models are a useful tool to extrapolate understanding and insight from one specific characterization to different conditions, for example, various battery designs and load situations.

Characterizing batteries is essentially estimating the parameters in electrochemical models. Broadly speaking, there are two approaches for this task. The first approach is to fit the models to experimental data. In this approach, a cost function is defined to quantify the difference between the model prediction and the experimental data. There are various types of cost functions, but the most commonly used one is the sum of weighted square errors as shown in equation (2.11). Then we find the best-fit model parameters by minimizing the cost function. Optimization tools are available in various commercial software packages or open-source libraries.

$$\text{cost} = \sum_{i=1}^{n} w_i (y_{\text{mod},\, i} - y_{\text{exp},\, i})^2 \tag{2.11}$$

In DC characterization methods, y is typically a cell voltage or an electrical current values under a known current or voltage profile, respectively, which are collected for n number of data. In AC methods, y is usually impedance including both real and imaginary components, though sometimes only one component is used. In impedance literature, this approach is also called complex nonlinear least squares (CNLS) fitting [37, 55, 56]. And, w_i is the weighting factor for the ith error component. For both DC and AC methods, one of the common weighting schemes is to scale the error component by the corresponding experimental value, $y_{\text{exp},\, i}$.

Another approach to estimate the model parameters for battery characterization is to reduce the models to an explicit algebraic equation that can be populated from a certain part of the experimental data. In general, such a reduced equation can be expressed by

$$\theta_j = f\left(y_{\text{exp},\, k},\, y_{\text{exp},\, l},\, \ldots;\, \theta_m,\, \theta_n,\, \ldots\right) \tag{2.12}$$

where θ_j is a parameter to estimate. The $y_{\text{exp},\, k}$, $y_{\text{exp},\, l}$ and so on are certain components of experimental data, while the θ_m, θ_n and so on are some of known parameters. There are various methods to reduce models, which are typically combined to derive a reduced equation. Though it really depends on which parameters to estimate and which data are available, one of the most common and effective methods is to exploit the separation of time scales discussed in the Introduction and to focus on a specific time regime where only one or a few relevant kinetic mechanisms dominate the battery dynamics. For example, to estimate the lithium-ion diffusivity in active materials, one can only look at the relaxation after a sufficiently long time where most of the other kinetic mechanisms reach their steady state. Reducing the models can be also achieved by assuming ideal kinetic constitute relationships or homogeneous rates throughout the porous electrodes (e.g., single-particle model). These approaches, however, should be used with caution only when the assumptions are valid in a given system.

2.3 DC electrochemical techniques

2.3.1 Open circuit voltage

Open circuit voltage (OCV) is the difference of electrical potential between the cathode and anode in the context of disconnected-circuit [57]. Because electrical potential of electrodes is stabilized without current flow, OCV value reflects the Gibbs free energy in a state of thermodynamic equilibrium. Knowing the OCV of a lithium-ion battery is, therefore, helpful to understand thermodynamic state of the lithium-ion active materials in the electrodes. In addition, OCV also provides the information regarding charge transfer in electrode materials, short-circuit occurrences, self-discharge, and SoC of the battery [57, 58]. On the other hand, the voltage during operation of the cell is deviated from this theoretical value, as the system moves into a non-equilibrium state with continuous flow of electrical current.

As an example, the voltage profile of graphite || $LiNi_xMn_yCo_{1-x-y}O_2$ cell in figure 2.6 shows how OCV and operating voltage vary during cell operation [59]. The overpotential between OCV (OCV_{CH} and OCV_{DCH}) and operating voltage (V_{cell_CH} and V_{cell_DCH}) is observed during both charging (CH) and discharging (DCH). In general, the overpotential in lithium-ion battery is observed due to various reasons such as cell aging, high-current rate, cell resistance, etc.

2.3.2 Conductivity

The electrical conduction is one of the most essential material properties that needs to be evaluated for most battery cell components, especially including solid active materials (cathode/anode). Under Ohm's law (equation (2.13)), the electrical current, I (amperes), through a conducting medium is proportional to the voltage, V (volts), where the resistance, R (ohms), is independent of the current as the constant of proportionality.

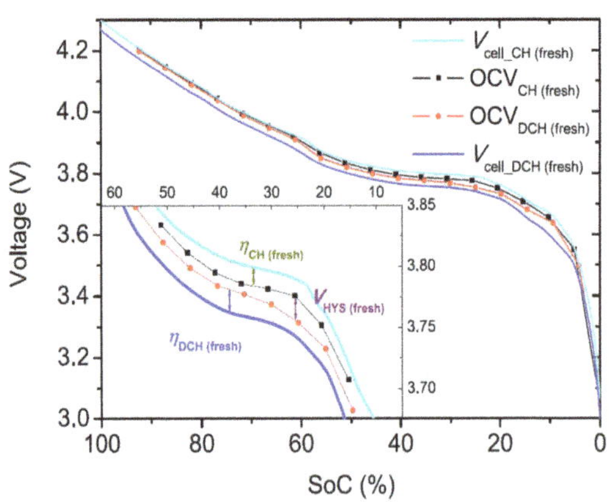

Figure 2.6. Charge (CH) and discharge (DCH) cell voltages of graphite || $LiNi_xMn_yCo_{1-x-y}O_2$ cells and OCV corresponding to a fresh cell. The inset shows how the overpotential and voltage hysteresis are obtained. Reproduced from [59], copyright (2019) with permission of Springer.

$$I = \frac{V}{R} \tag{2.13}$$

The resistance R is influenced by the specimen configuration while the resistivity ρ (Ohm-meters, Ω-m) is a material's intrinsic property, independent of the specimen geometry (equation (2.14)). Here, l is the distance between two end-points of voltage measurement, and A is the cross-sectional area of conducting path. The electrical conductivity σ is simply the reciprocal of the resistivity as expressed in equation (2.15).

$$\rho = \frac{RA}{l} = \frac{VA}{Il} \tag{2.14}$$

$$\sigma = \frac{1}{\rho} \tag{2.15}$$

In the aspect of physical phenomena in material or system, the electric current is the motion of electrically charged carriers induced by an externally applied electric field. The electronic conduction is indicative of the flow of electrons/holes, while the ionic conduction is related to a net motion of charged ions under given electrical bias. Both cathode and anode electrodes in batteries are required to ensure electronic and ionic conduction for efficient electrochemical reaction. The pathways of each conduction and corresponding contribution are illustrated in figure 2.2. The electronic conduction is related to the electronic charge transport/transfer passing through conducting agent carbons and active materials and from the current collector (aluminum or copper). The ionic conduction is directly influenced by the conductivity of electrolyte medium, charge transfer at the interface of electrolyte/active materials, and lastly the bulk solid diffusion of charge carriers in the active materials.

2.3.3 Transport and transference number

While the ionic conductivity and its measurement methods for the electrolyte simply provide information about the ability of ions to move from one electrode to another, those methods do not tell how the electric charge is carried within the electrolyte. Here, the transport number, t_+, is defined as the contribution of electric charge carried by the cation to overall electric charge. This is expressed by equation (2.16) as below.

$$t_+ = \frac{i_+}{i_+ + i_-} = \frac{i_+}{i} \tag{2.16}$$

where i, i_+, and i_- are total current, the partial currents carried by cations and anions, respectively [60]. The transport number can be readily obtained by various methods including alternating current impedance, DC polarization, Tubandt's method, Hittorf's method, and pulsed field gradient nuclear magnetic resonance (NMR) [61, 62].

However, application of transport number is only valid in dilute solution. Under the circumstances of concentrated solution, there can be a fast dynamic equilibria for ion association between cation Li^+ and anion A^-, generating [LiA], $[Li_2A]^+$, $[LiA_2]^-$, etc, and these reactions significantly influence the transport number. For example, when the cation forms a complex ion with multiple anions, the resultant complex ion could move to either cathode or anode depending on its electric charge. In this regard, experimentally achievable quantity is the difference between the fluxes of the cation containing species directed towards the cathode and anode, i.e. the net number of moles of the cation being transferred in one direction. Here, the transference number T_+ can be defined as the net number of faradays of charge carried by the cation associated species toward the cathode, during the passage of one faraday of charge across the cell. The most common experimental method to measure this type of transference number is called the Bruce–Vincent method [60]. The method utilizes symmetric Li/Li cell polarized by a small potential (around 10 mV) and measures initial and steady state current [63]. Then the transference number can be calculated by

$$T_+ = \frac{I^s(\Delta V - I^0 R_1^{\,0})}{I^0(\Delta V - I^S R_1^{\,S})} \tag{2.17}$$

where I^0 and I^s are initial and steady-state current, respectively; $R_1^{\,0}$ and $R_1^{\,S}$ are initial and steady state resistance, respectively; ΔV is the potential applied to the cell.

2.3.4 Linear sweep and cyclic voltammetry

Voltammetry is one of the most widely and commonly used techniques in the category of electroanalytical methods. The principle of voltammetry is to measure the current while the voltage of working electrode is varied with a scan/sweep rate, ν (mV s^{-1}). One can obtain the useful information from voltammetry technique including the oxidation/reduction reaction peaks, capacitive current, reversible/irreversible reaction, quantitative analysis of reaction rates and identification of species [64]. Based on the principle of voltammetry, one can extend it to cyclic voltammetry (CV), linear sweep voltammetry (LSV), chrono-amperometry (CA), chrono-potentiometry and stripping voltammetry (SV). Among them, CV and LSV have been widely used in the lithium-ion battery field to evaluate the active materials as well as electrolyte stability. Figure 2.7(a) illustrates the difference in voltage versus time profile for LSV and CV, indicating that CV is based on the principle of LSV with additional backward sweeping voltage and repeating cycles.

With LSV analysis, a cell may undergo the oxidation or reduction within the applied voltage range. Therefore, for instance, the electrochemical instability of electrolyte can be investigated by monitoring significant current surge at certain voltage. As a specific example, figure 2.7(b) shows one study for the electrochemical stability window of the electrolyte with or without additives. Here, the oxidative current peak at around 5.0 V (Li/Li^+) for the electrolyte with 2 wt% of vinylene carbonate (VC) additive could be indicative of worse electrochemical stability than the imide-based additives [65]. On the other hand, the factors affecting the shape of

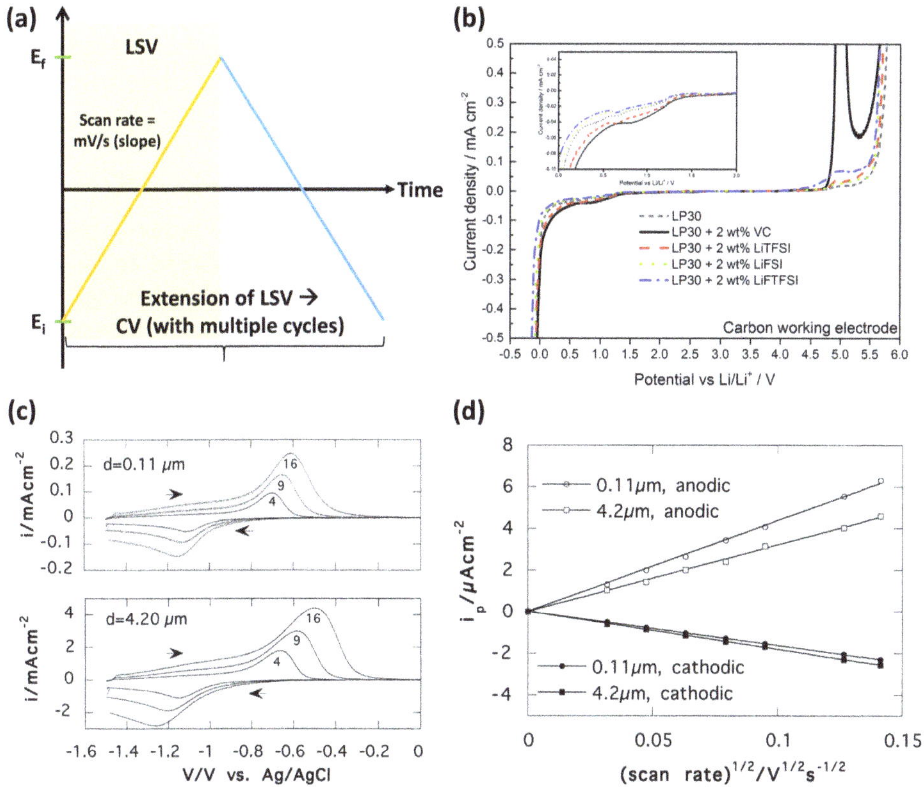

Figure 2.7. (a) Schematic illustration of linear sweep voltammetry (LSV) and cyclic voltammetry (CV) technique's profiles, (b) LSV profiles showing effect of additive introduction (vinylene carbonate and lithium imide salts) into graphite||$LiFePO_4$ cells. Reproduced from [65], copyright (2018) with permission from Elsevier. (c) CV results of two electrode thickness of TiO_2 electrode and (d) the corrected cathodic and anodic peak currents are plotted vs the square root of the scan rate. Reproduced with permission [67], copyright (1997) American Chemical Society.

CV-curves are a useful indicator to estimate the electrochemical performances, which are related to the particle size of active materials, electrolyte concentration, test temperature, electrode thickness and scan rate [64]. For example, as changing the sweep scan rate (ν), one can estimate the reasonable diffusion coefficients (D), by using the so-called Randles–Ševčík equation as below, of which different forms can be derived with consideration of electrode geometry and reaction reversibility [66].

$$i_p = 0.4463nFAC\left(\frac{nF\nu D}{RT}\right)^{1/2} \tag{2.18}$$

where i_p is a peak current, n is number of electrons transferred during the redox reaction of interest, A is electrode area, F is the Faraday constant, C is concentration of redox species, R is gas constant, and T is temperature. Figure 2.7(c) shows the results of CV-study with various scan rates (4, 9, and 16 mV s^{-1}) to investigate the kinetic characteristics of a TiO_2 based electrode as a function of electrode thickness.

The chemical diffusion coefficient for lithium ions in TiO_2 active materials can be calculated by using the slopes of the corrected cathodic/anodic peak current from figure 2.7(d) [67]. In that study, the thinner TiO_2 electrode was found to exhibit apparently higher lithium diffusion coefficient.

2.3.5 Constant current (galvanostatic) and potential (potentiostatic) method

Constant current (CC) method observes the voltage responses of batteries under CC condition. This is one of most frequently used methods to know the capacity, capacity reversibility, resistance, rate-capability of the electrodes. A value of the CC is generally defined by C-rate of a battery. A C-rate is a measure of the current rate at which a battery is completely charged with respect to its theoretical capacity. For example, a 1-C rate indicates the charge current that charges the entire battery in 1 h. In other words, for a battery with a capacity of 5 Ah, a 1-C rate equates to current of 5 A.

The period of each charging and discharging process with CC can be limited by either cutoff voltage or constant capacity. When cutoff voltage is applied, continuous charging and discharging is conducted with a given voltage range. This method allows one to achieve the quantity of electrical charge within the fixed voltage range over time (cycle counts). The period of (dis)charging also can be limited by constant capacity cutoff. The constant capacity cutoff is often used to test the anode materials that have a long flat voltage plateau, in order to avoid an undesired change in electric charge even by a small change in voltage.

Figure 2.8(a) shows the voltage response of graphite || $LiNi_{0.5}Mn_{0.3}Co_{0.2}O_2$ cell over time measured by CC method in the voltage range of 3 to 4.1V. The CC of 4 mA (1-C rate) was applied during the charging, and the CC of 2 mA (0.5-C rate) was applied during the discharging. In addition, plotting voltage response over capacity (figure 2.8(b)) is also often used to compare delivered capacity and voltage decay of cells from cycle to cycle. This voltage profile well reveals the capacity decrease and the overpotential increase of cells as cycle counts increase. For further understanding, the voltage profile (voltage over capacity with constant current) can be converted to the differential capacity curve (dQ/dV), as shown in figure 2.8(c). The dQ/dV curves are useful to track the characteristic features of the cell's

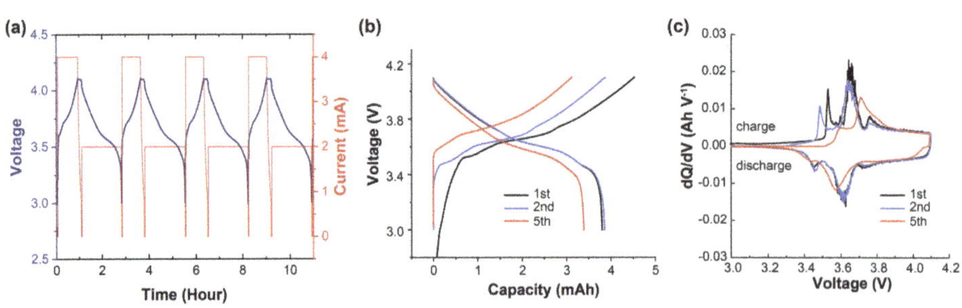

Figure 2.8. (a) The voltage response of graphite || $LiNi_{0.5}Mn_{0.3}Co_{0.2}O_2$ cell over time with CC condition. (b) The voltage response of graphite || $LiNi_{0.5}Mn_{0.3}Co_{0.2}O_2$ cell over capacity in specific cycle number. (c) The dQ/dV curve converted from figure 2.8(b).

electrochemical system over cycle life. For example, each peak in *dQ/dV* curve indicates phase change of electrode materials driven by electrochemical reactions. Identifying and comparing those peaks in *dQ/dV* curve enable us to better understand the underlying thermodynamics (e.g., electrochemical reaction associated with each peak) and kinetics (e.g., peak shift over cycle counts) of the cell system.

On the other hand, charging process with the constant voltage method forces a cell to be under a specific voltage value. As the constant voltage does not control the current, the method could bring unnecessarily high current to the cell depending on the applied voltage value. Thus, constant voltage is rarely used by itself [58], but is combined after constant current method. This particular method is termed the constant current—constant voltage (CCCV) method. Figure 2.9 shows an example of charging process with CCCV. The CC of 4 mA is applied until the cell reaches 4.1 V. Then the cell is held at 4.1 V until the current is stabilized at near zero. The current gradient during constant voltage is due to the Li-ion concentration gradient within the electrode materials. While the surface of electrode materials is fully charged after the cell reaches 4.1 V, the inside of electrode materials could not be fully charged, especially with high current. The CCCV method prevents this issue, as the constant voltage period allows entire electrode materials to reach close to thermodynamic equilibrium. The method also maximizes the energy storage capacity of batteries.

2.3.6 Galvanostatic/potentiostatic intermittent titration technique

With Weppner and Huggins's development of useful electroanalytical approaches for metallic alloys system, GITT (Galvanostatic Intermittent Titration Technique) and PITT (Potentiostatic Intermittent Titration Technique) were introduced in the 1970s and have been applied to study of the diffusion kinetics of guest ions in ion-insertion electrodes [57, 68, 69]. The controlled variables and measured one for both GITT and PITT are illustrated in figure 2.10, providing intuitive understanding for the difference of each other [70]. GITT is the constant current method measuring changes in OCV, while, in PITT, the constant voltage is applied at each potential

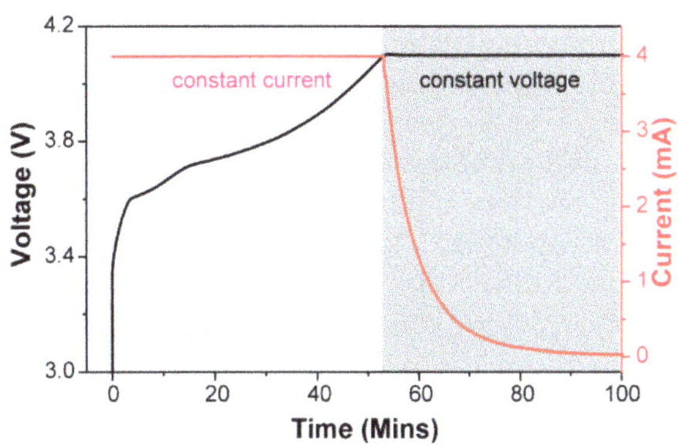

Figure 2.9. The voltage response of graphite || $LiNi_{0.5}Mn_{0.3}Co_{0.2}O_2$ cell during charging with CCCV method.

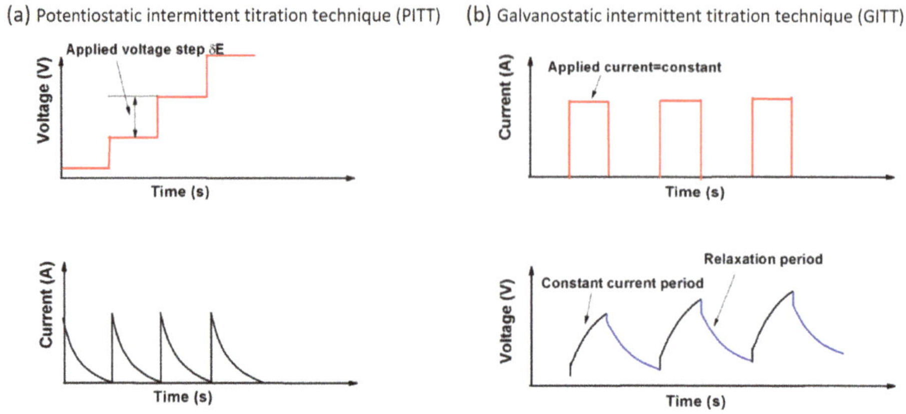

Figure 2.10. The controlled variables and the type of response of the relevant dependent variables are shown for (a) PITT and (b) GITT. Reproduced with permission from [70], copyright (2017) American Chemical Society.

step to measure current changes. From careful measurement of both GITT and PITT, diffusion coefficient can be calculated and the detailed derivation and experiment are discussed below.

The typical measurement of PITT is carried with the cell composed of metallic lithium as counter and reference electrodes, electrolyte and working electrode (cathode or anode materials of interest). After recording the OCV of the fabricated cell, the measurement starts with application of positive or negative potential increment (V_x) during a certain period of time, followed by relaxation time when the cell is switched off. This combination of a step increase or decrease in potential and relaxation phase is continued until the cell voltage reaches the desired upper and lower cut-off voltage for delithiation and lithiation, respectively. The accompanying current signal can be described with D_{PITT} in equation (2.19) [68, 71]

$$i = \frac{2FS(C_t - C_{t=0})D_{\mathrm{PITT}}}{L}\exp\left(-\frac{\pi^2 D_{\mathrm{PITT}} t}{4L^2}\right) \tag{2.19}$$

where F is the Faraday constant, S is electrode surface area, L is the characteristic length of the electrode material, and $(C_t - C_{t=0})$ is difference of the surface concentration of lithium ions at given time t from the one at the beginning of the potential pulse. Then, using equation (2.20), the diffusion coefficient D_{PITT} can be extracted by examining the slope of the $ln(i)$ curve as a function of time (t) for each potential pulse. As an example, Zheng Li *et al* reported the PITT study of $LiNi_{0.4}Mn_{0.4}Co_{0.2}O_2$, and extracted D_{PITT} values are shown in figure 2.11(c) and (d) [72].

$$D_{\mathrm{PITT}} = \frac{dln(i)}{dt}\frac{4L^2}{\pi^2} \tag{2.20}$$

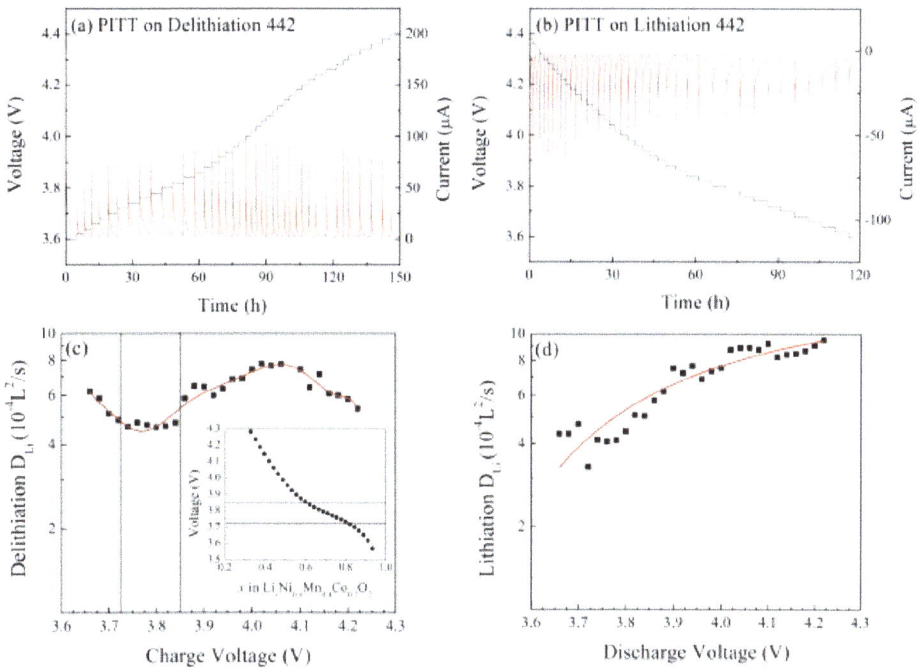

Figure 2.11. The voltage steps $V(t)$ and corresponding current decay $I(t)$ under each constant voltage and the calculated D_{PITT} upon charge (a, c) and discharge (b, d). Reproduced from [72], copyright (2014) with permission from Elsevier.

While sharing the same purpose of PITT technique, GITT is a constant current based method. The GITT consists of a series of current pulses and following relaxation time. During the latter, the OCV is monitored while no current passes through the cell. From GITT, one can calculate the diffusion coefficient of lithium, D_{GITT}, by using equation (2.21) [68].

$$D_{\mathrm{GITT}} = \frac{4}{\pi}\left(\frac{iV_m}{Z_A FS}\right)^2\left[\frac{(dE/d\delta)}{(dE/d\sqrt{t})}\right]^2 \tag{2.21}$$

where i is the current, V_m is the electrode's molar volume, Z_{A} is the charge number, F is the Faraday constant, S is the electrode/electrolyte contact area, $dE/d\delta$ is the slope of the coulometric titration curve, and $dE/d\sqrt{t}$ is the slope of the linearized plot of the potential during the current pulse of duration t(s). When the linearity of $dE/d\sqrt{t}$ is valid with the sufficiently small currents, the simplified equation, shown in equation (2.22), can be used.

$$D_{\mathrm{GITT}} = \frac{4}{\pi\tau}\left(\frac{n_m V_m}{S}\right)^2\left(\frac{\Delta E_S}{\Delta E_t}\right)^2 \tag{2.22}$$

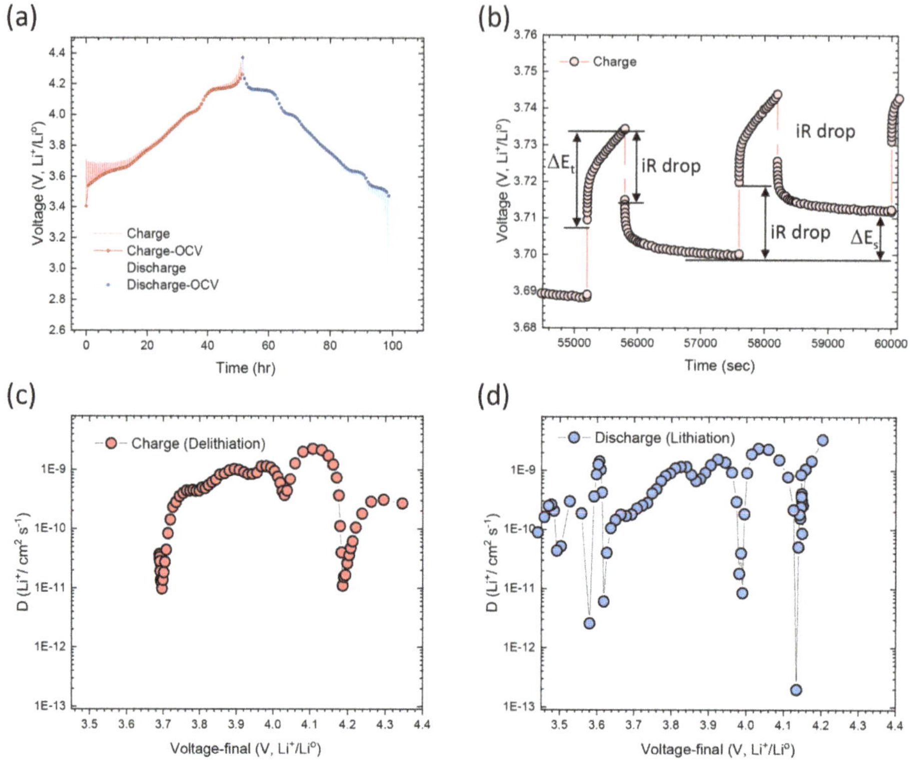

Figure 2.12. (a) Galvanostatic intermittent titration curve as a function of time during the charge and discharge pulses and (b) two charge steps. Solid lines with black dots in (a) represent OCV. The *iR* drop, ΔE_s and ΔE_t are shown in (b). The calculated chemical diffusion coefficient D_{GITT} during (c) delithiation and (d) lithiation for Li||$LiNiO_2$ cell.

where τ is the duration of the current pulse, n_m is the number of mole, ΔE_t is the voltage change during the constant current pulse, and ΔE_s is the steady-state voltage change. Here, ΔE_s and ΔE_t are carefully taken by eliminating *iR* drop indicated in figure 2.12(b). As a result of GITT measurement on Li||$LiNiO_2$ cell, figures 2.12(c) and (d) show the diffusion coefficient, D_{GITT}, calculated using the simplified equation (2.22).

Aurbach *et al* studied the comparison between GITT and PITT using thin composite graphite electrode and evaluated the limitations for each technique [73]. In their study, they concluded that GITT is superior compared to PITT due to the latter's insufficiently low amplitude of the voltage step. In addition, for GITT, short pulses with short enough relaxation time would enable one to decrease errors in calculating diffusion coefficient by eliminating parasitic background current contribution. However, one has to be discreet to choose which technique can be used in order to avoid an over/under-estimation of diffusion kinetic parameters.

2.3.7 Hybrid pulse power characterization (HPPC) test

The hybrid pulse power characterization (HPPC) test is a specifically designed test profile to elicit dynamic power capability of lithium-ion battery for an electric vehicle (EV) at given SoC and aging status [74, 75]. The HPPC test profile primarily includes discharge and regen current pulses which result in voltage drops to a cell. Then the resistance of the cell can be obtained by Ohm's law as in the equation below:

$$R_{\text{pulse}} = \frac{\Delta V}{\Delta I} \tag{2.23}$$

where R_{pulse} is the pulse resistance, ΔV is the voltage drop, and ΔI is the applied current. Figure 2.13(a) shows a schematic of voltage response to discharging current pulse. The voltage drop, ΔV is observed due to the pulse current ΔI during pulse length, Δt_p. As the HPPC test is intended to derive a cell resistance at specific SoC, the pulse length should be kept as short as ~ 10 s. This is to avoid solid-state diffusion in lithium-active materials, which could affect the surface stoichiometry of materials leading to changes in OCV of the cell. In a circumstance of OCV affected by longer pulse length, it is inappropriate to designate R_{pulse} to a single SoC point [75].

Figure 2.13(b) shows the area specific impedance (ASI) as a function of voltage obtained from HPPC profile. HPPC profiles were repeated three times at the cycle number of 7th, 30th, and 53rd. The pulse current was repeated to obtain the ASI values at every 10% of SoC from 10% to 90%. Detailed HPPC profiles used in literature can be found elsewhere [76, 77]. The achieved ASI profile provides the cell resistance at the specific SoC and voltage, enabling one to determine dynamic power capability of the cell over the device's usable voltage range.

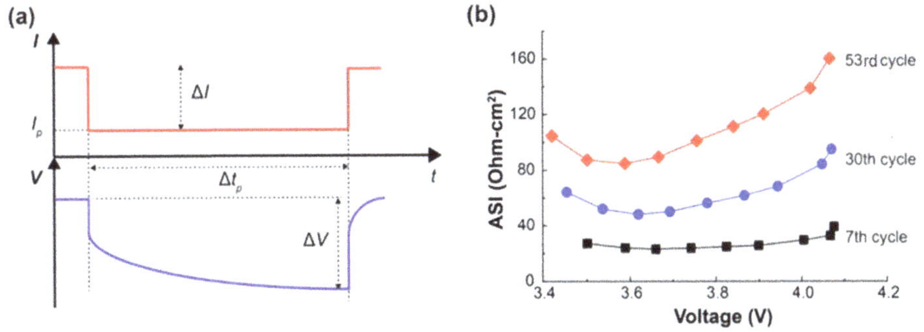

Figure 2.13. (a) Schematic of voltage response to pulse current. (b) The area specific impedance (ASI) as a function of cell voltage for graphite || $LiNi_{0.94}Co_{0.06}O_2$ cell.

2.4 AC electrochemical impedance spectroscopy

2.4.1 Principle of electrochemical impedance spectroscopy

Electrochemical impedance spectroscopy (EIS) has been considered a powerful tool to investigate various types of electrochemical systems including metal corrosion, batteries, photovoltaic cells, fuel cells, and electro/photocatalysts [78–81]. As an alternating current (AC) technique, EIS studies frequency dependent response of electrochemical systems, in principle, enabling one to differentiate between resistive and capacitive processes [82, 83]. This is based on their distinct response to alternating input signal; the former is independent on modulation frequency of input source, while the latter is inversely proportional. Contrary to most direct current (DC) techniques, which often impose large and consistent perturbation on the system, EIS utilizes a small magnitude modulation input and observes how the system responds at the vicinity of a steady state. Not only minimizing irreversible alteration of the system, therefore, EIS allows one to evaluate the response as a linearized current–potential behavior near equilibrium and, in turn, to examine kinetic processes with simplified theoretical consideration. Furthermore, in EIS, a small AC excitation with a wide range of frequency can be superimposed on the top of a fixed DC perturbation, linking steady-state characterization with a transient-state based one. This feature is especially useful for battery characterization, because, for example, redox reaction kinetics and charge/mass transport can be dependent on SoC of electrodes.

In typical EIS measurements, sinusoidal waveform of electrical voltage, $E(t) = E_0\sin(\omega t)$, with a small amplitude (E_0 ~ 10–50 mV) is applied at a controlled frequency. Here, ω is an angular frequency, which is defined as $\omega = 2\pi f$ with the conventional frequency, f, in Hz. The magnitude of AC voltage perturbation satisfying linearity assumption depends on the systems of interest and needs to be determined carefully. And, as aforementioned, AC perturbation can be added on the top of a fixed DC value; $E(t) = E_{\text{fixed}} + E_0\sin(\omega t)$. Then, corresponding current response, $I(t) = I_0\sin(\omega t + \varphi)$, is measured, which can be offset by a phase angle (φ). Here, the values of I_0, and especially φ depend on kinetic processes in the system, which will be further discussed later. Lastly, impedance is defined as a generalized concept of resistance in Ohm's law, $Z(\omega)=E(t)/I(t)$, in units of ohms (Ω). In most cases, impedance is expressed as a vector quantity in terms of complex notation ($j = -1$), as given in equation (2.24) and described in figure 2.14. The Z_{Re} and Z_{Im} represent the real and imaginary components of impedance, respectively. The magnitude or modulus of impedance is defined as $|Z| = \sqrt{(Z_{Re})^2 + (Z_{Im})^2}$, and the phase angle is given by $\tan\varphi = Z_{Im}/Z_{Re}$.

$$Z(\omega) = Z_{Re} + j \cdot Z_{Im} = |Z|(\cos\varphi + j \cdot \sin\varphi) \tag{2.24}$$

2.4.2 Equivalent circuit models for electrochemical systems

As mentioned in the preceding section in this chapter, equivalent circuit modeling has been found to be useful to analyze various physical phenomena observed in

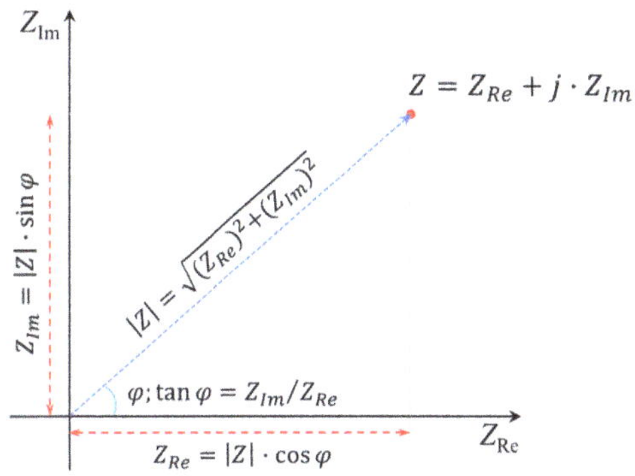

Figure 2.14. The graphic representation of an impedance vector.

electrochemical systems, by fitting a model to EIS data. This is based on the observation that components in an electrical circuit exhibit electrical behavior analogous to physical process in electrochemical systems. The fundamental circuit elements, frequently used for EIS data analysis, include resistor, capacitor, constant phase element, inductor, and diffusion element, as listed in table 2.2. It is worthwhile, however, to point out that physical relevancy of a circuit model needs to be carefully examined with supporting characterization of the system of interest. This is because there can exist many different circuits which could provide good fits to the one EIS data.

First, resistance is used to describe Faradaic processes at interfaces and some of non-Faradaic processes like transport of electronic/ionic charge species via conducting media, though the latter can be better treated with diffusion element in many cases. The primary examples include the polarization resistance (R_p)[1] associated with electron transfer at the electrode/electrolyte interface and the solution resistance (R_s) of a liquid electrolyte. The impedance of a pure resistor (Z_R) is simply its resistance (R) with $\varphi = 0$ and, in turn, is independent of frequency.

On the other hand, the non-Faradaic processes such as accumulation of charge species at interfaces can be modeled with capacitance. For example, the charging process occurred at the double-layer, which is formed as ions from electrolyte adsorb onto charged electrode surface, can be ideally represented as pure capacitance (C_d).[2] The impedance of a pure capacitor (Z_C) is inversely proportional to ω as $Z_C = 1/(j\omega C)$ with only an imaginary impedance component, and the phase difference (φ) between the voltage and the current is 90°.

The use of these elements can be exemplified at a simplest circuit shown in the inset of figure 2.15(a), called a Randles circuit [84], which is often employed to capture the basic description of the electrochemical reaction at the electrified electrode/electrolyte interface. Here, the parallel combination of resistive (R_p) and

[1] The polarization resistance (R_p) becomes the charge-transfer resistance (R_{ct}) at the standard electrode potential.

[2] The electrochemical double-layer capacitance associated with ideally polarizable liquid electrodes (e.g., mercury) can be considered a pure capacitor.

Table 2.2. Circuit elements used to model electrochemical impedance data.

Circuit element	Symbol	Mathematical expression
Resistor (R)		$Z_R = R$
Capacitor (C)	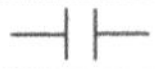	$Z_C = 1/(j\omega C)$
Constant Phase Element (CPE)		$Z_{CPE} = 1/((j\omega)^{\beta} Q)$
Inductor (L)		$Z_L = j\omega L$
Warburg diffusion (W)[a]	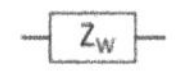	$Z_W = \frac{A_W}{\sqrt{\omega}}(1 - j)$ $A_W = \frac{RT}{n^2F^2A\sqrt{2}}\left(\frac{1}{C_O\sqrt{D_O}} + \frac{1}{C_R\sqrt{D_R}}\right)$

[a]Impedance only for the case of semi-infinite linear diffusion is included in the table.

capacitive (C_d) elements is introduced, given that the total current flow at the electrified interface is the sum of both Faradaic and non-Faradaic processes through the working interface. And, one more resistive element is included in series to represent the solution resistance (R_s); in practice, other contributions related to electrical lead or contact resistances can be combined. The total impedance of this equivalent circuit is given as [85]

$$Z(\omega) = R_S + \frac{R_P}{1 + R_P C_d(j\omega)} \tag{2.25}$$

Figure 2.15 shows two popular ways of visualizing EIS data: (a) the Nyquist plot (complex plane plot[3]) and (b) the Bode-modulus/-phase angle plots. Here, the simulated EIS data ($R_s = 10\ \Omega$, $R_p = 100\ \Omega$, $C_d = 100\ \mu$F for the Randles circuit) is used as an example. In the Nyquist plot, similar to figure 2.14, impedance is displayed as a vector; the Z_{Re} (Z') and Z_{Im} (Z'') for different ω values are plotted on the *x*- and *y*-axes, respectively. As the convention in electrochemistry community, negative values of Z_{Im} are plotted at the first quadrant of a graph.[4] This type of plot allows a quick assessment of physical characteristics of reaction kinetics occurring in the system, based on their distinctive graphical representation.[5] For example, in figure 2.15(a), a semicircle comes from a parallel combination of a resistor and a capacitor. However, without additional annotation, frequency dependence of impedance is hidden in a plot itself. On the other hand, the Bode plots present $|Z|$ and φ on the *y*-axis as a function of modulation frequency *f* on the *x*-axis (typically, a logarithmic scale for $|Z|$ and *f*).

[3] Complex plane plot is more accurate terminology, but Nyquist plot is more frequently used in electrochemistry literature. Nyquist plot is used to evaluate the stability of a system with feedback in automatic control and signal processing.

[4] This is because, in many cases, electrochemical systems engage with capacitance which has a negative imaginary impedance component.

[5] Here, it is worthwhile to point out that the unit length of *x*- and *y*-axes needs to be identical in a Nyquist plot. Otherwise, a plot can be deformed that would mislead interpretation.

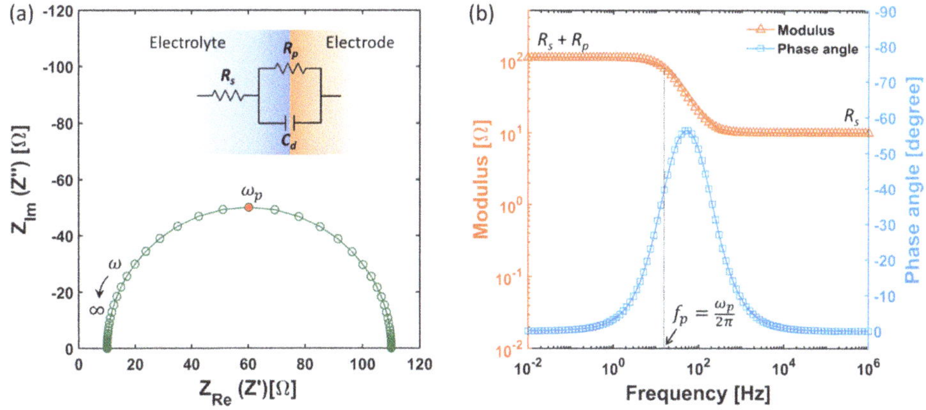

Figure 2.15. (a) Nyquist and (b) Bode-modulus/-phase angle plots displaying the simulated EIS data with the equivalent circuit in the inset. Values used to generate simulated data: $R_s = 10\ \Omega$, $R_p = 100\ \Omega$, and $C_d = 10^2\ \mu$F. The angular peak frequency (ω_p) associated with characteristic time constant and corresponding frequency (f_p) are shown in the plots.

Let us further examine the Randles circuit used as an example and its graphical representation shown in figure 2.15. At high frequency ($\omega \rightarrow \infty$), Z_C approaches 0, so that $Z(\omega)$ is dominated by non-Faradaic process, R_s, as indicated in equation (2.25) and both plots. As frequency decreases and consequently Z_C increases, more Faradaic component at the interface (R_p) is involved, and, therefore, $Z(\omega)$ becomes larger. In this frequency regime, the characteristic semicircle is formed in the Nyquist plot, and $|Z|$ increases and φ exhibits non-zero values in the Bode-phase angle plot. Lastly, at low frequency limit ($\omega \rightarrow 0$), $Z_C \rightarrow \infty$. As a result, Faradaic process is prevailing at the interface, and $Z(\omega)$ becomes the sum of R_p and R_s. In fact, this low frequency limit can be connected to the experimental condition of other DC-based characterization techniques such as chronoamperometry (holding potential for a long period of time). For the case of an ideal semicircle (in other words, an ideal capacitive component), its diameter is given by R_p and the capacitance can be obtained from the peak frequency (ω_p) in the Nyquist plot, as indicated in figure 2.15(a), with the relation of $\omega_p = 1/(RC) = 1/\tau$ where τ is the characteristic time constant for the parallel circuit.

In practical experiments, there exists deviation from ideality, and other circuit elements need to be introduced to properly describe electrochemical systems with an equivalent circuit. For example, a double-layer capacitor on a solid electrode does not behave like a pure capacitor because of chemical inhomogeneities, structural irregularities, or roughness in the electrode surface. Impedance associated with a non-ideal capacitance can be treated with constant phase elements (CPE), which is defined by $Z_{CPE} = 1/((j\omega)^{\beta}Q)$ where Q represents a non-ideal capacitance in units of F·s$^{\beta-1}$ and β indicates an ideality ranging from 0 to 1 (for a pure capacitor, $\beta = 1$) [86]. As demonstrated in figure 2.16 with modeled data, a non-ideal capacitance, replacing a pure capacitance in the Randles circuit with the CPE having $\beta < 1$,

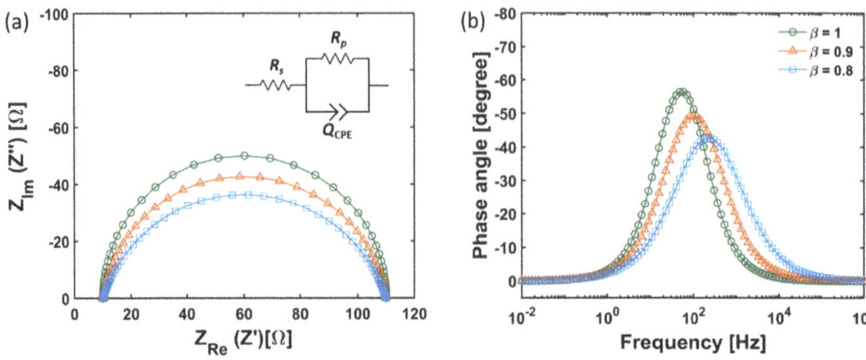

Figure 2.16. (a) Nyquist and (b) Bode-phase angle plots displaying the simulated EIS data with the equivalent circuit in the inset. Values used to generate simulated data: $R_s = 10\ \Omega$, $R_p = 100\ \Omega$, and $Q_{CPE} = 10^2\ \mu F \cdot s^{\beta-1}$.

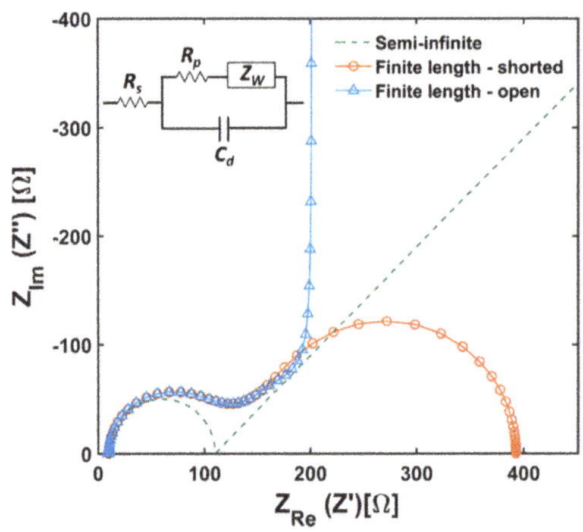

Figure 2.17. Nyquist plots displaying the simulated EIS data with the equivalent circuit in the inset with different types of the Warburg element (Z_W). Values used to generate simulated data: $R_s = 10\ \Omega$, $R_p = 100\ \Omega$, $C_d = 10^2\ \mu F$, $\sigma = 100\ \Omega \cdot s^{-1/2}$, and $L/\sqrt{D} = 2$.

results in a depressed semicircle in the Nyquist plot and a broadened peak in the Bode-phase angle plot.

Mass transport of mobile charge carriers through an electrode and an electrolyte is often modeled with the Warburg impedance (Z_W) [87]. The frequently used equivalent circuit is shown at the inset in figure 2.17. Here, a Warburg impedance element is introduced in series with R_p in the Randles circuit, in order to describe diffusion of electroactive species toward an electrode–electrolyte interface where the redox reaction of interest occurs. Warburg impedance can also be used to explain movement of spectator species such as counterions if their concentration, in turn, charge balance, becomes critical for determining overall reaction kinetics [88]. The Warburg impedance, in the case of a semi-infinite linear diffusion, is given by

$$Z_W = \frac{A_W}{\sqrt{\omega}}(1 - j) \tag{2.26}$$

$$A_W = \frac{RT}{n^2F^2A\sqrt{2}}\left(\frac{1}{C_O\sqrt{D_O}} + \frac{1}{C_R\sqrt{D_R}}\right) \tag{2.27}$$

where A_W is the Warburg coefficient which describes diffusion-related resistance as functions of involved charges (n), surface area of an electrode (A), and concentration (C_O, C_R) and diffusion coefficients (D_O, D_R) of oxidized and reduced species; other terms have their usual meanings. As indicated by its mathematical expression ($Z_W \propto 1/\sqrt{\omega}$), the Warburg impedance increases as frequency decreases (in other words, diffusion distance of charged species becomes longer). On the Nyquist plot, as shown in figure 2.17, the semi-infinite Warburg impedance is characterized with a diagonal line with the slop of 1 ($\varphi = 45°$), originated in the same value but the opposite sign of its real and imaginary parts. Therefore, the overall impedance shows a semicircle from parallel coupling of R_p and C_d at high frequencies followed by a straight line from Z_W at low frequencies. When a diffusion layer has finite thickness (δ), which is often found in many practical cases (e.g., diffusion of charged ions within porous channel or electron/hole transport within electrode), the Warburg impedance is given by

$$W_S = \frac{A_W\sqrt{2}}{\sqrt{j\omega}}\tanh\left[\frac{\delta}{\sqrt{D}}\sqrt{j\omega}\right] \tag{2.28}$$

$$W_O = \frac{A_W\sqrt{2}}{\sqrt{j\omega}}\coth\left[\frac{\delta}{\sqrt{D}}\sqrt{j\omega}\right] \tag{2.29}$$

depending on a boundary condition. Here, the finite length transmissive and reflective impedance elements (W_S and W_O) describe shorted and open boundaries at the electrode–electrolyte interfaces for zero and infinite charge transfer resistance, respectively. Their graphical representation in the Nyquist plot is shown in figure 2.17 [89]. The characteristic diffusion lifetime can be obtained from $\tau = (\delta/\sqrt{D})^2$ where D is the diffusion coefficient of the mobile charge species.

As explained at the Introduction, most practical batteries employ porous electrode structures to ensure large interfacial areas and, in turn, enhanced electrochemical reaction kinetics. The impedance of porous film electrodes, prepared on a conductive substrate and immersed in a liquid electrolyte, has been commonly evaluated by using transmission line model, as illustrated in figure 2.18(a) [43, 90]. This model includes serial and parallel connection of circuit elements to describe AC response at different regions of the film. Impedance associated with transport of charged species through the electrolyte within the pore channel and the electrode's solid phase is modeled by χ_1 and χ_2, respectively; here, L is the length of the pore. The element ζ describes electrical current flow via Faradaic and non-Faradaic processes occurring at the electrode/electrolyte interface in the pore inner wall.

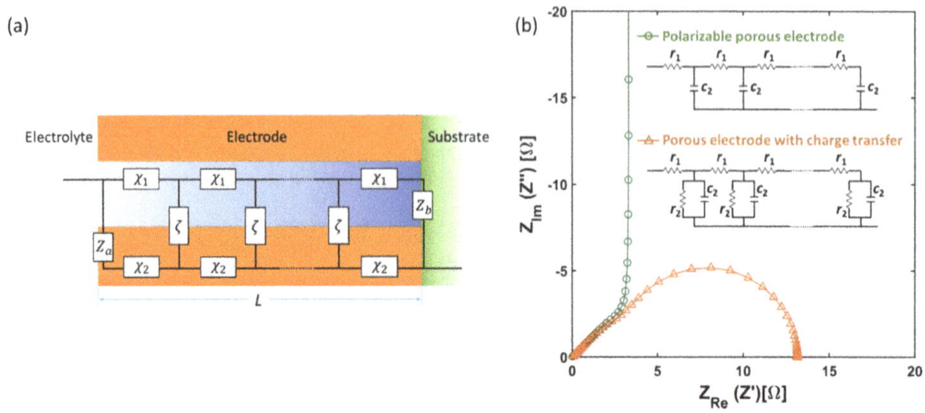

Figure 2.18. (a) Schematic representation of a generic transmission line model describing elemental reactions in a porous electrode. The used terms are discussed in the text. (b) Nyquist plots displaying the simulated EIS data with the transmission line models in the inset. Values used to generate simulated data: $r_1 = r_2 = 10\ \Omega$ and $c_2 = 1$ mF.

Lastly, Z_a and Z_b are impedance related to boundary condition for two interfaces, namely, electrolyte/electrode's top surface and electrolyte/substrate, respectively. Figure 2.18(b) illustrates two representative models with corresponding simulated Nyquist plots. One is a polarizable porous electrode showing capacitive behavior ($\zeta = c_2$) at the pore's internal surface, while the other is the case when charge transfer reaction occurs at the same interface (ζ is now replaced by the parallel combination of resistive and capacitive elements). Both models start with a line of slope 1, the symbolic feature associated with charge diffusion, at the high frequency regime. Then, the former appears as a vertical line at lower frequency due to capacitance in the double layer. On the other hand, the latter is characterized by a semicircle arc with a finite resistance in the DC limit.

Lastly, in many cases of practical experiments, it is necessary to add an inductor in an equivalent circuit, in order to take account of inductance which can be originated from close placement of metallic contacts or electrical leads. The impedance of an inductive element is given by $Z_L = j\omega L$. Contrary to capacitive impedance (Z_C), Z_L is proportional to ω and it has only positive imaginary part, appearing as a straight line on the lower quadrant of the Nyquist plot (figure 2.19).

2.4.3 Reliability of impedance data

The valid measurement and analysis of impedance spectra require four conditions listed below [91, 92].

- Linearity: In general, electrochemical systems are non-linear, following the Butler–Volmer behavior, i.e., an exponential relationship between voltage and current. On the other hand, electrical systems, applied to model EIS response, are linear (Ohm's law). Therefore, the amplitude of applied AC signal must be small enough to ensure a linear relation between perturbation and response during the entire measurement.

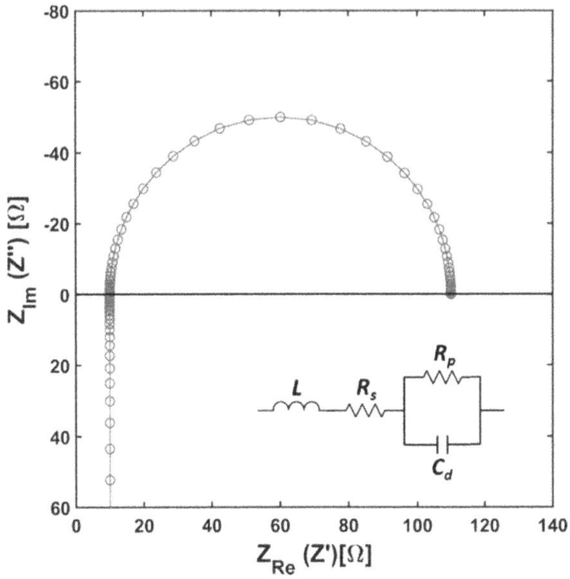

Figure 2.19. Nyquist plot displaying the simulated EIS data with the equivalent circuit in the inset. Values used to generate simulated data: L = 10 μH, R_s = 10 Ω, R_p = 100 Ω, and C_d = 10^2 μF.

- Causality: The response measured from the system must be directly originated from the applied AC perturbation.
- Stability: The system must be returned to its original state without significant alteration after the applied AC perturbation is eliminated.
- Finiteness: The measured impedance must be a finite value at the whole range of measurement frequency.

Careful design and practice for EIS measurements, therefore, is necessary to fulfill the above-mentioned conditions so that one can avoid misinterpretations of EIS data and consequently wrong assignment of electrochemical processes. The Kramers–Kronig (K–K) relations have been used to assess the validity of impedance data. With the EIS data satisfying all four conditions, the K–K relations allow one to obtain the imaginary impedance of the complex function, that is independent of the involved system or physical processes, from its real part and vice versa, as shown in equations (2.30) and (2.31). Therefore, for the valid EIS data, the computed impedance components should be close to the experimentally obtained ones. The methods for carrying out the K–K relations (integration from zero to infinity) to practically feasible data collected over a finite range of frequency can be found in the literature [93, 94].

$$Z_{Im}(\omega) = -\left(\frac{2\omega}{\pi}\right)\int_0^{\infty} \frac{Z_{Re}(x) - Z_{Re}(\omega)}{x^2 - \omega^2}dx \tag{2.30}$$

$$Z_{Re}(\omega) = Z_{Re}(\infty) + \left(\frac{2}{\pi}\right)\int_0^\infty \frac{xZ_{Im}(x) - \omega Z_{Im}(\omega)}{x^2 - \omega^2}dx \tag{2.31}$$

2.4.4 Application to battery characterization

EIS has been widely employed to characterize materials' fundamental properties used in batteries and their change/interaction with adjacent components during battery operation. As performed in frequency domain, in principle, EIS allows one to de-convolute reaction kinetics of elemental processes present in the system of interest, within a relatively short period of measurement time. In addition, EIS can be performed under dynamic operating condition without disassembly of lab-scale cells or practical devices [95]. This aspect is beneficial not only to avoid any unnecessary degradation of sensitive materials upon exposure to oxygen and moisture but also to consider EIS a promising diagnosis tool in a battery management system.

Important properties of materials, which can be obtained by EIS, include parameters related to movement of charged species. To be specific, ionic conductivity and relative permittivity of an electrolyte can be extracted from R_s and C_d, respectively, while lithium or electronic charge diffusion coefficients in a solid electrode can be studied with relevant resistive or Warburg components [96–98]. On the other hand, interfacial reactions, which play a critical role for determining battery performance, have been investigated by using EIS [99]. For instance, impedance characteristics originating from SEI layers formed at electrode–electrolyte interfaces can be modeled by introducing a parallel combination of resistive and capacitive elements as shown in the inset of figure 2.20 [100, 101]. Therefore, by inspecting the related EIS feature (the semicircle at high frequency regime for this example) as functions of cycle number, additives in electrolyte, surface coating, and

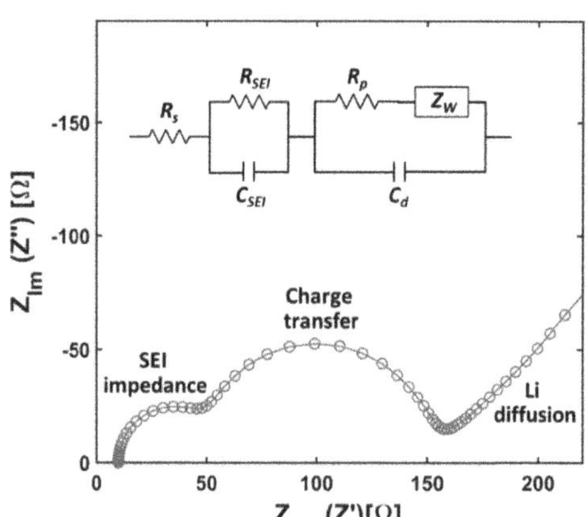

Figure 2.20. Nyquist plot of an equivalent circuit example, shown in the inset, to model SEI formation in a Li ion battery half-cell system.

so on, impact of SEI formation on overall cell impedance and battery cycling efficiency can be systematically studied. As the last example, the SoC and the SoH of operating dynamic battery systems have been estimated by correlating them with change in parameters extracted from equivalent circuit analysis, including, for instance, charge-transfer resistance and series resistance, respectively. More detailed examples of EIS application to batteries characterization are available in the literature [78, 79, 102–104].

Appendix A. List of symbols in table 2.1

Symbols

a_s	specific surface area of active material, m^{-1}
A	cell cooling surface, m^2
$c_{s,i}$	lithium concentration in the active material, mol m^{-3} (i = a, c)
$c_{e,i}$	lithium concentration in the electrolyte phase, mol m^{-3}
C_p	lumped specific heat capacity, J $(kg.K)^{-1}$
$D_{s,i}$	diffusion coefficient in the active material, m^2 s^{-1} (i = a, c)
D_e	diffusion coefficient in the electrolyte phase, m^2 s^{-1}
E_a	activation energy, eV
F	Faraday's constant, 96 485 C mol^{-1}
$f_{\pm}$	mean molar activity of the salt
h	convective heat transfer coefficient, W $(m^2 \cdot K)^{-1}$
i_0''	exchange current density at the surface of active material, A m^{-2}
i''	local current density by redox reaction at surface of active material, A m^{-2}
k_i	rate constant of redox reaction at surface of active material, (A m^{-2})/(C mol^{-1})/(mol $m^{-3})^{1.5}$, (i = a, c)
l_a, l_s, l_c	anode, separator and cathode thickness, m
m	mass of the cell, kg
p	Bruggeman exponent coefficient
$\dot{q}$	heat generation rate (per volume), W m^{-3}
R	universal gas constant, 8.314 J $(mol.K)^{-1}$
r	radial position in the 1D spherical particle domain, m
R_a, R_c	radius of active material, m
T	absolute temperature, K
t	time, s
t_+^0	lithium-ion transference number with respect to solvent velocity
U	open circuit voltage, V
x	distance from current collector/anode interface, m
V	cell volume, m^3
V_{cell}	cell voltage, V

Greek symbols

α_a, α_c	anodic and cathodic transfer coefficients, 0.5
ε_e	electrolyte volume fraction
κ	ionic conductivity, S m^{-1}

η	overpotential, V, (i =a,c)
ρ	cell density, kg m^{-3}
σ	electronic conductivity, S m^{-1}
$\phi_{s,\,i}$	solid phase electrical potential, V (i =a,c)
ϕ_e	electrolyte phase electrical potential, V (i =a,c)

Superscripts and subscripts

a	anode
c	cathode
eff	effective value
max	maximum
ohm	ohmic
ref	reference
rev	reversible
rxn	reaction
s	solid phase
s/e	solid/electrolyte interface
tot	total
e	electrolyte phase
+	positive current collector
−	negative current collector

References

[1] Nitta N, Wu F, Lee J T and Yushin G 2015 Li-ion battery materials: present and future *Mater. Today* **18** 252–64

[2] Xu J, Lin F, Doeff M M and Tong W 2017 A review of Ni-based layered oxides for rechargeable Li-ion batteries *J. Mater. Chem.* A **5** 874–901

[3] Mao C *et al* 2018 Selecting the best graphite for long-life, high-energy Li-ion batteries *J. Electrochem. Soc.* **165** A1837–45

[4] Marom R, Amalraj S F, Leifer N, Jacob D and Aurbach D 2011 A review of advanced and practical lithium battery materials *J. Mater. Chem.* **21** 9938–54

[5] Bareno J, Shkrob I A, Gilbert J A, Klett M and Abraham D P 2017 Capacity fade and its mitigation in Li-ion cells with silicon-graphite electrodes *J. Phys. Chem.* C **121** 20640–9

[6] Cao P F *et al* 2018 Effect of binder architecture on the performance of silicon/graphite composite anodes for lithium ion batteries *ACS Appl. Mater. Inter.* **10** 3470–8

[7] Xu K 2004 Nonaqueous liquid electrolytes for lithium-based rechargeable batteries *Chem. Rev.* **104** 4303–17

[8] Orikasa Y *et al* 2016 Ionic conduction in lithium ion battery composite electrode governs cross-sectional reaction distribution *Sci. Rep.* **6** 26382

[9] Krewer U *et al* 2018 Review—dynamic models of li-ion batteries for diagnosis and operation: a review and perspective *J. Electrochem. Soc.* **165** A3656–73

[10] Energy, U.S.D.O. 2017 Basic research needs for next generation electrical energy storage. Report of The Office of Basic Energy Sciences Workshop on Energy Storage, March 27–29, 2017

[11] Shi S Q *et al* 2012 Direct calculation of Li-ion transport in the solid electrolyte interphase *J. Am. Chem. Soc.* **134** 15476–87

[12] Wang Y X, Nakamura S, Ue M and Balbuena P B 2001 Theoretical studies to understand surface chemistry on carbon anodes for lithium-ion batteries: reduction mechanisms of ethylene carbonate *J. Am. Chem. Soc.* **123** 11708–18

[13] Ganesh P, Kent P R C and Jiang D E 2012 Solid-electrolyte interphase formation and electrolyte reduction at Li-ion battery graphite anodes: insights from first-principles molecular dynamics *J. Phys. Chem.* C **116** 24476–81

[14] Leung K and Budzien J L 2010 Ab initio molecular dynamics simulations of the initial stages of solid-electrolyte interphase formation on lithium ion battery graphitic anodes *Phys. Chem. Chem. Phys.* **12** 6583–86

[15] Methekar R N, Northrop P W C, Chen K J, Braatz R D and Subramanian V R 2011 Kinetic Monte Carlo simulation of surface heterogeneity in graphite anodes for lithium-ion batteries: passive layer formation *J. Electrochem. Soc.* **158** A363–70

[16] Doyle M, Fuller T F and Newman J 1993 Modeling of galvanostatic charge and discharge of the lithium/polymer/insertion cell *J. Electrochem. Soc.* **140** 1526–33

[17] Newman J and Tiedemann W 1975 Porous-electrode theory with battery applications *AIChE J.* **21** 25–41

[18] Thomas K E, Newman J and Darling R M 2002 *Advances in Lithium-Ion Batteries* ed W A van Schalkwijk and B Scrosati (Cham: Springer) pp 345–92

[19] Andre D *et al* 2011 Characterization of high-power lithium-ion batteries by electrochemical impedance spectroscopy. II: modelling *J. Power Sources* **196** 5349–56

[20] Gao P, Zhang C F and Wen G W 2015 Equivalent circuit model analysis on electrochemical impedance spectroscopy of lithium metal batteries *J. Power Sources* **294** 67–74

[21] Moss P L, Au G, Plichta E J and Zheng J P 2008 An electrical circuit for modeling the dynamic response of Li-ion polymer batteries *J. Electrochem. Soc.* **155** A986–94

[22] Doyle M, Fuller T F and Newman J 1993 Modeling of galvanostatic charge and discharge of the lithium polymer insertion cell *J. Electrochem. Soc.* **140** 1526–33

[23] Doyle M and Newman J 1995 Modeling the performance of rechargeable lithium-based cells—design correlations for limiting cases *J. Power Sources* **54** 46–51

[24] Doyle M, Newman J, Gozdz A S, Schmutz C N and Tarascon J M 1996 Comparison of modeling predictions with experimental data from plastic lithium ion cells *J. Electrochem. Soc.* **143** 1890–903

[25] Guo M, Sikha G and White R E 2011 Single-particle model for a lithium-ion cell: thermal behavior *J. Electrochem. Soc.* **158** S11–1

[26] Ning G and Popov B N 2004 Cycle life modeling of lithium-ion batteries *J. Electrochem. Soc.* **151** A1584–91

[27] Dubarry M, Vuillaume N and Liaw B Y 2009 From single cell model to battery pack simulation for Li-ion batteries *J. Power Sources* **186** 500–7

[28] Feng T H, Yang L, Zhao X W, Zhang H D and Qiang J X 2015 Online identification of lithium-ion battery parameters based on an improved equivalent-circuit model and its implementation on battery state-of-power prediction *J. Power Sources* **281** 192–203

[29] Cai L and White R E 2009 Reduction of model order based on proper orthogonal decomposition for lithium-ion battery simulations *J. Electrochem. Soc.* **156** A154–61

[30] Han X B, Ouyang M G, Lu L G and Li J Q 2015 Simplification of physics-based electrochemical model for lithium ion battery on electric vehicle. Part I: diffusion simplification and single particle model *J. Power Sources* **278** 802–13

[31] Subramanian V R, Boovaragavan V and Diwakar V D 2007 Toward real-time simulation of physics based lithium-ion battery models *Electrochem Solid St* **10** A255–60

[32] Westerhoff U, Kurbach K, Lienesch F and Kurrat M 2016 Analysis of lithium-ion battery models based on electrochemical impedance spectroscopy *Energy Technol-Ger* **4** 1620–30

[33] Effendy S, Song J H and Bazant M Z 2020 Analysis, design, and generalization of electrochemical impedance spectroscopy (EIS) inversion algorithms *J. Electrochem. Soc.* **167** 106508

[34] Illig J, Ender M, Weber A and Ivers-Tiffee E 2015 Modeling graphite anodes with serial and transmission line models *J. Power Sources* **282** 335–47

[35] Schmidt J P, Berg P, Schonleber M, Weber A and Ivers-Tiffee E 2013 The distribution of relaxation times as basis for generalized time-domain models for Li-ion batteries *J. Power Sources* **221** 70–7

[36] Schmidt J P *et al* 2011 Studies on $LiFePO_4$ as cathode material using impedance spectroscopy *J. Power Sources* **196** 5342–8

[37] Boukamp B A 2020 Distribution (function) of relaxation times, successor to complex nonlinear least squares analysis of electrochemical impedance spectroscopy? *J. Phys.: Energy* **2** 042001

[38] Song J and Bazant M Z 2018 Electrochemical impedance imaging via the distribution of diffusion times *Phys. Rev. Lett.* **120** 116001

[39] Osaka T, Momma T, Mukoyama D and Nara H 2012 Proposal of novel equivalent circuit for electrochemical impedance analysis of commercially available lithium ion battery *J. Power Sources* **205** 483–6

[40] Song J and Bazant M Z 2014 Electrochemical impedance of a battery electrode with anisotropic active particles *Electrochim. Acta* **131** 214–27

[41] Song J H and Bazant M Z 2013 Effects of nanoparticle geometry and size distribution on diffusion impedance of battery electrodes *J. Electrochem. Soc.* **160** A15–24

[42] Abarbanel D W, Nelson K J and Dahn J R 2016 Exploring impedance growth in high voltage NMC/graphite Li-Ion cells using a transmission line model *J. Electrochem. Soc.* **163** A522–9

[43] Bisquert J 2000 Influence of the boundaries in the impedance of porous film electrodes *Phys. Chem. Chem. Phys.* **2** 4185–92

[44] Jamnik J and Maier J 2001 Generalised equivalent circuits for mass and charge transport: chemical capacitance and its implications *Phys. Chem. Chem. Phys.* **3** 1668–78

[45] Newman J, Thomas K E, Hafezi H and Wheeler D R 2003 Modeling of lithium-ion batteries *J. Power Sources* **119–21** 838–43

[46] Gu W B and Wang C Y 2000 Thermal-electrochemical modeling of battery systems *J. Electrochem. Soc.* **147** 2910–22

[47] Srinivasan V and Wang C Y 2003 Analysis of electrochemical and thermal behavior of Li-ion cells *J. Electrochem. Soc.* **150** A98–106

[48] Kwon K H, Shin C B, Kang T H and Kim C S 2006 A two-dimensional modeling of a lithium-polymer battery *J. Power Sources* **163** 151–7

[49] Kim U S, Yi J, Shin C B, Han T and Park S 2011 Modeling the dependence of the discharge behavior of a lithium-ion battery on the environmental temperature *J. Electrochem. Soc.* **158** A611–8

[50] Kim G H, Smith K, Lee K J, Santhanagopalan S and Pesaran A 2011 Multi-domain modeling of lithium-ion batteries encompassing multi-physics in varied length scales *J. Electrochem. Soc.* **158** A955–69

[51] Guo M, Kim G H and White R E 2013 A three-dimensional multi-physics model for a Li-ion battery *J. Power Sources* **240** 80–94

[52] Lee K J, Smith K, Pesaran A and Kim G H 2013 Three dimensional thermal-, electrical-, and electrochemical-coupled model for cylindrical wound large format lithium-ion batteries *J. Power Sources* **241** 20–32

[53] Kim H K, Kim C J, Kim C W and Lee K J 2018 Numerical analysis of accelerated degradation in large lithium-ion batteries *Comput. Chem. Eng.* **112** 82–91

[54] Kim H K, Choi J H and Lee K J 2019 A numerical study of the effects of cell formats on the cycle life of lithium ion batteries *J. Electrochem. Soc.* **166** A1769–78

[55] Macdonald J R 1990 Impedance spectroscopy—old problems and new developments *Electrochim. Acta* **35** 1483–92

[56] Macdonald J R, Schoonman J and Lehnen A P 1982 Applicability and power of complex nonlinear least squares for the analysis of impedance and admittance data *J. Electroanal. Chem. Interfacial Electrochem.* **131** 77–95

[57] Huggins R 2008 *Advanced Batteries: Materials Science Aspects* (Cham: Springer)

[58] Park J-K 2012 *Principles and Applications of Lithium Secondary Batteries* (New York: Wiley)

[59] Ovejas V J and Cuadras A 2019 Effects of cycling on lithium-ion battery hysteresis and overvoltage *Sci. Rep.* **9** 14875

[60] Evans J, Vincent C A and Bruce P G 1987 Electrochemical measurement of transference numbers in polymer electrolytes *Polymer* **28** 2324–8

[61] MacInnes D A 1961 *The Principles of Electrochemistry* (New York: Dover)

[62] Bockris J O M and Reddy A K N 1970 *Modern Electrochemistry.* vol 1 (New York: Plenum)

[63] Mindemark J, Lacey M J, Bowden T and Brandell D 2018 Beyond PEO—alternative host materials for Li^+-conducting solid polymer electrolytes *Prog. Polym. Sci.* **81** 114–43

[64] Yang X and Rogach A L 2019 Electrochemical techniques in battery research: a tutorial for nonelectrochemists *Adv. Energy Mater.* **9** 1900747

[65] Sharova V *et al* 2018 Comparative study of imide-based Li salts as electrolyte additives for Li-ion batteries *J. Power Sources* **375** 43–52

[66] Bard A J and Faulkner L R 2001 *Electrochemical Methods—Fundamentals and Applications* (New York: Wiley) pp 226–60

[67] Lindström H *et al* 1997 Li^+ Ion Insertion in TiO_2 (Anatase). 2. voltammetry on nanoporous films *J. Phys. Chem.* B **101** 7717–22

[68] Wen C J, Boukamp B A, Huggins R A and Weppner W 1979 Thermodynamic and mass transport properties of 'LiAl' *J. Electrochem. Soc.* **126** 2258–66

[69] Weppner W and Huggins R A 1977 Determination of the kinetic parameters of mixed-conducting electrodes and application to the system Li_3Sb *J. Electrochem. Soc.* **124** 1569–78

[70] Zhu Y, Gao T, Fan X, Han F and Wang C 2017 Electrochemical techniques for intercalation electrode materials in rechargeable batteries *Acc. Chem. Res.* **50** 1022–31

[71] Xie J *et al* 2009 Li-ion diffusion kinetics in $LiFePO_4$ thin film prepared by radio frequency magnetron sputtering *Electrochim. Acta* **54** 4631–7
[72] Li Z *et al* 2014 Towards understanding the rate capability of layered transition metal oxides $LiNi_yMn_yCo_{1-2y}O_2$ *J. Power Sources* **268** 106–12
[73] Markevich E, Levi M D and Aurbach D 2005 Comparison between potentiostatic and galvanostatic intermittent titration techniques for determination of chemical diffusion coefficients in ion-insertion electrodes *J. Electroanal. Chem.* **580** 231–7
[74] Consortium USAB 1996 Electric vehicle battery test procedures manual *USABC*
[75] Barai A *et al* 2019 A comparison of methodologies for the non-invasive characterisation of commercial Li-ion cells *Prog. Energy Combust. Sci.* **72** 1–31
[76] Long B R *et al* 2016 Enabling high-energy, high-voltage lithium-ion cells: standardization of coin-cell assembly, electrochemical testing, and evaluation of full cells *J. Electrochem. Soc.* **163** A2999–3009
[77] Tornheim A, Sahore R, He M, Croy J R and Zhang Z 2018 Preformed anodes for high-voltage lithium-ion battery performance: fluorinated electrolytes, crosstalk, and the origins of impedance rise *J. Electrochem. Soc.* **165** A3360
[78] Qu D *et al* 2018 Electrochemical impedance and its applications in energy-storage systems *Small Methods* **2** 1700342
[79] Bredar A R C, Chown A L, Burton A R and Farnum B H 2020 Electrochemical impedance spectroscopy of metal oxide electrodes for energy applications *ACS Appl. Energy Mater.* **3** 66–98
[80] García-Antón J *et al* 2014 Improvement of the electrochemical behaviour of Zn-electroplated steel using regenerated Cr (III) passivation baths *Chem. Eng. Sci.* **111** 402–9
[81] Lai W and Haile S M 2005 Impedance spectroscopy as a tool for chemical and electrochemical analysis of mixed conductors: a case study of Ceria *J. Am. Ceram. Soc.* **88** 2979–97
[82] Bard A J and Faulkner L R 2001 *Electrochemical Methods—Fundamentals and Applications* (New York: Wiley) pp 368–416
[83] Macdonald J R and Johnson W B 1987 *Impedance Spectroscopy—Emphasizing Solid Materials and Systems* ed J R Macdonald (New York: Wiley) pp 1–26
[84] Randles J E B 1947 Kinetics of rapid electrode reactions *Discuss. Faraday Soc.* **1** 11–9
[85] Lasia A 2014 *Electrochemical Impedance Spectroscopy and its Applications* (Cham: Springer) pp 7–66
[86] Lasia A 2014 *Electrochemical Impedance Spectroscopy and its Applications* (Cham: Springer) pp 177–202
[87] Warburg E 1901 Ueber die Polarisationscapacität des Platins *Ann. Phys.* **311** 125–35
[88] Thomas M G S R, Bruce P G and Goodenough J B 1985 AC impedance analysis of polycrystalline insertion electrodes: application to $Li_{1-x}CoO_2$ *J. Electrochem. Soc.* **132** 1521–8
[89] Bisquert J 2002 Theory of the impedance of electron diffusion and recombination in a thin layer *J. Phys. Chem.* B **106** 325–33
[90] de Levie R 1963 On porous electrodes in electrolyte solutions: I. Capacitance effects *Electrochim. Acta* **8** 751–80
[91] Macdonald D D, Sikora E and Engelhardt G 1998 Characterizing electrochemical systems in the frequency domain *Electrochim. Acta* **43** 87–107

[92] Lasia A 2014 *Electrochemical Impedance Spectroscopy and its Applications* (Cham: Springer) pp 271–300

[93] Agarwal P, Orazem M E and Garcia-Rubio L H 1996 The influence of error structure on interpretation of impedance spectra *Electrochim. Acta* **41** 1017–22

[94] Boukamp B A 1995 A Linear Kronig-Kramers transform test for immittance data validation *J. Electrochem. Soc.* **142** 1885–94

[95] Lohmann N, Weßkamp P, Haußmann P, Melbert J and Musch T 2015 Electrochemical impedance spectroscopy for lithium-ion cells: test equipment and procedures for aging and fast characterization in time and frequency domain *J. Power Sources* **273** 613–23

[96] Park J-K 2012 *Principles and Applications of Lithium Secondary Batteries* (Weinheim: Wiley-VCH Verlag & Co. KGaA) pp 231–61

[97] Ho C, Raistrick I D and Huggins R A 1980 Application of A-C techniques to the study of lithium diffusion in tungsten trioxide thin films *J. Electrochem. Soc.* **127** 343–50

[98] Xia H, Lu L and Ceder G 2006 Li diffusion in $LiCoO_2$ thin films prepared by pulsed laser deposition *J. Power Sources* **159** 1422–7

[99] Mei B-A *et al* 2018 Physical interpretations of electrochemical impedance spectroscopy of redox active electrodes for electrical energy storage *J. Phys. Chem.* C **122** 24499–511

[100] Aurbach D 2000 Review of selected electrode–solution interactions which determine the performance of Li and Li ion batteries *J. Power Sources* **89** 206–18

[101] Zhang W *et al* 2018 Degradation mechanisms at the $Li_{10}GeP_2S_{12}/LiCoO_2$ cathode interface in an all-solid-state lithium-ion battery *Acs. Appl. Mater. Inter.* **10** 22226–36

[102] Osaka T, Mukoyama D and Nara H 2015 Review—development of diagnostic process for commercially available batteries, especially lithium ion battery, by electrochemical impedance spectroscopy *J. Electrochem. Soc.* **162** A2529–37

[103] Choi W, Shin H-C, Kim J M, Choi J-Y and Yoon W-S 2020 Modeling and applications of electrochemical impedance spectroscopy (EIS) for lithium-ion batteries *J. Electrochem. Sci. Technol.* **11** 1–13

[104] Aurbach D, Levi M D and Levi E 2008 A review on the solid-state ionics of electrochemical intercalation processes: how to interpret properly their electrochemical response *Solid State Ionics* **179** 742–51

Chapter 3

Synchrotron-based battery imaging with multi-modal x-ray signals

Guibin Zan, Piero Pianetta and Yijin Liu

3.1 Introduction

Synchrotrons generate x-ray beams with high brilliance and a broad spectrum, which interact with matter in different ways and, therefore, are sensitive to different material properties. Synchrotron-based experimental techniques have been broadly applied to battery research, formulating a comprehensive understanding of the behaviors of battery materials at different electrochemical states and/or under different operating conditions. Given the structural hierarchy and the chemical heterogeneity in batteries, there is a growing interest in using multi-modal synchrotron-based probes to conduct multi-scale and high-dimensional imaging of batteries with structural, compositional, chemical, and morphological sensitivities. In this chapter, we review state-of-the-art developments in synchrotron-based battery research with the focused theme of highlighting the structural and chemical complexities in batteries. We elaborate on different ways of probing spectroscopic and scattering signals using synchrotron beams in a spatially resolved manner in order to image batteries using different contrast mechanisms. We will also discuss specific case studies that not only benefit from the cutting-edge synchrotron techniques but also leverage advanced computational developments.

3.2 General synchrotron techniques: microscopy, spectroscopy, and scattering

3.2.1 Overview of synchrotron radiation and its interaction with matter

Synchrotron radiation is the electromagnetic radiation generated when charged relativistic particles such as electrons are forced into a curved path by externally applied magnetic fields. Thanks to its intensity, brilliance, and broad coverage of the electromagnetic spectrum, synchrotron radiation has been identified as a powerful

doi:10.1088/978-0-7503-2682-7ch3

tool for a wide range of research disciplines. To effectively utilize synchrotron radiation for research purposes, dedicated synchrotron facilities have been developed and deployed across the world. In a synchrotron facility, bunches of electrons are accelerated to a final energy of about 1.5–8 GeV, close to the speed of light, and circulated around a storage ring that is defined by a series of magnets [1]. Bending magnets are the most common device for creating a single curved electron trajectory. As the electrons are deflected through the magnetic field, they give off electromagnetic radiation that is guided using the mirrors and other optics in the beamlines to the end-station for specific experiments (see schematic illustration in figure 3.1(A)). To optimize for the x-ray yield, different types of magnetic structures called insertion devices, e.g., wigglers and undulators, are also used. These have been broadly adopted in the third-generation synchrotron facilities. In an insertion device, the electrons follow a periodic trajectory determined by the particular periodic magnet structure, and the radiation emitted at each turn is superimposed. The synchrotron radiation from a bending magnet covers a wide and continuous

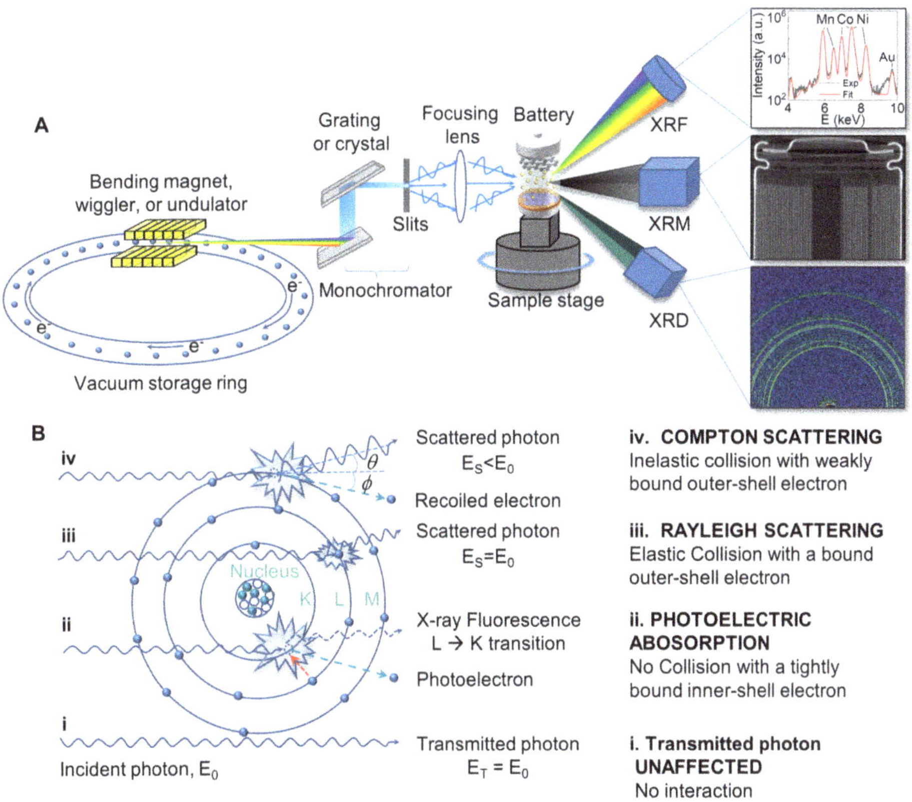

Figure 3.1. Schematic illustration of a synchrotron beamline is shown in panel (A) with multi-modal signal detection highlighted on the right. XRF, XRM, and XRD are acronyms for x-ray fluorescence, x-ray microscopy, and x-ray diffraction, respectively. Panel (B) illustrates four different types of x-ray-to-matter interactions.

spectrum and features a relatively larger beam emittance compared with that from a wiggler or undulator. Over the past few decades, tremendous scientific advancements have been facilitated by the continuous improvement of the synchrotron light sources' brightness, incorporation of precision instruments, advances in x-ray focusing optics, and development of innovative experiment techniques. An in-depth discussion of the physics of synchrotron radiation is outside the scope of this chapter and we refer to more comprehensive descriptions of synchrotron radiation in the literature [1, 2].

Synchrotron-generated x-rays are extensively used in a broad range of research fields because they interact with the sample in several different ways and, therefore, carry useful information associated with different sample properties (see illustration in figure 3.1(B)). Generally speaking, there are four types of matter–beam interactions when x-rays transmit through the material: (i) given the strong penetration capability, a significant portion of the hard x-ray photons do not interact with material and are unaffected as they transmit through; (ii) photoelectric absorption results in total removal of the incident x-ray photon with energy greater than the binding energy of the electron in its shell, with the excess energy distributed to the kinetic energy of the photoelectron; (iii) Rayleigh scattering is the interaction of the x-ray photon with an electron (or whole atom), in which no energy is exchanged and a small angular change in direction occurs; (iv) Compton scattering interactions occur with essentially unbound electrons, with transfer of energy shared between recoiled electron and scattered photon [2, 3]. Based on the interactions between the x-ray and the matter, a number of different experimental methods have been developed to study the material with structural, compositional, chemical, and morphological sensitivities [4]. These synchrotron-based experimental techniques are often combined, offering an unprecedented understanding of the investigated samples from different perspectives in a synergistic manner. In the next section, we will elaborate on this point with more details that are relevant to battery research.

3.2.2 The interplay among lattice, electronic structure, and micromorphology in battery

Batteries are designed to be structurally hierarchical and chemically heterogeneous. To fulfill the batteries' functionality as an energy reservoir, different battery components need to execute their respective roles in a synergistic fashion upon battery operation. In fact, heterogeneity in physical structure, defects in the atomic lattice, inhomogeneities in composition are considered as the key factors for a battery's capacity fade and cathode degradation [4]. Structural and chemical heterogeneities can easily lead to inhomogeneous current distribution, which could be responsible for both acute and chronic capacity loss. For a systematic modeling and description of the aging process of the battery, an accurate and faithful experimental observation is critical. The experimental data can be utilized to formulate an in-depth understanding of degradation mechanisms, which could inform the electrode design and improve the battery lifespan. In this section, we narrow down the topic of our discussion to the battery cathode, which is the most significant limiting factor for further improving the battery energy density. We will cover some anode studies in the later part of this chapter. The battery cathode is often made of many micron-sized particles of active material that are embedded in a porous carbon and binder matrix, which provide mechanical support as well as different conducting pathways for lithium ions and electrons. As the battery is cycled, the lattice of the active cathode undergoes repeated and often anisotropic expansion and contraction, which is accompanied by cationic (transition metal cations) and/or anionic (oxygen anions) redox reactions for energy storage and release. The micromorphology evolves upon battery cycling and, in turn, modulates how the reactions proceed. This is a highly complicated process and these factors intertwine across different time and length scales.

As illustrated in figures 3.2(A) to 3.2(D), the three-dimensional morphology can be revealed using x-ray microscopy [5]; the cathode material's lattice structure can be probed using x-ray diffraction [6]; the charge transfer and redox reactions can be fingerprinted using x-ray spectroscopy [7, 8]. In the example we showed in figure 3.2(A), we show the comparison of the morphological characteristics in nickel manganese cobalt (NMC) cathode secondary particles as they are cycled under different rates. More severe damage is observed upon fast charging, which can be attributed to the escalated heterogeneity in local current density and in phase

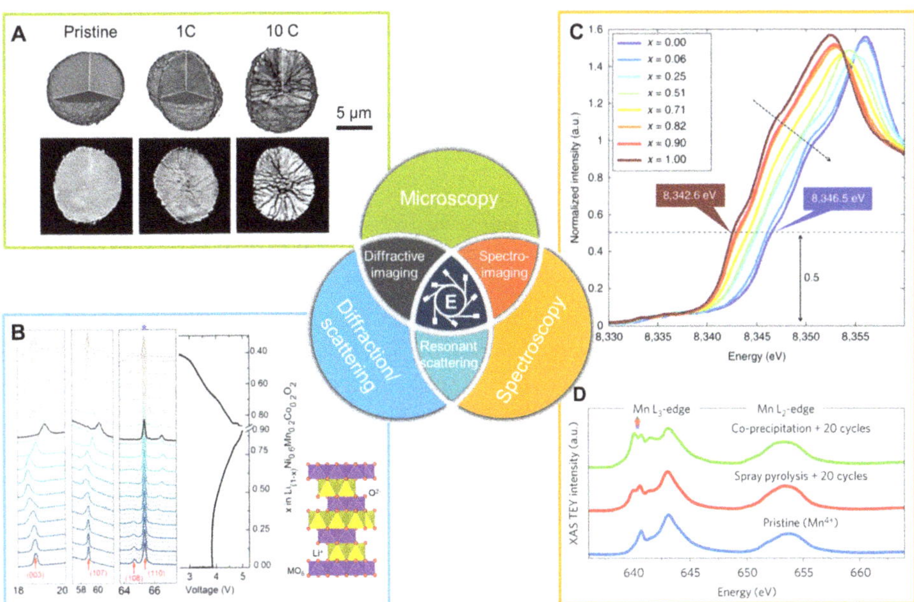

Figure 3.2. Panels (A) to (D) highlight a few case studies that utilize different synchrotron techniques to probe different material properties. Panel (E) schematically illustrates the three different synchrotron-based characterization techniques including diffraction/scattering, spectroscopy, and microscopy. The hybrid experimental approaches are also illustrated. (A) Reprinted with permission from [5]. Copyright 2018 Elsevier. (B) Reprinted with permission from [6]. Copyright 2018 IOP Publishing. (C) Reprinted with permission from [7]. Copyright 2017 Springer Nature. (D) Reprinted with permission from [8]. Copyright 2016 Springer Nature.

separation under fast charging conditions. In figure 3.2(B), the evolution of selected diffraction peaks of an NMC cathode is monitored as the battery is cycled, showing the anisotropic lattice deformation as a function of the state-of-charge. In figures 3.2(C) and 3.2(D), we present two examples of spectroscopic studies on battery cathodes in hard and soft x-ray regimes, respectively. The hard x-ray near edge structure (XANES, see figure 3.2(C)), on the other hand, offers a bulk averaged signal and reveals the gradual oxidation of the Ni cation in a spinel $LiMn_{1.5}Ni_{0.5}O_4$ cathode material upon chemical delithiation. The soft x-ray absorption spectroscopy with a total electron yield (TEY) signal detection (figure 3.2(D)) has a very shallow penetration depth (~5 nm), which results in a high sensitivity to the surface chemistry and reveals the electrochemical-cycling-induced changes of Mn's valence state for the NMC cathodes synthesized by different methods.

These experimental techniques can be further utilized in a hybrid fashion (see illustration in figure 3.2(E)). For example, by conducting the x-ray scattering experiment at a selected x-ray energy that resonates with the absorption edge of an element of interest, which is termed resonant scattering, one could have an elemental specific scattering signal that highlights the specific role of a targeted element. Diffractive imaging and spectro-imaging are both broadly applied to the battery research as they carry spatially resolved information regarding the local lattice arrangements/deformations and the valence/charge states, respectively [4]. Given the importance of structural and chemical heterogeneity in batteries, in this chapter we will discuss the synchrotron-based microscopy techniques in different imaging modalities that could offer structural, compositional, and chemical sensitivities. In the next section, we will review and compare several synchrotron imaging methods from a technical perspective and with battery case studies.

3.3 Synchrotron multi-modal microscopy for battery research

3.3.1 An overview of the synchrotron-based imaging approaches

X-rays have been broadly used for imaging since they were discovered more than a century ago. Synchrotron x-ray microscopy offers the capability of directly visualizing battery materials in a noninvasive manner due to their exceptional penetration capability. It becomes even more powerful when it is coupled with the computed tomography (CT) technique, in which projection images of the specimen are acquired at different viewing angles [9, 10]. With different configurations of the x-ray imaging setup, x-ray CT can offer spatial resolution at different scales, provide different contrast mechanisms, and even record the dynamically evolving structure (four-dimensional tomography). Coupled with advanced quantification and statistical analysis methods, many different attributes of the sample can be revealed by the multi-scale, multi-modal, and multi-dimensional imaging experiments.

Depending on the data acquisition approach, the implementation of x-ray micro/nano imaging using synchrotron x-ray sources can be categorized into two groups: the full-field imaging methods and the scanning imaging methods. In full-field imaging methods, a full image is directly recorded using an area detector in a single shot (figures 3.3(A), 3.3(B), 3.3(D) and 3.3(E)). On the other hand, scanning imaging

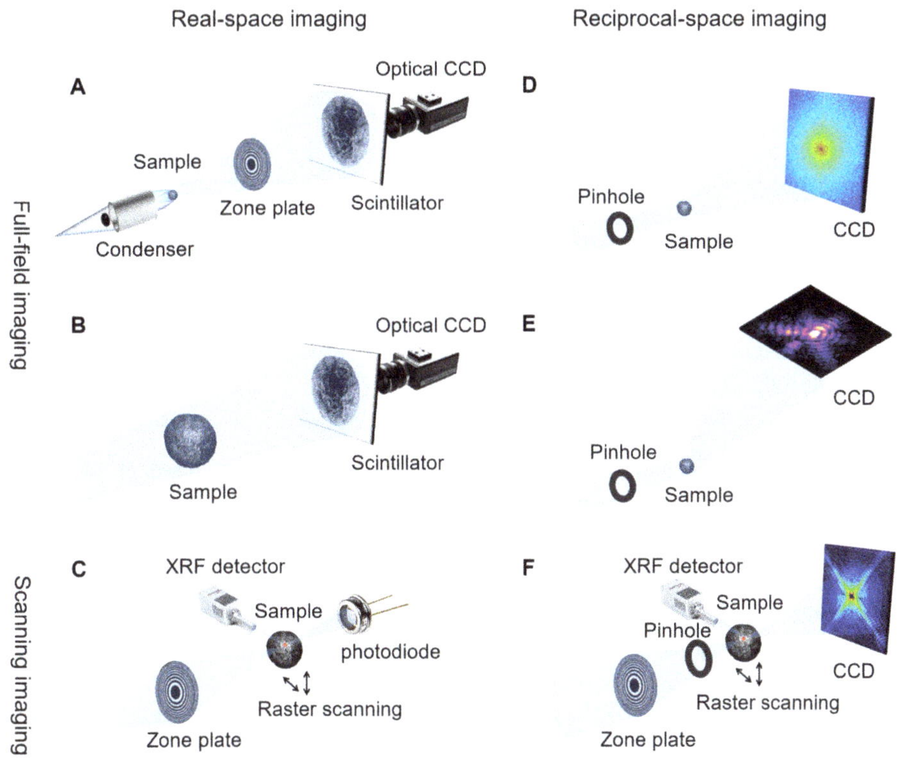

Figure 3.3. Schematic illustration of different synchrotron-based imaging modalities: (A) nano-resolution transmission x-ray microscopy (Nano-TXM); (B) micro-transmission x-ray microscopy (Micro-TXM); (C) scanning transmission x-ray microscopy (STXM); (D) coherent diffraction imaging (CDI); (E) Bragg coherent diffraction imaging (BCDI); (F) ptychography. Panels (A) to (C) are real-space imaging methods. Panels (D) to (F) are reciprocal-space imaging methods. Panels (A), (B), (D), and (E) are full-field imaging approaches. Panels (C) and (F) are scanning imaging approaches. (A–F) Reprinted with permission from [4]. Copyright 2017 American Chemical Society. Note: CCD, charge-coupled device.

methods build up an image pixel-by-pixel through a mechanical raster scan of the sample (figures 3.3(C) and 3.3(F)). We illustrate several popular synchrotron imaging approaches in figure 3.3. While the comparison of these methods is covered in greater detail in the literature [1, 4, 11–14], here we make some general remarks. The full-field imaging methods are generally fast and are often utilized to study the dynamic evolution that requires a relatively high temporal resolution. The scanning imaging methods, on the other hand, are relatively slow due to the required mechanical scan. In the scanning imaging methods, multiple detectors can work in parallel (see illustration in figures 3.3(C) and 3.3(F)), offering a high sensitivity to different aspects of the sample property. Another way of categorizing the x-ray imaging methods is by the imaging systems and the corresponding principles of the information extraction. When a real space image is directly acquired as raw data, we refer to these approaches as real space imaging methods (figures 3.3(A), 3.3(B), and 3.3(C)). On the other hand, the sample's structural information can be encoded in the diffraction pattern using the coherence property of the incoming x-ray beam, these methods are referred to as reciprocal space imaging methods (figures 3.3(D), 3.3(E) and 3.3(F)) [4]. The real-space imaging method provides an image of the sample in a straightforward manner, in contrast, the reciprocal-space imaging method requires sophisticated algorithms to reconstruct an image from the diffraction patterns. This process, termed phase retrieval, is nontrivial, we refer to existing comprehensive review articles in the literature for more details [15]. We summarize typical specifications of these synchrotron imaging methods in table 3.1. The pros and cons for each of these imaging modalities determine how they will be utilized for battery research. This will be elaborated through the selected case studies that will be included in our discussion below.

3.3.2 Battery research using synchrotron-based microscopy with absorption, phase, and scattering contrasts

Before discussing synchrotron imaging methods with more advanced contrast mechanisms, e.g., diffractive imaging and spectro-microscopy, in this section we review some of the synchrotron x-ray microscopic studies of batteries using absorption, phase, and scattering signals. These approaches have been broadly applied in a wide range of research fields, e.g., bio-medical research. The recent adoption of these methods for battery research have opened vast opportunities as we will elaborate below.

3.3.2.1 Imaging heavy-metal-based battery anode with absorption contrast

The research for high-performance anode materials constitutes one of the frontier challenges in developing the next-generation lithium-ion batteries, in particular for addressing the requirements in large-scale applications, e.g., electric vehicles and power grids. Conversion-type transition-metal compounds are regarded as promising anode candidates because of their high theoretical specific capacities. However, these materials undergo substantial volume change upon electrochemical battery cycling, which leads to particle fracturing and structural disintegration of the electrode. This phenomenon severely hinders the adoption of these materials in practical applications and, therefore, an in-depth understanding of the conversion-type battery anode's chemomechanical interplay is vitally important.

To visualize the structural transformation of the conversion-type battery anode upon battery operation, synchrotron-based x-ray microscopy with absorption contrast is an ideal tool. Here we discuss, in figures 3.4(A) to 3.4(E), an *operando* micro-tomographic study of SnO upon electrochemical cycling [16]. In this work, Ebner *et al* utilized the TOMCAT beamline at the Swiss Light Source to conduct time-resolved micro-tomography on an operating battery in a coin-type half-cell configuration. With a nominal spatial resolution at the sub-micrometer level, many particles can be visualized in a single scan (figure 3.4(A)). This work chose to study the battery at a rather slow charging rate and, therefore, tomograms were recorded every 15 min. A 2D virtual slice over two selected particles was visualized

Table 3.1. Synchrotron x-ray imaging techniques for battery research.

Techniques	Imaging modality	Energy range	Spatial resolution	Time resolution	Sample thickness	Advantages	Limitations
Nano-TXM	2D/3D morphology; electron density (phase contrast); element; chemistry	5–14 keV (hard x-ray); 200–5000 eV (soft x-ray)	20–30 nm (hard x-ray); sub-10 – 10nm (soft x-ray)	A few seconds to a few minutes	Tens of nanometers to tens of micrometers	High time resolution; chemical analysis; atmosphere condition	Weak image contrast for light elements for hard x-ray TXM
Micro-TXM	2D/3D morphology and structure imaging	Up to 60 keV	sub-μm– tens of micrometers	A few seconds to a few minutes	Micrometers to centimeters	High time resolution; large field of view	Low resolution; weak contrast for light elements
STXM	Element (including light elements)	<2 keV	12–40 nm	Hundreds of seconds to a few hours	Tens of nanometers to hundreds of nanometers	Chemical analysis; light-element mapping	Thin sample; vacuum chamber
CDI	2D/3D morphology, electron density	200 eV–15 keV	Several nanometers	A few seconds	Hundreds of nanometers to tens of micrometers	Higher spatial and temporal resolution	Small field of view
BCDI	Strain mapping; defects or distortions; 3D image	5 keV–12 keV	sub-10–10 nm	A few minutes to a few hours	Hundreds of nanometers to tens of micrometers	Strain and structural analysis	Crystalline structure sample required
Ptychography	2D/3D morphology; electron density; Element; chemistry	200 eV–15 keV	sub-5–10 nm	A few minutes to a few hours	Tens of micrometers	Trace element detection (down to ppm atomic concentration)	Low time resolution

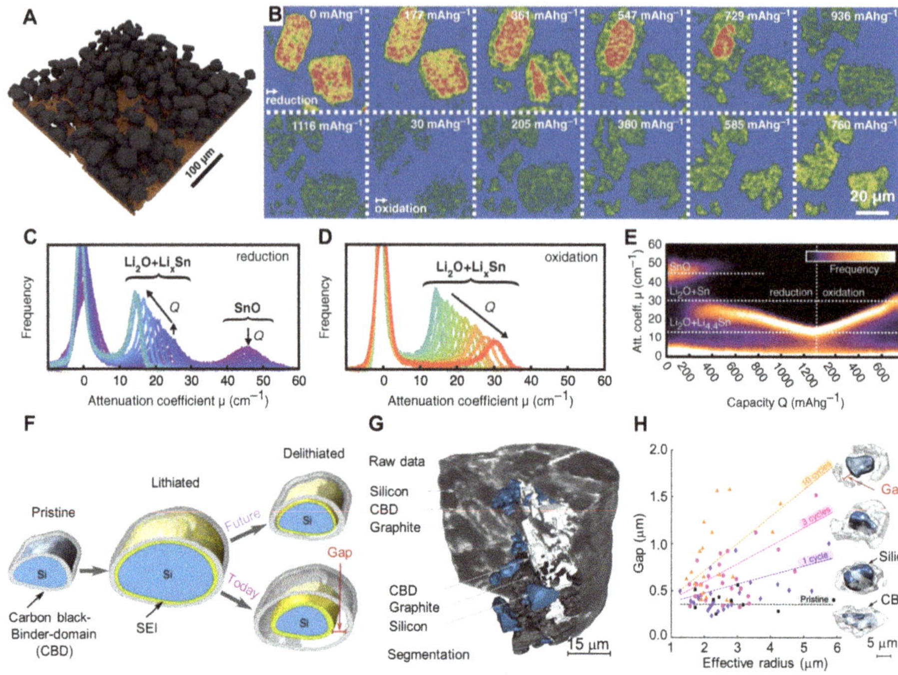

Figure 3.4. Synchrotron-based micro- and nano- tomographic studies on conversion-type battery anode materials. Panels (A) to (E) is an *operando* micro-tomographic study of SnO anode. Panel (F) is an illustration of the degradation mechanism in silicon electrode involving the detachment of Si particle from the CBD matrix. Panel (G) shows the tomographic data of a pristine silicon–graphite electrode that is used for multi-phase segmentation. Panel (H) shows the averaged gap between the surface of the silicon particle and the corresponding carbon black-binder domain as a function of the respective particle size and cycling history. (A–E) Reprinted with permission from [16]. Copyright 2013 American Association for the Advancement of Science. (F–H) Reprinted with permission from [20]. Copyright 2018 Springer Nature.

throughout the whole charging-discharging cycle (figure 3.4(B)) with the probability distribution of the attenuation coefficient quantified at different states of charge (figures 3.4(C), 3.4(D), and 3.4(E)). It shows that the absorption coefficient, which is correlated with the density of the material, can be utilized to fingerprint different chemical species in the SnO anode because of the dramatic volume expansion and contraction upon cycling. It is worth mentioning that the temporal resolution of synchrotron-based micro-tomography can reach ~1 s level [17, 18], which opens up the possibility of dynamic battery imaging under fast charging conditions. Rapid synchrotron micron-resolution radiography has even been utilized to image a thermal runaway process [19]. These x-ray imaging studies that require very high frame rate are not feasible without the intense x-ray beam from a synchrotron facility.

With the development of x-ray optics, e.g., the Fresnel zone plate, the spatial resolution of synchrotron-based x-ray absorption contrast tomography has been pushed down to the level of tens-of-nanometers. To visualize and quantify the degradation mechanism including the detachment of the carbon black-binder network from the active Si anode material, Müller *et al* performed the absorption-based transmission x-ray microscopy (TXM) experiments to image the carbon–silicon composite electrodes (figure 3.4(F)) at the 32-ID-C beamline of the Advanced Photon Source at the Argonne National Laboratory [20]. Figure 3.4(G) shows the reconstructed tomogram of a pristine sample, in which the four-phase segmentation (the silicon particles, the carbon black-binder domain, the graphite particles, and the pore space) is demonstrated. This imaging experiment was conducted for four different electrodes at different cycle states: the pristine state (black squares), after one cycle (purple diamonds), after three cycles (magenta circles), and after ten cycles (orange triangles). The results demonstrate that the gaps increase with increasing silicon effective particle radius and increasing cycle number (figure 3.4(H)), featuring an interesting dependence on the electrode's local morphology and electrochemical history. Weker *et al* utilized the transmission x-ray microscope at beamline 6–2c at the Stanford Synchrotron Radiation Lightsource to capture germanium (Ge) anode particles in a functioning battery pouch cell [21]. They reported significant size dependence on the cycling characteristics of Ge particles. Small particles appear to be more structurally robust as only Ge particles with diameters larger than a few microns develop cracks upon cycling. They also observed that small Ge particles experience volume expansion before their larger counterparts, but they rapidly lose electrical contact and become deactivated in the subsequent battery operation. For more detailed interpretation of the nano-tomography results, refer to the original publication [21].

3.3.2.2 Imaging battery electrode with x-ray phase contrast

As demonstrated in the conversion-type anode studies discussed above, it is clear that attenuation-based synchrotron tomography is a powerful tool for battery research. This imaging modality, however, is not effective when materials that consist of low-Z elements are of interest. Most of the low-Z materials are nearly equally transparent to hard x-rays, resulting in extremely poor absorption contrast. In a commercial lithium-ion battery, for example, the carbon/binder matrix and the pore structure in the cathode, the polymer separator, the graphite anode, the Li dendrites, and the liquid electrolyte are all low-Z materials with very significant scientific importance and technological implications. X-ray imaging of these battery components with absorption contrast has not been very successful. One could still improve the image quality by increasing the x-ray exposure time in order to accumulate the signal. This approach, however, has a rather poor dose efficiency and often causes radiation damage that jeopardizes the study in a different way.

An alternative approach to address this challenge is to conduct x-ray phase contrast imaging (XPCI) [22, 23], which generates image contrast stemming from the x-ray phase shift induced by the sample. The strength of attenuation and phase shift of a certain material is determined by its complex refractive index, $n = 1 - \delta + i\beta$, where δ describes the phase shift and β determines the attenuation [2]. The phase term (δ) is typically about three orders of magnitude higher than the attenuation term (β) at an x-ray energy between 5 and 120 keV for the low-Z materials [24]. This suggests that XPCI could offer a significantly increased contrast for imaging low-Z

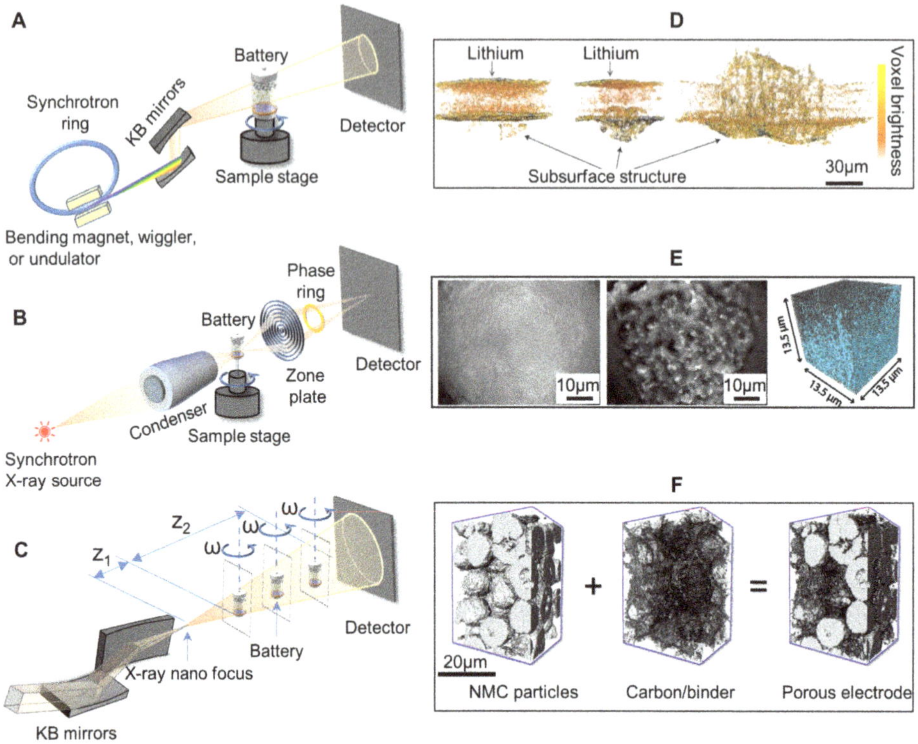

Figure 3.5. The left column contains schematic illustrations for three different XPCI methods, including (A) propagation-based in-line XPCI, (B) zone plate and phase ring-based Zernike-type XPCI, and (C) holography phase contrast. The right column contains three case studies on battery electrode materials using the respective methods on the left. (D) Reprinted with permission from [29]. Copyright 2014 Springer Nature. (E) Reprinted with permission from [32]. Copyright 2020 American Chemical Society. (F) Reprinted with permission from [35]. Copyright 2020 Springer Nature.

samples. Another inherent advantage of XPCI for imaging soft materials is that δ decreases at a rate that is much slower than that of β ($1/E^2$, as compared to $1/E^4$ in the hard x-ray region) with increasing x-ray energy [25, 26]. These advantages have attracted broad interest and have motivated a number of battery studies using XPCI.

There are many different experimental configurations designed to achieve XPCI and specific phase retrieval algorithms have been developed to serve these XPCI methods. In this chapter, we focus on three different approaches used in XPCI studies of battery materials using synchrotron-based hard x-ray illumination: (1) propagation-based in-line XPCI, (2) zone plate and phase ring-based Zernike-type XPCI, (3) holographic phase contrast.

Figure 3.5 A outlines a simple and popular XPCI method known as the propagation-based in-line XPCI, which is remarkably similar to a micro-tomography setup except that the detector is placed at a distance downstream of the sample. This method was pioneered by Nugent [27] and Wilkins [28] in the 1990s. It is quite robust in the sense that it requires very minimum modification to a standard synchrotron imaging setup and with no additional components. This method can also work with x-rays of a rather broad bandwidth and, therefore, can make better use of the intense polychromatic x-ray beam from a synchrotron source. The in-line propagation induces a 'halo' on the interface between two materials of different refractive indices. This edge enhancement effect can be directly utilized for outlining the sample morphology, but a phase retrieval procedure is often necessary for a more quantitative analysis. Harry *et al* utilized synchrotron hard x-ray phase-contrast microtomography at beamline 8.3.2 at the Advanced Light Source to image symmetric lithium–polymer–lithium cells that are cycled at 90 °C [29]. They show that during the early stage of dendrite development, the bulk of the dendritic structure lies within the electrode, underneath the polymer/electrode interface. Monochromatic hard x-rays with energies in the 22–25 keV range were chosen for this experiment due to the desired penetration depth, which, in turn, results in very low absorption contrast. By conducting tomographic reconstruction of the phase contrast images, they visualize the evolution of the dendrite growth in the cell. Their data suggests that there are always some subsurface structures within the lithium electrode under the dendritic structure emanating from the electrode/electrolyte interface (figure 3.5(D)). This study suggests that the sub-surface microstructure design for the Li metal electrode is a viable approach for suppressing the development of the Li dendrite, which is an intensively studied topic in this field.

Another x-ray phase contrast approach that is compatible with TXM leverages the developments by Zernike in the optical regime (figure 3.5(B)). TXM utilizes a Fresnel zone plate as its objective lens, which typically magnifies the image by a factor of ~50 before the x-rays are converted into visible light by a scintillator. Zernike-type x-ray phase contrast imaging with a TXM setup achieves phase contrast by inserting a phase ring at the back focal plane of the objective Fresnel zone plate such that the zero-order beam is modulated by the phase plate, while the higher-order photons are not. The Zernike-type x-ray phase contrast also features an edge enhancement effect [30]. A quantitative phase retrieval is rather challenging despite some efforts in this field [31]. Su *et al* utilized the TXM setup at ID-32-C beamline at the Advanced Photon Source to image the 3D morphology of a Li–O_2 battery electrode that consists of porous carbon and Li_2O_2 [32]. As shown in figure 3.5(E), the phase contrast image demonstrates a dramatically enhanced image contrast in comparison with the absorption image. In addition to utilizing the phase contrast modality, the authors implemented a machine learning approach for image segmentation, which facilitates the further quantification analysis more efficiently. By analyzing the cathode morphology and the spatial arrangement of the Li_2O_2 phase, the authors derived the potential implications in terms of carbon surface passivation and pore-clogging. This approach can also be utilized to image the residual Li_2O_2 and other byproducts after a designed cycling process, which would offer insights into the structure-functionality-degradation interplay at the micro-to-nano-scales.

The x-ray phase contrast imaging can also be carried with a hard x-ray nanoprobe setup by utilizing the nano-sized focal spot as a secondary source and placing the sample downstream of the focal point (figure 3.5(C)) [33, 34]. The sample can be moved between the focal spot and the detector. This approach not only modulates the propagation-induced phase effect, which is needed for a precise phase retrieval, but also effectively adjusts the magnification factor, $(Z_2 + Z_1)/Z_1$ (see annotation in figure 3.5(C)). This method has been broadly utilized for biological studies. Recently, it has been applied to imaging composite battery cathode electrodes with both heavy active materials, e.g., NMC, and light carbon and binder matrix (figure 3.5(F)) [35]. Yang *et al* utilized this method to image a thick cathode electrode and revealed the depth-dependent particle cracking profile in the electrode that is cycled for 10 times. This phenomenon was attributed to the cell polarization effect and is supported by finite element modeling results [36]. Jiang *et al* further quantified the NMC particles' detachment from the carbon/binder matrix [35]. We will discuss this further in a later section in this chapter.

3.3.2.3 Battery imaging with small angle scattering contrast

X-ray scattering occurs as certain x-ray photons depart from their original direction as they interact with the electrons in a material. By observing the interference pattern and by analyzing the characteristics of all scattered waves, the information related to the electron density distribution inside the sample can be unveiled. As a result, x-ray scattering techniques have become a vital method to study the structures of the substance. Small-angle x-ray scattering (SAXS) is one of the most versatile tools to characterize nanoscale structures and dimensions with a considerable small scale of 1–100 nm in a variety of sample types. Synchrotron-based SAXS has been widely adopted to probe liquid electrolytes, including the liquid structures in a water-in-salt electrolyte, the characterization of solvation structures, and redox active electrolytes. We refer to a recent review article for more comprehensive discussion of the SAXS application in battery research [37]. By conducting x-ray scattering measurements with a focused x-ray beam, one could harvest this information in a spatially resolved manner and build up 2D and 3D images through a point-to-point scan. In particular, when the incident beam is of sufficient spatial coherence, ptychographic reconstruction can reveal fine structural details in the sample at a scale that is smaller than the size of the x-ray beam. We refer to a recent study by Tsai *et al* [38], in which the authors utilized the cSAXS beamline (X12SA) at the Swiss Light Source to conduct 3D ptychographic imaging of lithium-rich NMC particles at different states of charge and to correlate the measured electron density with the lattice constants that are measured using a micro diffraction technique. It is worth pointing out that the ptychography technique utilizes the coherent small angle scattering signal for phase retrieval. The scattering power of the sample largely determines the ultimate resolution of this technique. Therefore, we discuss this approach in this section instead of doing so in the previous section dedicated to the phase-contrast battery imaging.

3.3.3 Synchrotron-based diffractive imaging for battery research

When an incident monochromatic synchrotron beam interacts with a battery particle that has good crystallinity, Bragg diffraction occurs in specific angles that satisfy the Bragg condition [39]. The position and the shape of the Bragg diffraction peaks carry useful information regarding the material's lattice configuration, which is important to battery performance. Experimental techniques have been developed to harness this effect for imaging the sample with sensitivity to its lattice arrangements, distortions, and defects.

Hong *et al* utilized the Hard x-ray Nanoprobe (HXN) beamline 3-ID of the National Synchrotron Light Source II at Brookhaven National Laboratory to conduct diffractive imaging on $LiCoO_2$ (LCO) cathode particles with trace element doping [40]. This experiment was carried out using a nanoprobe setup in a raster-scanning-style data acquisition approach (figure 3.6(A)). Unlike the NMC secondary cathode particles, the LCO is usually in the form of single-crystal micron-sized particles. As the nano-focused beam illuminates an LCO particle, several different Bragg peaks that arise from different crystal planes can be observed at different angles. To locate the target Bragg peak, the authors first rotated the sample crystal over a 180-degree range with the diffraction patterns recorded for peak indexing. Then they choose to monitor the (101) peak, which has both the in-plane and out-of-plane structural contributions, as the particle is raster scanned. If the imaged particle is truly single crystal without any lattice imperfections, there shall be no change in the Bragg peak during the raster scan. Instead, motions of the monitored Bragg peak can be observed and were attributed to three different types of lattice distortion: bending, twisting, and d-spacing heterogeneity (figures 3.6(B) and 3.6(C)). The intriguing finding of this work is that the doping-induced pre-existing lattice defects in the LCO particle led to an enhanced lattice stability with a suppressed O_3 to H_{1-3} phase transition as this material is charged to the voltage range of 4.40–4.60 V.

When the incident beam has a good spatial coherence, the intensity pattern around a Bragg peak can be used to conduct phase retrieval. This approach is known as Bragg coherent diffractive imaging (BCDI) and has been applied to the study of layered oxide cathode materials under operating conditions by Singer *et al* [41] using the ID-34 beamline at the Advanced Photon Source and the P10 beamline at PETRA III. In this experiment, the sample was rotated, and 3D phase maps were reconstructed as the battery was charged to high voltage. Dislocation lattice defects can be visualized using this method and the formation and evolution of dislocation networks inside this particle were followed as a function of the state of charge. While this method is very powerful, the BCDI experiment has its complexity, and it is challenging to combine the BCDI technique with the other imaging modalities for a correlative study. This is because, in the BCDI study of battery materials, usually many particles are simultaneously illuminated by the x-ray beam and a few selected Bragg peaks are chosen for further investigation. It is often not possible to identify which particle gives rise to the imaged Bragg peak.

3.3.4 X-ray spectro-microscopy for battery research

Another major advantage provided by synchrotron radiation is its tunability over a wide energy range, which then allows incorporation of compositional and chemical sensitivities into the imaging data. Thus, by exploiting the x-ray absorption edges of

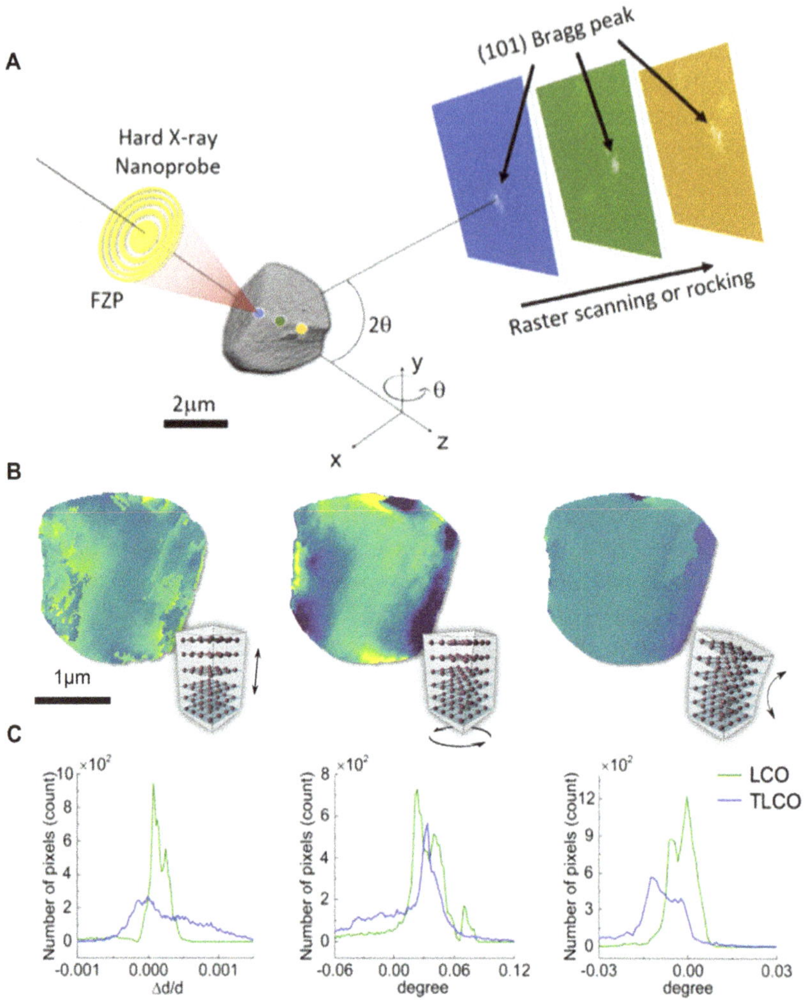

Figure 3.6. Panel (A) shows a schematic illustration of the synchrotron-based hard x-ray nanoprobe setup for raster imaging with XRD signal. Panel (B) shows the spatial distributions of different lattice distortion modes in a Ti-doped LCO particle. Panel (C) shows the comparisons between a Ti-doped LCO particle and a bare LCO particle, highlighting the trace-doping-induced lattice deformations that are indicated by the broadening of the curves. (A–C) Reprinted with permission from [40]. Copyright 2020 Elsevier.

the different materials, this approach essentially combines the x-ray spectroscopy with x-ray microscopy and, therefore, is termed x-ray spectro-microscopy [42, 43]. Spectro-microscopy experiments can be carried out using different imaging configurations in either the full-field imaging mode [42] or the scanning imaging mode [44], both of which are compatible with the tomography technique for unraveling the elemental and/or chemical distributions in 3D. This method has been broadly adopted for battery research. In this section, we discuss the mesoscale x-ray spectro-microscopic study of battery cathode particles [45], featuring the heterogeneity in composition and local state of charge at the particle level.

3.3.4.1 Synchrotron-based x-ray spectro-imaging with compositional sensitivity
There is a broad interest in and strong desire that battery cathode materials maintain their structural robustness against the undesirable side reactions that occur when the cathodes are cycled to high cut-off voltages. While high-voltage cycling is an effective strategy for improving the battery energy density, it induces local phase transformations, surface passivation, particle cracking, and oxygen release, all of which are harmful to the battery's state of health. In a study by Lin *et al* [8] superior stability was demonstrated in the $LiNi_{0.4}Mn_{0.4}Co_{0.2}O_2$ (NMC-442) cathode materials that are synthesized through a spray pyrolysis approach. The authors turned to energy-resolved full-field TXM for investigating the underlying mechanism. Nano-resolution spectro-tomography was conducted for resolving the spatial distribution of the three transition metal elements (Ni, Co, and Mn) in the NMC-442 spherical particles. More specifically, x-ray nano-tomography was conducted at six different energies (at 8344, 8324, 7720, 7700, 6550 and 6530 eV), above and below the K-edges of Ni, Co, and Mn, respectively. After the tomographic reconstruction, the elemental distributions and associations can be directly visualized (figure 3.7(A)) and quantified in a depth-dependent plot (figure 3.7(B)). Their data indicates that the particle surface is relatively rich in manganese and poorer in cobalt and nickel relative to the nominal NMC-442 composition. This depth-dependent metal segregation effect results in a more robust particle surface and, subsequently, improves the high-voltage cycling performance for the NMC-442 cathode synthesized through the spray pyrolysis method. The nano-resolution full-field spectro-tomography technique played a significant role in this study.

Full-field spectro-tomography provides rather efficient 3D imaging capabilities with compositional sensitivity. However, this method has limited sensitivity and can only resolve elements with relatively high local concentrations. This acceptable for studying the spatial distribution of Ni, Mn, and Co in an NMC-442 particle, but it is not feasible to image the trace elements using this method. Trace amounts of foreign element doping in cathode materials is a prevailing and promising strategy to improve the practical discharge capacity with an acceptable level of cycle reversibility. It is also useful to point out that the doping concentrations adopted in industrial production are exceedingly lower than those in laboratory-scale experiments and that co-doping with multiple elements is often carried out in practice with expectation for a synergistic effect among multiple dopants. However, the functional roles of these trace dopants are not well-understood and the optimization of the synthesis procedure including the doping process is often carried out by trial-and-error. To investigate the fundamental mechanisms, the detection of trace elements with high sensitivity is of pivotal importance but is a daunting challenge.

In a recent study, Zhang *et al* reported the study of $LiCoO_2$ (LCO) with trace amounts of Ti–Mg–Al co-doping (~0.1 wt% for each dopant) (TMA-LCO) using nano-resolution fluorescence 3D mapping [46]. This material exhibits an improved structural stability at a high charging voltage of 4.6 V. The scientific goal of their study is to understand the roles of these dopants and the authors turned to 3D x-ray fluorescence imaging for resolving the spatial distribution of the low-concentration

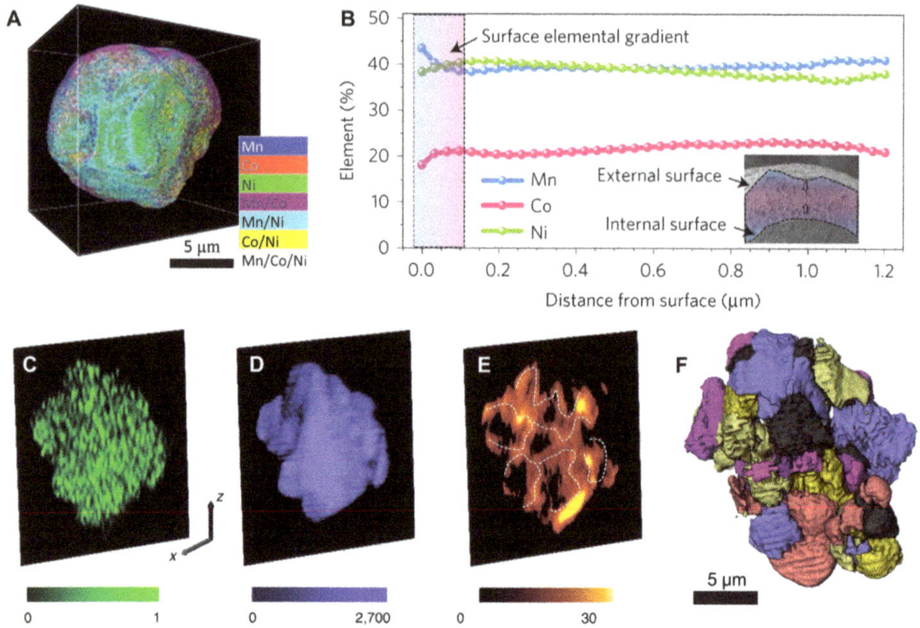

Figure 3.7. Synchrotron-based spectro-tomography on battery cathode materials with compositional sensitivity. Panel (A) is a 3D rendering of the elemental association map for an NMC cathode particle synthesized through spray pyrolysis approach. The depth-dependent elemental distribution for this particle is shown in panel (B), highlighting a Mn-rich surface. Panels (C) to (E) show the distribution of Al, Co, and Ti over a virtual slice through the center of a trace doped $LiCoO_2$ cathode particle. The particle is broken down to several domains that are separated by Ti-segregated grain boundaries. Panel (F) shows the sub-particle domain separated by the Ti-rich boundaries. (A–B) Reprinted with permission from [8]. Copyright 2016 Springer Nature. (C–F) Reprinted with permission from [46]. Copyright 2019 Springer Nature.

dopants at the mesoscale. The x-ray fluorescence signals fingerprint the elemental concentrations with remarkably high sensitivity (down to parts-per-million, ppm). The authors conducted fluorescence mapping using a hard x-ray nanoprobe setup with a nano-focused beam of 50 nm. The tomographic scan was conducted on a total of 51 different viewing angles from −75° to 75°, and an iterative algorithm was employed to improve the reconstruction quality. As shown in figures 3.7(C) to 3.7(E), the spatial arrangements of Al, Co and Ti over a virtual 2D slice throughout the center of the particle have distinct features. It appears that Al is distributed across the entire particle, with a small variation in its concentration. In contrast, a large degree of Ti segregation was observed in figure 3.7(E). The Ti-rich phase constitutes a complex interconnected network (as highlighted by the dashed lines in figure 3.7(E)), which divides the $LiCoO_2$ particle into several subdomains, as shown in figure 3.7(F). These experimental observations hint at different functional roles of these co-existing dopants. While Mg and Al atoms can be incorporated into the LCO lattice, affecting the phase transition behavior in the cycling process, even a trace amount of Ti would spontaneously segregate, forming Ti-rich grain boundaries that modify the microstructure of the particle and promotes the mechanical robustness of the LCO cathode. We highlight here again that the fluorescence mapping approach using a synchrotron nanoprobe beamline can resolve the distribution of trace elements with well-balanced spatial resolution and compositional sensitivity. It is applicable to a wide range of battery research, in which trace element doping is adopted.

3.3.4.2 Synchrotron-based x-ray spectro-imaging with chemical sensitivity
In addition to imaging, the battery material with compositional sensitivity, imaging with chemical sensitive probes is also particularly important for this field. In particular, the battery electrochemical cycling is accompanied by wanted and/or unwanted redox reactions. Spatially resolving the oxidation states of selected elements of interest can help the researchers to understand the reaction heterogeneity and kinetics, which are directly related with the battery performance and degradation.

The absorption coefficient of a certain element exhibits a steep jump as a function of x-ray energy at its absorption edge, whereas it is a relatively smooth curve away from the absorption edge. Moreover, the absorption edges of different elements are often well-separated in energy. When combined with x-ray absorption spectroscopy (XAS) [47, 48], the x-ray imaging approach becomes a powerful tool for studying chemical heterogeneity. This experimental approach can be demonstrated through collecting a series of x-ray transmission images as the x-ray energy is tuned across the absorption edge of a targeted element. Here we discuss an *operando* spectro-imaging study of an industry-relevant LCO cathode material by Xu *et al* [49]. The authors designed and constructed a pouch cell setup that is compatible with the transmission x-ray microscope at beamline 6–2c at the Stanford Synchrotron Radiation Lightsource. They used this cell to monitor the LCO particle's behavior under different cycling conditions. Before the spectro-imaging experiment, the spectroscopic fingerprints of the two principal chemical compounds, the LCO at a fully charged state (at 4.6 V) and a fully discharged state (at 3.0 V), were measured using the conventional bulk-averaged XAS technique. These two signature spectra were then applied to a linear combination fitting of the *in situ* XANES imaging data. The evolution of the phase transformation in the studied LCO particles exhibits a clear cycling-rate-dependence, as shown in figure 3.8(A). In addition to the mesoscale chemical complexity (within an individual particle), in a follow-up study by the same group, they analyzed the spectro-imaging data from over 100 LCO particles (figures 3.8(B) and 3.8(D)) and captured a particle in the form of unanticipated metallic Co (figures 3.8(B) and 3.8(C)). This is attributed to Co cation dissolution from the cathode and precipitation on anode side, which is a side reaction that has been long speculated without direct evidence. We will discuss this case study further in the next section.

Extending into the third spatial dimension, Wang *et al* developed a capillary battery cell and conducted spectro-tomographic imaging of a lithium iron phosphate ($LiFePO_4$) cathode particle as the cell is electrochemically operated [50]. This five-dimensional (5D) (that is, three spatial dimensions, energy, and time) dataset directly reveals that the propagation of the reaction front changes from an

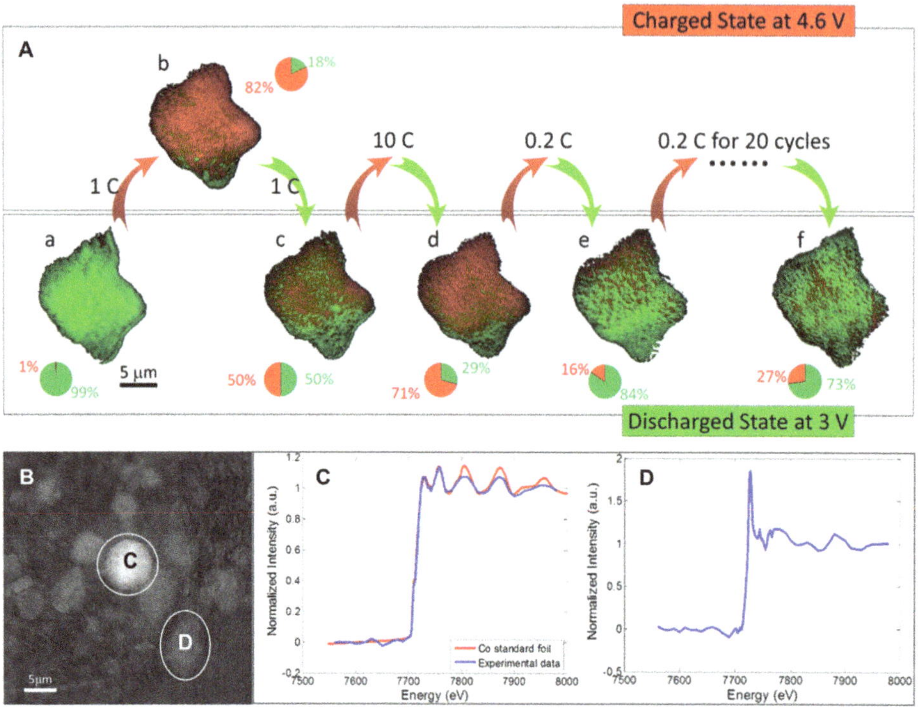

Figure 3.8. Panel (A) shows an *operando* spectro-microscopy study of a randomly selected $LiCoO_2$ cathode particle inside a functioning pouch cell. Panel (B) shows a follow-up study of over 100 particles in the same cell, capturing different spectroscopic fingerprints of different particles imaged (panels C and D). (A–D) Reprinted with permission from [49]. Copyright 2017 American Chemical Society.

anisotropic to an isotropic manner during the delithiation process. This is another excellent example of a spectro-microscopy study of battery materials and it highlights the high-resolution and high-dimensionality of the synchrotron spectro-imaging data. This also necessitates the use of more advanced computational methods, which we will cover in the next section.

3.4 Data science approach for synchrotron-based battery research

The high-resolution and high-dimensional synchrotron-based multi-modal imaging techniques open up remarkable scientific opportunities. At the same time, they bring enormous challenges in data reduction, analysis, and interpretation. In a modern synchrotron-based high-throughput experiment, the data is generated at unprecedented rate. For example, in a full-field spectro-microscopy experiment, nearly four million spectra (2048 × 2048 pixels, typically with some empty regions on the image) can be generated over the course of 10–15 min [42, 43]. This is at a data rate of several orders of magnitude higher than that of a conventional synchrotron spectroscopic scan. In the spectro-microscopy dataset, every spectrum is associated with a spatial unit (i.e., a pixel in 2D or a voxel in 3D), representing the corresponding local chemical characteristics, which are often inhomogeneous, yet functionally important. The full-field spectro-microscopy technique benefits from the use of an area detector, which is now being adopted in many other types of synchrotron experiments. The big data challenge is becoming a common theme for synchrotron-based techniques and a data science approach is becoming a necessity. To get the most out of the advances in synchrotron-based experimental techniques, it is crucial to leverage cutting-edge computing methods that can effectively extract scientifically important information from the large-scale data. In this section, we discuss a few battery studies (from our group) that benefit from the synergy between the synchrotron-based experimental techniques and advanced computational approaches.

As discussed in the previous section, Xu *et al* conducted an *operando* spectro-imaging study of the LCO cathode in a pouch cell setup [49]. While the dynamic evolution of the chemical heterogeneity within a single LCO particle reveals a strong rate-dependent behavior, the statistical representativeness of this type of study can raise questions. After all, the entire cell consists of a large number of particles, a localized observation could miss some other localized chemical evens that are functionally important. To address this concern, Zhang *et al* performed a synchrotron-based *in situ* spectro-microscopic study on hundreds of active LCO cathode particles in the same cell [51]. They scanned over many different fields of views over the cell after the designed electrochemical cycling process. In order to reduce the quantity and complexity of the spectroscopic data, they first extracted several critical data attributes that serve as the key descriptors of the spectroscopic fingerprint. Correlation analysis of these data attributes was carried out to ensure that they were independent so that over-emphasis or under-emphasis of certain spectroscopic features can be avoided. These descriptors were then fed into an unsupervised machine learning algorithm for classification analysis, which then identified certain particles with unanticipated spectroscopic characteristics. The authors concluded their study with the discovery of two unanticipated minority chemical phases, over-lithiated LCO and metallic Co, both of which provide direct evidence and insights into the side reactions that occur during the fast charge-discharge cycles. In this work, domain-knowledge-based principal component analysis was implemented to guide the extraction of spectroscopic data attributes. This approach significantly improved the robustness and the accuracy of the conventional clustering algorithm.

Another illustrative example is the study of a $LiNi_{0.6}Mn_{0.2}Co_{0.2}O_2$ (NMC-622) cathode material by Mao *et al.* The authors implemented a hybrid supervised and unsupervised machine learning approach to map out the spatial distribution of the chemical outliers in the NMC-622 particles that were subjected to a high-voltage cycling process [52]. Although the clustering algorithm is adopted from the literature, the authors carried out the implementation in a way that is informed by their electrochemical treatment of the sample. The algorithm illustrated in figure 3.9(A) features a training stage and an interference stage. They first used an unsupervised clustering algorithm to categorize the spectro-microscopic dataset on mildly cycled NMC-622 particles (cycled between 3 V and 4.4 V) into two subsets, e.g., clusters 2# and 3# in figure 3.9(B). These two clusters stand for two slightly different spectroscopic signatures associated with different oxidation states of Ni, in

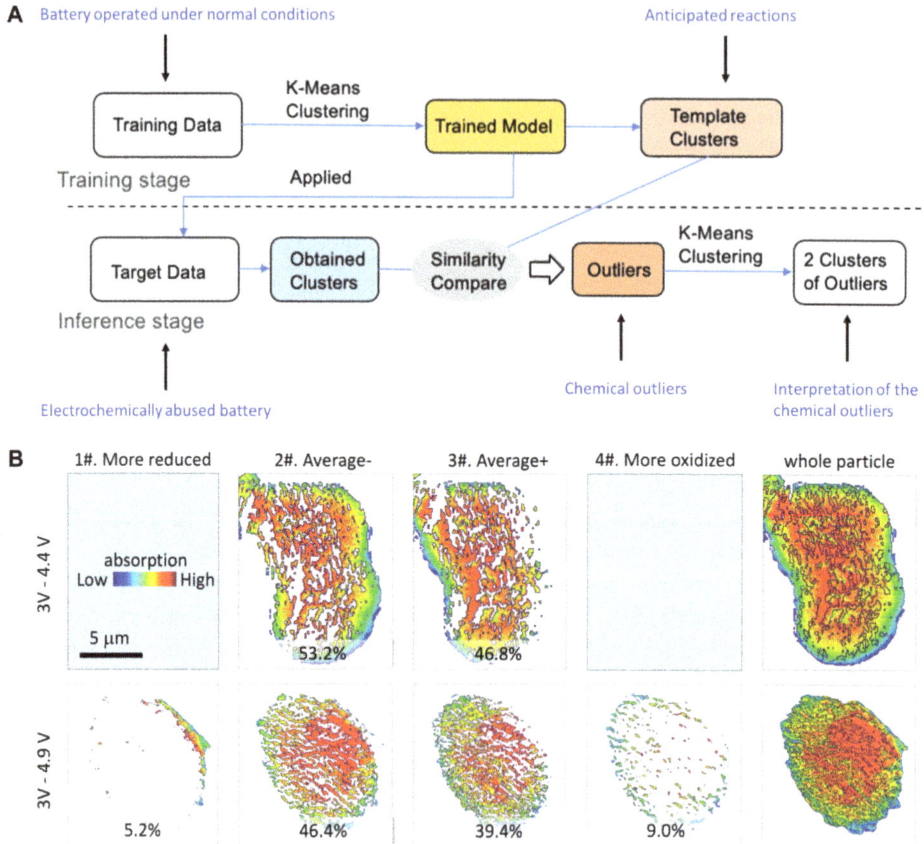

Figure 3.9. Panel (A) illustrates the hybrid supervised and unsupervised data classification algorithm for analyzing large scale spectro-microscopic data on cycled NMC-622 cathode material. Panel (B) shows the comparison between clustering results for the particles cycled under different conditions. While anticipated chemical heterogeneity is observed in the 3–4.4 V cycled particle, the 3–4.9 V cycled particle features two types of distinct chemical outliers (1# and 4#). (B) Reprinted with permission from [52]. Copyright 2019 John Wiley and Sons.

which 2# denotes as more-reduced average oxidation state (average–) and 3# denotes as more-oxidized average oxidation state (average+). The non-uniform distribution of Ni oxidation states reveals the heterogeneity of the Li distribution within the particle. This is because the Ni redox reaction is the main mechanism for charge compensation during electrochemical cycling of NMC-622. The two initial clustering results then serve as template clusters to facilitate a supervised machine learning procedure for further study of those particles that were cycled more aggressively (between 3 and 4.9 V). The spectra with adequate similarity to the clusters 2# and 3# are assigned to the corresponding groups, otherwise the others that have a considerable discrepancy are labeled as outliers. To further understand the chemical state of those outliers, another round of unsupervised clustering was conducted to divide them into two group of outlier clusters, the more-reduced-than-average (cluster 1#) and the more-oxidized–than-average (cluster 4#) groups, as demonstrated in the bottom row of figure 3.9(B).

This work not only shows the sub-particle-level chemical heterogeneity, but also highlights the identification of undesirable secondary phases at the mesoscale in the NMC-622 particle that was aggressively cycled. The advanced computational approach was indispensable in this study. The spatial distributions and the spectroscopic fingerprints of the chemical outliers (clusters 1# and 4#) indicate the co-existence of surface degradation (1#) and the local deactivation of isolated domains (4#) at the mesoscale when the NMC-622 material was deeply delithiated.

In addition to the machine-learning-based classification analysis of the spectro-imaging data, advanced algorithms in computer vision have also been leveraged for analyzing the synchrotron-based nano-tomographic data on battery materials. The visualization of the composite electrode microstructure is invaluable to understand how individual battery particles are charged and discharged in a lithium-ion battery. The conventional absorption-based contrast mechanism suffers from the limited capability of imaging the weak-absorbing matter, such as the carbon/binder domain (CBD) in the electrode. Jiang *et al* carried out a high-resolution hard x-ray nano-tomography based upon x-ray phase contrast modality to address this frontier challenge at the beamline ID16A-NI at the European Synchrotron Radiation Facility [35]. By using this 3D non-invasive virtual imaging with quantitative phase contrast, the carbon matrix, void, binder, and active particles in a Ni-rich $LiNi_{0.8}Mn_{0.1}Co_{0.1}O_2$ (NMC-811) composite cathode were clearly resolved with exceptional image contrast, as shown in figure 3.5(F).

In order to avoid the potentially biased characterization results and to ensure representativeness, statistical analysis on a large number of particles is pivotal for a comprehensive investigation when it comes to the highly complex composite electrode. The segmentation and identification of every single particle in the reconstructed 3D volume is labor-intensive and tedious in the conventional manual approach. Although the exceptional image contrast makes it relatively easy to distinguish NMC-811 from the surrounding matrix, identification of individual particles can be challenging, particularly for the severely damaged particles. This is because some of the particles have severely fractured into detached fragments. To associate these fragments that break away from the same particle requires some degree of 'intelligence' and, therefore, often requires human intervention in the process. Without a proper particle identification/association process, the subsequent statistical analysis could falsely conclude on many small particles, which are just broken fragments in reality.

In order to tackle this issue, a machine learning model was designed to conduct the identification and quantification of over 650 NMC-811 particles automatically with superior accuracy and efficiency. Figure 3.10(A) illustrates the high-level workflow. The authors trained a machine-learning model by providing a training set of nano-tomographic slices from a similar electrode that were marked with manual annotations. The concept of fragment association is passed on to the model in this training process without being formulated explicitly. Then the trained model was applied to process every slice of the targeted volume and the generated 2D

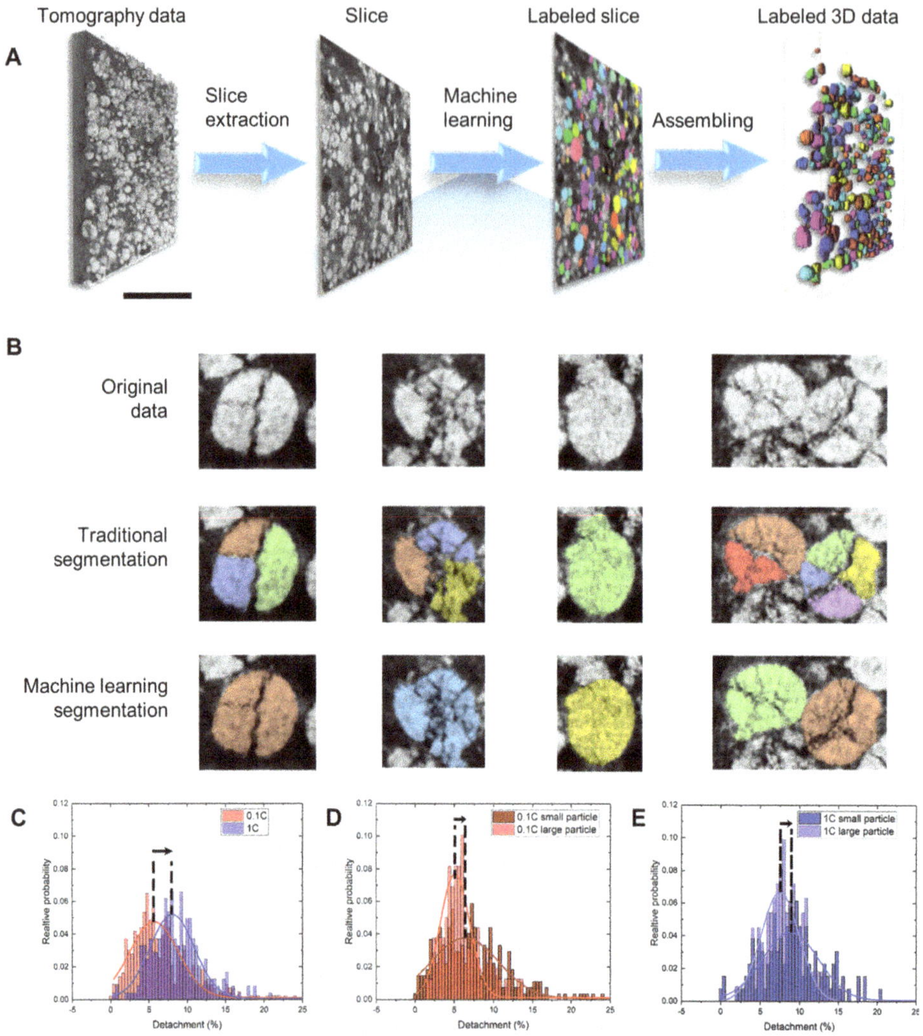

Figure 3.10. Panel (A) illustrates a machine-learning-assisted NMC particle identification approach. Panel (B) compares the segmentation results between the traditional method and the machine-learning approach. Panel (C) shows the statistical analysis of the degree of particle detachment from the carbon/binder matrix in electrodes that were cycled under different rates. The particle size effect is also elucidated in panels (D) and (E). (A–E) Reprinted with permission from [35]. Copyright 2020 Springer Nature.

masks were merged to reconstruct the 3D particles afterwards. Figure 3.10(B) exhibits the comparison of the segmentation results on a few representative particles using the conventional method (watershed and separation algorithms) and the developed machine learning approach. It is clear that the conventional segmentation method failed to distinguish different particles, while the machine learning approach outperformed the conventional approach, highlighting a significantly improved performance and robustness.

Based on the machine-learning-assisted segmentation result, the particles' detachment from the carbon/binder matrix was quantified for every individual NMC particle in the imaged electrodes that were cycled with different electrochemical protocols (0.1C and 1C). They analyzed over 650 NMC particles in total and compared the probability distributions of the degree of particle detachment between the slow (0.1 C, red) and fast (1 C, blue) cycled particles (see figure 3.10(C)). It clearly displayed a shift of the peak, indicating that the particles experienced more severe detachment when cycled at faster charging rate. Such an effect could be part of the reason for the rapid capacity decay associated with the fast-charging conditions. To understand the size effect on the particle detachment, particles were divided into two groups (big and small) based on their respective volumes. The relative probability distribution of the particle detachment in the small and the big groups are compared for the slow (0.1 C, figure 3.10(D)) and fast (1 C, figure 3.10(E)) cycled electrodes, respectively. It is interesting to observe that, in both cases, small particles exhibit a more broadly scattered probability distribution with the peak position slightly shifted to a higher value. Through the machine-learning-assisted analysis, this work revealed that the degree of particle detachment from the carbon/binder matrix is highly dependent on the charging rate and the particle size and this result was concluded with statistical significance.

3.5 Future directions of synchrotron-based battery research

In this chapter, we discussed synchrotron-based experimental tools that have been broadly adopted for battery research. Because of the scientific and technical importance in the structural and chemical heterogeneity in batteries, we elaborated on the synchrotron-based multi-modal and high-dimensional imaging approach. These experimental modalities, respectively, carry valuable information regarding different aspects of the battery material, e.g., the lattice structure, the valence state, and the micro-morphological arrangement, which collectively determine the ultimate functionality of the battery. It is worth mentioning that there is an increasing effort to harness these experimental probes in a correlative and synergistic manner. Examples include (1) the combination of soft x-ray nano-probe and hard x-ray TXM to evaluate the surface-to-bulk mutual modulation in an NMC particle [53]; (2) the correlation of ptychographic reconstructed electron density with the micro diffraction signal to reveal that the morphological degradation in NMC could have radial dependency and secondary particle size dependency [38]; (3) utilizing spectro-tomography and soft x-ray resonant inelastic x-ray scattering (RIXS) to elucidate the interplay between the cationic and anionic redox reaction in Li-rich cathode material [54].

With the projected major developments in the next-generation synchrotron facilities [1, 55], we can foresee emerging opportunities and challenges for synchrotron-based battery research. On the one hand, brighter synchrotron beam could potentially improve the signal-to-noise ratio of the measurement, promote the spatial and temporal resolutions, enhance the sensitivity to material with extremely low concentrations. These advantages, on the other hand, come with the concerns of radiation damage, which could jeopardize the study, in particular for the *in situ* and *operando* experiments. Efforts will be devoted to carefully addressing these concerns and to developing comprehensive strategies for dose mitigation and management. A good example is the recent RIXS study led by Yang *et al* [56]. These types of benchmarking efforts are valuable, in particular in the context of future battery research at the next-generation synchrotron facilities.

Finally, we discussed several case studies that leverage the modern developments in the computational methods for artificial intelligence, e.g., machine learning algorithms. Advanced data-mining approaches offer great opportunities to extract and analyze the important information from the large-scale experimental data in an effective and efficient way. This approach is also applicable to a broader range of research fields well-beyond the battery studies.

References

[1] Willmott P 2019 *An Introduction to Synchrotron Radiation: Techniques and Applications* (New York: Wiley)

[2] Als-Nielsen J and McMorrow D 2011 *Elements of Modern X-ray Physics* (New York: Wiley)

[3] Seibert J A and Boone J M 2005 X-ray imaging physics for nuclear medicine technologists. Part 2: x-ray interactions and image formation *J. Nucl. Med. Technol.* **33** 3–18

[4] Lin F *et al* 2017 Synchrotron x-ray analytical techniques for studying materials electrochemistry in rechargeable batteries *Chem. Rev.* **117** 13123–86

[5] Xia S *et al* 2018 Chemomechanical interplay of layered cathode materials undergoing fast charging in lithium batteries *Nano Energy* **53** 753–62

[6] Tian C *et al* 2018 Depth-dependent redox behavior of $LiNi_{0.6}Mn_{0.2}Co_{0.2}O_2$ *J. Electrochem. Soc.* **165** A696–704

[7] Kuppan S, Xu Y, Liu Y and Chen G 2017 Phase transformation mechanism in lithium manganese nickel oxide revealed by single-crystal hard x-ray microscopy *Nat. Commun.* **8** 14309

[8] Lin F *et al* 2016 Metal segregation in hierarchically structured cathode materials for high-energy lithium batteries *Nat. Energy* **1** 15004

[9] Liu Y, Kiss A M, Larsson D H, Yang F and Pianetta P 2016 To get the most out of high resolution x-ray tomography: a review of the post-reconstruction analysis *Spectrochim. Acta* B **117** 29–41

[10] Bautz W, Kalender W and Godfrey N 2005 Hounsfield and his influence on radiology *Radiologe* **45** 350–5

[11] Wang L, Wang J and Zuo P 2018 Probing battery electrochemistry with In operando synchrotron x-ray imaging techniques *Small Methods* **2** 1700293

[12] Cao C *et al* 2019 Emerging x-ray imaging technologies for energy materials *Mater. Today.* **34** 132–47

[13] Yuan Q, Deng B, Guan Y, Zhang K and Liu Y 2019 Novel developments and applications of nanoscale synchrotron radiation microscopy *Physics (N.Y.)* **48** 205–18

[14] Jacobsen C 2020 *X-ray Microscopy* (Cambridge: Cambridge University Press)

[15] Miao J, Ishikawa T, Robinson I K and Murnane M M 2015 Beyond crystallography: diffractive imaging using coherent x-ray light sources *Science* **348** 530–5

[16] Ebner M, Marone F, Stampanoni M and Wood V 2013 Visualization and quantification of electrochemical and mechanical degradation in Li ion batteries *Science* **342** 716–20

[17] Walker S M *et al* 2014 *In vivo* time-resolved microtomography reveals the mechanics of the blowfly flight motor *PLoS Biol.* **12** e1001823
[18] Mokso R *et al* 2015 Four-dimensional *in vivo* x-ray microscopy with projection-guided gating *Sci. Rep.* **5** 1–6
[19] Finegan D P *et al* 2015 In-operando high-speed tomography of lithium-ion batteries during thermal runaway *Nat. Commun.* **6** 6924
[20] Müller S *et al* 2018 Quantification and modeling of mechanical degradation in lithium-ion batteries based on nanoscale imaging *Nat. Commun.* **9** 2340
[21] Weker J N *et al* 2014 *In situ* nanotomography and operando transmission x-ray microscopy of micron-sized Ge particles *Energy Environ. Sci.* **7** 2771–7
[22] Liu Y, Nelson J, Holzner C, Andrews J C and Pianetta P 2013 Recent advances in synchrotron-based hard x-ray phase contrast imaging *J. Phys. D: Appl. Phys.* **46** 494001
[23] Eastwood D *et al* 2014 The application of phase contrast x-ray techniques for imaging Li-ion battery electrodes *Nucl. Instrum. Methods Phys. Res.* B **324** 118–23
[24] Momose A and Fukuda J 1995 Phase-contrast radiographs of nonstained rat cerebellar specimen *Med. Phys.* **22** 375–9
[25] Wilkins S *et al* 2014 On the evolution and relative merits of hard x-ray phase-contrast imaging methods *Philos. Trans. R. Soc.* A **372** 20130021
[26] Pietsch P and Wood V 2017 X-ray tomography for lithium ion battery research: a practical guide *Annu. Rev. Mater. Sci.* **47** 451–79
[27] Nugent K, Gureyev T, Cookson D, Paganin D and Barnea Z 1996 Quantitative phase imaging using hard x-rays *Phys. Rev. Lett.* **77** 2961
[28] Wilkins S, Gureyev T E, Gao D, Pogany A and Stevenson A 1996 Phase-contrast imaging using polychromatic hard x-rays *Nature* **384** 335–8
[29] Harry K J, Hallinan D T, Parkinson D Y, MacDowell A A and Balsara N P 2014 Detection of subsurface structures underneath dendrites formed on cycled lithium metal electrodes *Nat. Mater.* **13** 69–73
[30] Schmahl G *et al* 1995 Phase contrast studies of biological specimens with the x-ray microscope at BESSY *Rev. Sci. Instrum.* **66** 1282–6
[31] Yang Y *et al* 2014 Approaching quantitative Zernike phase contrast in full-field transmission hard x-ray microscopy: Origin and reduction of artifacts *Appl. Phys. Lett.* **105** 094101
[32] Su Z *et al* 2020 x-ray nanocomputed tomography in zernike phase contrast for studying 3D morphology of Li–O_2 battery electrode *ACS Appl. Energy Mater.* **3** 4093–102
[33] Da Silva J C *et al* 2017 Efficient concentration of high-energy x-rays for diffraction-limited imaging resolution *Optica* **4** 492–5
[34] Cloetens P *et al* 1999 Holotomography: quantitative phase tomography with micrometer resolution using hard synchrotron radiation x-rays *Appl. Phys. Lett.* **75** 2912–4
[35] Jiang Z *et al* 2020 Machine-learning-revealed statistics of the particle-carbon/binder detachment in lithium-ion battery cathodes *Nat. Commun.* **11** 2310
[36] Yang Y *et al* 2019 Quantification of heterogeneous degradation in Li-ion batteries *Adv. Energy Mater.* **9** 1900674
[37] Qian K, Winans R E and Li T 2021 Insights into the nanostructure, solvation, and dynamics of liquid electrolytes through small-angle x-ray scattering *Adv. Energy Mater.* **11** 2002821
[38] Tsai E H *et al* 2019 Correlated x-ray 3D ptychography and diffraction microscopy visualize links between morphology and crystal structure of lithium-rich cathode materials *IScience* **11** 356–65

[39] Le Bail A, Cranswick L M and Madsen I 2008 *Powder Diffraction: Theory and Practice* (Cambridge: Royal Society of Chemistry)

[40] Hong Y *et al* 2020 Hierarchical defect engineering for $LiCoO_2$ through low-solubility trace element doping *Chem.* **6** 2759–69

[41] Singer A *et al* 2018 Nucleation of dislocations and their dynamics in layered oxide cathode materials during battery charging *Nat. Energy* **3** 641–7

[42] Meirer F *et al* 2011 Three-dimensional imaging of chemical phase transformations at the nanoscale with full-field transmission x-ray microscopy *J. Synchrotron Radiat.* **18** 773–81

[43] Liu Y *et al* 2012 3D elemental sensitive imaging using transmission x-ray microscopy *Anal. Bioanal. Chem.* **404** 1297–301

[44] Obst M and Schmid G 2014 3D chemical mapping: application of scanning transmission (soft) x-ray microscopy (STXM) in combination with angle-scan tomography in bio-, geo-, and environmental sciences *EM. Microscopy* (Berlin: Springer) pp 757–81

[45] Wei C *et al* 2018 Mesoscale battery science: the behavior of electrode particles caught on a multispectral x-ray camera *Acc. Chem. Res.* **51** 2484–92

[46] Zhang J *et al* 2019 Trace doping of multiple elements enables stable battery cycling of $LiCoO_2$ at 4.6 V *Nat. Energy* **4** 594–603

[47] Bunker G 2010 *Introduction to XAFS: A Practical Guide to X-ray Absorption Fine Structure Spectroscopy* 1st edn (Cambridge: Cambridge University Press)

[48] Koningsberger D and Prins R 1988 *X-ray Absorption: Principles, Applications, Techniques of EXAFS, SEXAFS, and XANES* (New York: Wiley-Interscience)

[49] Xu Y *et al* 2017 *In situ* visualization of state-of-charge heterogeneity within a $LiCoO_2$ particle that evolves upon cycling at different rates *ACS Energy Lett.* **2** 1240–5

[50] Wang J, Karen Chen-Wiegart Y-c, Eng C, Shen Q and Wang J 2016 Visualization of anisotropic-isotropic phase transformation dynamics in battery electrode particles *Nat. Commun.* **7** 12372

[51] Zhang K *et al* 2017 Finding a needle in the haystack: identification of functionally important minority phases in an operating battery *Nano Lett.* **17** 7782–8

[52] Mao Y *et al* 2019 High-voltage charging-induced strain, heterogeneity, and micro-cracks in secondary particles of a nickel-rich layered cathode material *Adv. Funct. Mater.* **29** 1900247

[53] Li S *et al* 2020 Mutual modulation between surface chemistry and bulk microstructure within secondary particles of nickel-rich layered oxides *Nat. Commun.* **11** 4433

[54] Zhang J *et al* 2020 Depth-dependent valence stratification driven by oxygen redox in lithium-rich layered oxide *Nat. Commun.* **11** 6342

[55] Eriksson M, Van der Veen J F and Quitmann C 2014 Diffraction-limited storage rings–a window to the science of tomorrow *J. Synchrotron Radiat.* **21** 837–42

[56] Li Q *et al* 2021 Could irradiation introduce oxidized oxygen signals in resonant inelastic x-ray scattering of battery electrodes? *J. Phys. Chem. Lett.* **12** 1138–43

IOP Publishing

Batteries
Materials principles and characterization methods
Chen Liao

Chapter 4

Vibrational spectroscopy for batteries

Bertrand J Tremolet de Villers and Sang-Don Han

4.1 Fundamental principles and methods

Vibrational spectroscopy is one of the most important avenues for understanding and quantifying the interfacial and bulk chemical and electrochemical reactions occurring in a battery during operation. Fourier transform infrared (FTIR) and Raman spectroscopy are attractive because of their versatility, and they are used extensively by the battery community in their *ex situ* form. Furthermore, FTIR and Raman modules are ubiquitous across research laboratories because of their relatively low initial cost, reliability, and ease of use. They can also be used to study gases, liquids, and solids. Both techniques provide chemical identification of samples, and each has its own rich history in the advancement of energy storage materials and technologies.

In this chapter, the fundamental principles and measurement methods of FTIR and Raman scattering, the most-commonly available IR spectroscopy, are presented. Then, the importance of vibrational spectroscopy to various aspects of rechargeable battery research is discussed, with several examples illustrating the usefulness of the techniques. Moreover, special consideration is given to recent efforts involving *in situ* electrospectroscopic analysis during battery cycling. This provides additional insight into the chemical and electrochemical mechanisms underlying the battery performance that is largely inaccessible with traditional *ex situ* measurements. Of course, neither technique is perfect on its own, some of their biggest limitations are acknowledged here, but they are complimentary to each other and to other advanced characterization tools such as optical and atomic microscopy, x-ray-based spectroscopies, electrochemical and analytical evaluations. Finally, the authors provide their perspective on the continuing evolution of vibrational spectroscopy in electrochemical energy storage research.

doi:10.1088/978-0-7503-2682-7ch4

4.1.1 Fourier-transform infrared spectroscopy (FTIR)

As the name implies, Fourier-transform infrared spectroscopy (FTIR) involves the absorption of infrared (IR) radiation by a molecule. Infrared is the region of the electromagnetic spectrum encompassing energies just below the visible region. Furthermore, FTIR spectroscopy is usually performed in one of three IR sub-regions—near-IR from 12500–4000 cm^{-1} (0.8–2.5 μm), mid-IR from 4000–400 cm^{-1} (2.5–25 μm), and far-IR from 400–10 cm^{-1} (25–1000 μm). Whereas ultraviolet and visible light can excite an electron within an atom or molecule, IR light excites a molecular bond vibration [1]. The absorption frequency of a bond vibration is specific to the atoms participating in the bond, thus FTIR provides specific chemical identification. For example, carbon–hydrogen bond stretching excitations occur at 3300–2850 cm^{-1} and carbon–oxygen double bond (carbonyl) stretching excitations occur at 1850–1650 cm^{-1}. Tables listing the vibration excitation modes and IR absorption energies for many different chemical functional groups are readily available [2].

Conventionally in FTIR, the absorption frequency or wavenumber has units of inverse centimeter (cm^{-1}) and derives from the fact that the technique, which will be discussed in more detail below, relies on the physical translation of a mirror that is moving at a constant speed. Analogous to the temporal frequency that is the number of waves per time, the wavenumber is a spatial frequency that is the number of waves per distance. The temporal frequency and wavenumber are simply related by the constant speed of light in a vacuum, $\approx 3e^8$ m s^{-1}, and therefore wavenumber is the inverse of wavelength. For ease of conversion between wavenumber and wavelength, one should remember, wavelength (μm) = $1e^4$ μm cm^{-1}/wavenumber (cm^{-1}).

Most light absorption measurements rely on a beam of monochromatic light that passes through a sample. The wavelength of the light is then varied, usually with a prism or a diffraction grating. The intensity change of the light at each wavelength is then displayed as an absorption spectrum. For example, this is the most common method applied in a UV–vis spectrophotometer. FTIR works in a different manner in that a broad band of light shines through the sample and then onto a detector. Wavelength or frequency selection is achieved by passing the light through an interferometer before the sample. The interferometer usually consists of a beam splitter, a fixed mirror, and a moving mirror. Typically, the beam splitter directs 50% of the light intensity to each mirror. After reflecting from each mirror, the light beams return to the beam splitter and the light waves interfere with each other before passing onto a detector. Because the 'broadband' light consists of many different wavelengths at once, some wavelengths will constructively interfere while others will destructively interfere resulting in a complex interference pattern. After the moving mirror translates a short distance, the path length of the light to that mirror has changed slightly, resulting in a different interference pattern after the beam splitter. Measurements are taken at many discrete positions of the movable mirror, and then a Fourier transform is used to reconstruct the spatial spectrum, in micrometers or inverse centimeters [3].

There are four common configurations used in FTIR measurements: transmission, attenuated total internal reflection (ATR), specular reflection, and diffuse reflection infrared Fourier-transform (DRIFT). Figure 4.1 illustrates the concepts of each configuration. In transmission mode, the IR light passes directly through the sample (exactly one time) with the absorption path length determined by the sample thickness and angle of incidence. Transmission FTIR can be used to measure a solid pellet or film, a liquid usually sandwiched between two salt plates, or a gas in a sealed tube with IR windows [1, 3].

Due to the ease of the sample preparation and measurement, ATR-FTIR has become a very popular configuration for most laboratory FTIR spectrometers. In ATR-FTIR, the IR excitation light is first passed through a high refractive index IR-transparent crystal at a critical angle (or above) such that the light totally internally reflects at the crystal surface in contact with the sample. Due to the wave nature of the light and in accordance with Maxwell's equations in a dielectric medium, there is a surface wave that is created at the reflection boundary between the ATR crystal and sample [4]. This surface wave is known as an evanescent wave and propagates along the surface of the crystal. Its intensity decays exponentially perpendicular to the surface. Therefore, the ATR-FTIR measures the absorption only up to a few microns into a sample. The penetration depth (d) is defined as the distance from the interface where the intensity decays to $1/e$ of its maximum; and this depends on the

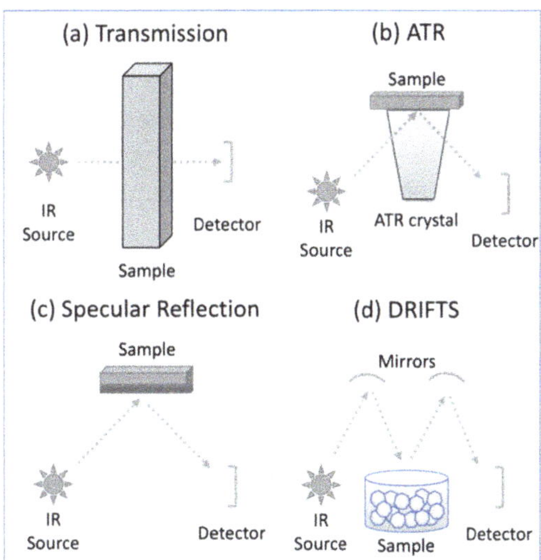

Figure 4.1. Conceptual illustrations of the four most common FTIR measurement configurations. (a) Transmission is amenable to measuring a solid pellet or film, a liquid sandwiched between two salt plates, or a gas in a sealed tube with IR windows. (b) Attenuated total internal reflection (ATR) is used to measure smooth solid samples and liquids and requires very little sample preparation. (c) Specular reflection mode is used for flat and shiny samples that are not easily measured by other configurations. (d) Diffuse reflection infrared Fourier-transform (DRIFT) requires meticulous sample preparation but is useful to measure solid powders.

refractive indices of the ATR crystal (n_1), the sample (n_2), the IR wavelength (λ), and the angle of incidence (θ) according to the equation,

$$d = \frac{\lambda}{2\pi\, n_1\sqrt{\sin^2\theta - (n_2/n_1)^2}}$$

The most common ATR crystals are diamond ($n_1 = 2.40$), zinc selenide ($n_1 = 2.43$), silicon ($n_1 = 3.42$), and germanium ($n_1 = 4.01$). As an example, figure 4.2 shows how the penetration depth varies with frequency, using a diamond ATR crystal, a sample with $n_2 = 1.4$ (typical of a carbonate-based liquid electrolyte for lithium-ion batteries), and $\theta = 45°$. Note that because the penetration depth varies by a factor of 10 across the spectrum and the sample refractive index is not actually constant with frequency, the band intensities in an ATR spectrum cannot be compared directly to the corresponding transmission spectrum without applying the appropriate corrections. ATR-FTIR is very useful to measure liquids and most types of solid samples as long as they can make good contact to the ATR crystal, they are not too rough (not IR scatterers), and their refractive index is not similar to that of the ATR crystal.

In specular reflection, the light is directly reflected from the sample surface and the reflection angle is equal to the incident angle. The sample must be flat and shiny, exhibiting little to no scattering (diffuse reflections). Light reflectance and transmittance at a surface are governed by the Fresnel equations [1]. For IR absorbing materials, the imaginary part of the wavelength-dependent complex refractive index can be significant which leads to very strong reflection bands in certain areas of the spectrum. As an example, the specular reflectance of a polystyrene film is shown in figure 4.3 (top). To compare with absorption spectra for chemical identification, it is necessary to first apply a mathematical Kramers–Kronig transform, figure 4.3 (bottom). Specular reflectance is mostly used on existing samples that cannot be measured with other FTIR configurations, for example, determining the surface composition of a painting or piece of art.

Sometimes reflection FTIR refers to a technique that is more accurately described as transflection spectroscopy. In this configuration, the sample is a material that has

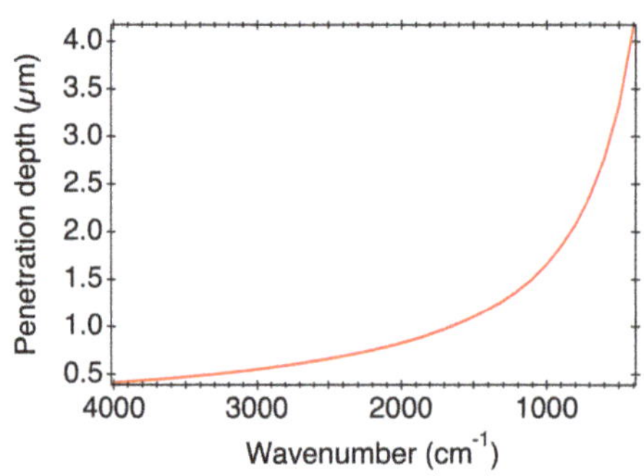

Figure 4.2. ATR-FTIR penetration depth varies with excitation light frequency. Here, the values of the refractive indices of the ATR crystal and the sample are 2.4 and 1.4, respectively, the angle of incidence is 45°.

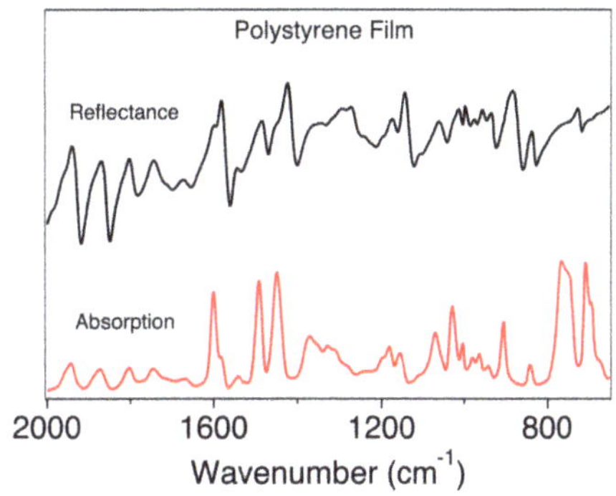

Figure 4.3. (a) Specular reflectance FTIR spectrum of a polystyrene film and (b) the absorption spectrum derived from a Kramers–Kronig transformation of the reflectance spectrum.

been deposited on a flat reflective surface such as a metal. The IR light penetrates the sample, then reflects off the substrate for a second pass through the sample. This method is akin to transmission mode but with a longer sampling pathlength to compensate for materials that have a weak absorption. In the special case when the sample thickness is much less than the wavelength of light and the incidence angle is kept small (grazing incidence), the polarization of the excitation light becomes important and can provide surface enhancement at the substrate to greatly amplify the absorption signal. This case is often called infrared reflection-absorption spectroscopy (IRRAS) or surface-enhanced infrared reflection-absorption spectroscopy (SEIRRAS) and can be used to study monolayer coatings on a metal.

The fourth commonly used FTIR configuration is diffuse reflection infrared Fourier-transform spectroscopy (DRIFTs). This method is used to measure scattering solids such as powders or films with rough surfaces [1]. Often finely-ground powder samples are diluted in a matrix of potassium bromide to increase the excitation depth into the sample which creates more scattered light for collection. It is important to note that within the sample, the excitation light not only reflects off particle surfaces, but also passes through the particles (refraction) and thus the collected spectrum looks somewhat like an absorption measurement. The Kubelka–Munk model is often applied to separate the scattering contribution from the absorption, assuming that the scattering coefficient is constant. In practice, the accuracy of a DRIFTs measurement depends greatly on careful sample preparation. However, if done correctly, DRIFTs can yield excellent qualitative and quantitative characterization of powders.

Finally, it is possible to achieve both chemical and spatial information by coupling IR measurements with microscopy or imaging. Transmission IR microscopy requires a thin (< 10 μm), smooth and level sample, therefore sample preparation can be challenging. More often, reflection IR microscopy is used. In this case, a micro-ATR crystal is integrated onto the microscope objective, and the principles of the FTIR measurement are the same as traditional ATR-mode. The spatial resolution of the microscope can be as small as a few microns and depends on the numerical aperture of the objective and the excitation wavelength used.

4.1.2 Raman spectroscopy

Raman spectroscopy is another type of vibrational spectroscopy, but instead of characterizing light absorption, it involves the measurement of inelastically-scattered light [5]. The scattering occurs when incident light interacts with the movement of a molecule's electron cloud (polarizability) during bond vibration. The energy difference between the incident and scattered light, known as the Raman shift, corresponds to the mid-infrared, thus both Raman and IR spectra of a material can look similar. Some molecular modes are Raman active, but not IR active and vice versa, for example carbon dioxide symmetric and asymmetric stretches, respectively. Other species may be observed by both techniques, and overall these spectroscopies are complementary rather than competitive [5, 6].

Raman spectra are measured using a high-intensity monochromatic light source, usually an ultraviolet, visible or near-infrared laser, and collecting, filtering and decomposing the scattered light by a dispersive element [5, 6]. There are three types of scattered radiation: Rayleigh, Raman Stokes, and Raman anti-Stokes scattering. In the energy-level diagram of Raman and Rayleigh scattering (figure 4.4) [7], the first pair of red and grey arrows depicts the energy change (ground electronic state to virtual energy state, $E = h\nu$) in the molecule when it interacts with an incident photon, and the Rayleigh scattering which results from elastic collisions between the photon and the molecule ($\Delta E = 0$). The energy changes that produce Stokes and anti-Stokes scattering are non-zero and in figure 4.4 correspond to the energy of the first vibrational level of the ground state ($\pm\Delta E$). Because ground-state excitation is much more probable, Stokes scattering is strongly favored over anti-Stokes. In addition, if a vibration is infrared active, the energy of its absorption would also be ΔE, so the Raman shift and infrared absorption energies are identical.

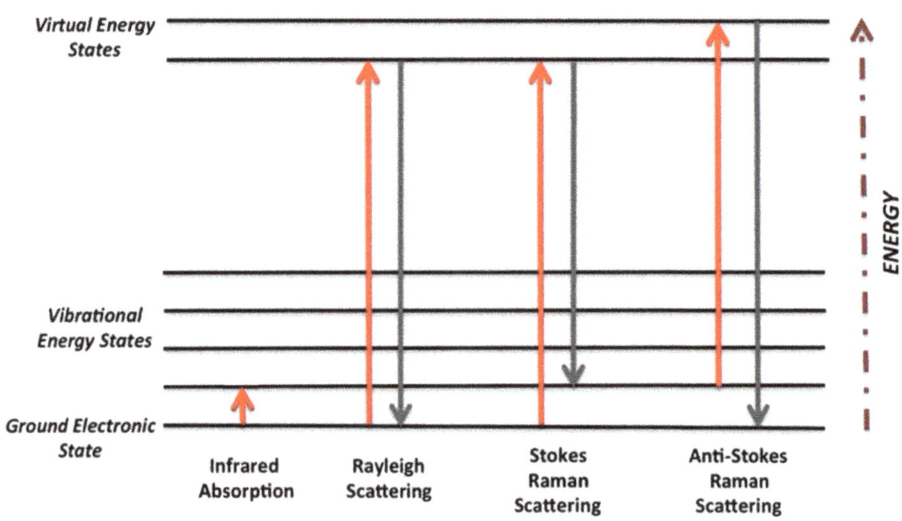

Figure 4.4. Origin of Rayleigh and Raman scattering compared to infrared absorption [7].

Raman spectra can be analyzed by the number of detectable bands and the wavenumbers, intensities, widths (usually the full width at half maximum, FWHM) and line shapes of the bands [3]. Specifically, the intensity of Raman peak depends upon the concentration of the active species, the polarizability of the molecule, the intensity of the laser source, as well as other factors. Rayleigh scattering is the most probable because the scattered and incident light have the same energy. It is often filtered out using narrow band filters. Generally, anti-Stokes scattering is less intense than the corresponding Stokes scattering, so usually only the Stokes part of a spectrum is used. In some cases, for example with strongly-fluorescing samples, anti-Stokes signals may be useful due to serious interference of the fluorescence with the observation of Stokes shifts.

Common laser sources for Raman spectroscopy include argon ion laser (488.0 or 514.5 nm), a krypton ion laser (530.9 or 647.1 nm), a helium/neon laser (632.8 nm), a diode laser (782 or 830 nm) and a neodymium-doped yttrium aluminum garnet (Nd: YAG) laser (1064 nm), but there are other choices as well [5]. Considering Raman intensities are proportional to the fourth power of the excitation frequency, ultraviolet lasers can provide more intense Raman signals compared with other sources, given the same input power. This advantage, however, may be nullified by possible photodecomposition caused by high-energy ultraviolet irradiation. On the other hand, near-infrared sources (e.g., diode laser and Nd:YAG) can be operated at much higher power without degrading the sample, and fluorescence is greatly reduced or nonexistent with these lasers.

In Raman microscopy, the achievable spatial resolution is primarily a function of the wavelength of the light and the microscope optics. The smallest resolutions that can be distinguished in microscopy are governed by the Abbe diffraction limit, given by:

$$d = \frac{1.22\,\lambda}{2n\,\sin\theta} = \frac{0.61\,\lambda}{\mathrm{NA}}$$

where NA is the numerical aperture of the objective, n is the refractive index of the imaging medium between the objective and the sample, and θ is half the angular aperture. To resolve two adjacent points completely, the separation required is $2d$ for a high-quality microscopy image [3]. For more information on the fundamental details on Raman microscopy including theory and instrumentation see, for example, references [3, 5].

4.2 Vibrational spectroscopy for batteries

The typical Li-ion battery comprises a positive and a negative electrode (cathode and anode, respectively) and a liquid nonaqueous electrolyte allowing Li^+ ion transport to each electrode. In addition, chemical and electrochemical reactions at the interface between the electrolyte and the electrode yield a passivation layer on the electrode surface, often referred to as the solid–electrolyte interphase (SEI). Over the decades of battery research, FTIR and Raman spectroscopies have proved to be invaluable tools to study and optimize the battery components and the SEIs.

Generally, FTIR and Raman spectroscopy is surface- and bulk-sensitive techniques, respectively, that provide local chemical and structural information of the SEI and near surface bulk electrodes. Spectroscopic methods are also used to analyze the liquid samples, such as identifying solution structures (i.e., ion solvation and ionic association interactions via the anion and/or solvent vibrational bands) in a bulk electrolyte and within an electrode–electrolyte interphase (EEI).

4.2.1 *Ex situ* FTIR and Raman

4.2.1.1 FTIR

The most prevalent application of the FTIR technique in battery research has been, and continues to be, towards the understanding of the chemical reactions occurring at the negative electrode surface. Seminal work investigating the reactivity of lithium metal anodes in nonaqueous electrolytes was carried out by A N Dey, G Nazri and R H Muller, E Peled, and D Aurbach *et al* [8–12] From FTIR measurements, D Aurbach *et al* identified lithium alky carbonates ($ROCO_2Li$) as major surface species resulting from the reduction of the propylene carbonate (PC) solvent, figure 4.5(a) [8]. The resulting film formed on the lithium surface acts as a passivation layer. This finding helped elucidate the reasons why lithium metal, which should be thermodynamically reactive with polar aprotic solvents, instead shows stable behavior. E Peled formulated a model that described the passivation layer

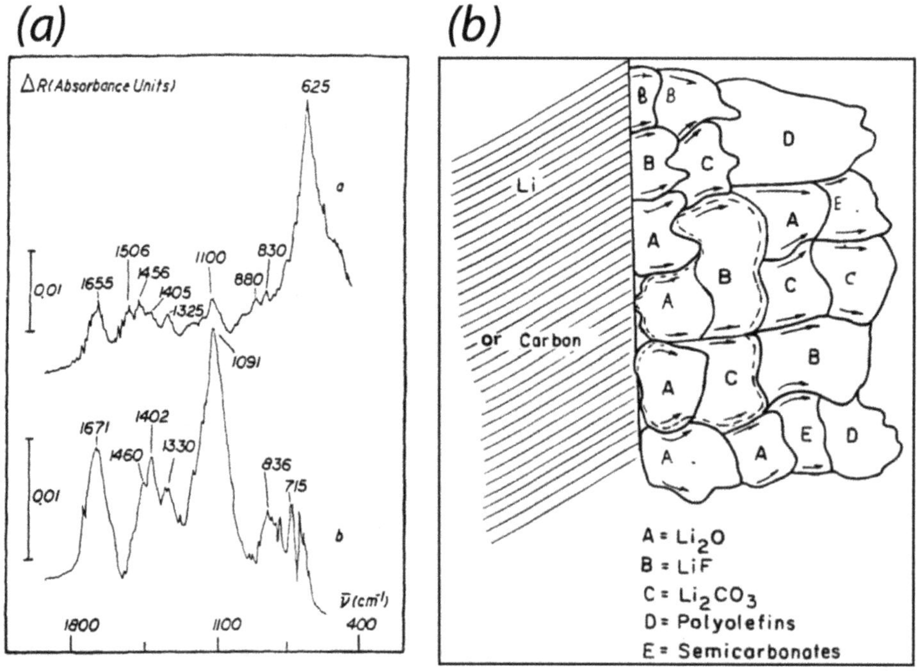

Figure 4.5. (a) FTIR spectra of lithium electrode surfaces. The sample potential was held at 0.2V vs. Li/Li^+ for 10 min followed by storage in PC solution for 24 h: (top) 0.2M $LiAsF_6$ in PC, (bottom) 0.2M $LiSO_3CF_3$ in PC [8]; (b) schematic presentation of a poly-hetero microphase SEI on lithium or carbon electrode [14].

as a sort of solid electrolyte that is ionically conductive (to Li^+ ions) but electrically insulating. He called his model the SEI model [12]. The idea that it is necessary for efficient and long lasting batteries to form a stable SEI at the electrode surface through which ions can pass but electrons cannot has been adopted beyond lithium metal and extended to carbon-based anodes like graphite. After extensive FTIR characterization, much of it carried out by D Aurbach *et al*, and combined with numerous other characterizations, the SEI on graphite is now known to consist of a denser inner layer made up mostly of inorganic species such as LiF, LiO_2, Li_2CO_3, and a more-porous outer shell containing organic species such as $ROCO_2Li$, semicarbonates, and polymers [13, 14]. A revised Peled model is shown in figure 4.5(b)[14]. Similar approaches continue to be applied to newer negative electrodes of interest such as silicon and tin alloying metals [15].

Vibrational spectroscopy is also useful to investigate liquid nonaqueous Li-ion battery electrolyte compositions. They generally comprise more than one solvent, one or more additives at low concentrations, and a Li salt that partially dissociates in the solution (solvation) [16]. The extent of Li^+ ion solvation by the solvent molecules is a critical property of the electrolyte [16–19]. On the one hand, Li^+ ions should be separated from their counter-anions by a partial shell of solvent molecules, allowing them to diffuse in the solution; but on the other hand, if the solvent molecules interact too strongly with the Li^+ ions, it will be difficult for the ions to transfer from the solution to the electrode during the battery charging and discharging. Because infrared vibrational modes are sensitive to the solution components' local bonding environments, there is a distinction between electrolyte species—solvent molecules or counter-ions—coordinated to nearby ions or surrounded only by other solvent molecules.

A commonly-used electrolyte is 1.2 M lithium hexafluorophosphate ($LiPF_6$) in a 3:7 weight ratio of ethylmethyl carbonate (EMC) and ethylene carbonate (EC) [20, 21]. The molecular structures of EMC and EC, coordinated to Li^+ ion and non-coordinating are shown in figure 4.6 along with the FTIR spectrum between 1350–1150 cm^{-1} where the solvent -C(O)-O- vibration mode excitation occurs. For example, an EMC solvent molecule in pure EMC has vibrational absorption at 1263 cm^{-1}, but the absorption energy shifts to 1306 cm^{-1} when EMC is sovating a Li^+ ion. By measuring the ratios of coordinated to pure solvent IR absorptions, the solvent coordination number to a Li^+ ion can be determined for a variety of solvents, salts, and concentration ranges [22–24].

With an appropriate reference database and by applying machine learning techniques, it has been shown recently that FTIR can be used to quickly determine the compositions of unknown electrolytes including the major solvent constituents and salt concentration [25]. At the heart of this technique is the understanding of the vibrational absorption behavior in an electrolyte by observing the Li^+-solvent coordination. This method also showcases the usefulness of FTIR compared to more complicated and more expensive analytical methods such as gas chromatography (figure 4.7).

Lithium transition-metal oxides and phosphates such as $LiNi_{1-x-y}Mn_xCo_yO_2$, $LiMn_xNi_yO_4$, and $LiFePO_4$ are commonly used as the positive electrode in Li-ion

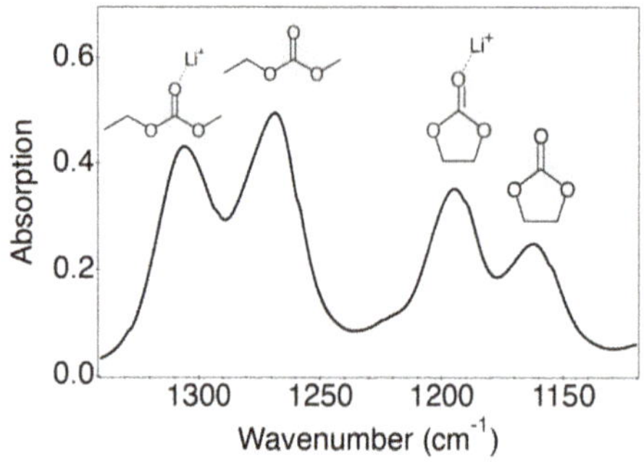

Figure 4.6. ATR-FTIR spectrum of electrolyte 1.2 M lithium hexafluorophosphate ($LiPF_6$) in a 3:7 weight ratio of ethylmethyl carbonate (EMC) and ethylene carbonate (EC). The peaks at 1306, 1263, 1198, and 1150 cm^{-1} correspond to EMC coordinating a Li^+ ion, EMC (non-coordinating), EC coordinating a Li^+ ion, and EC (non-coordinating), respectively. The grey shading indicates the excited vibrational mode.

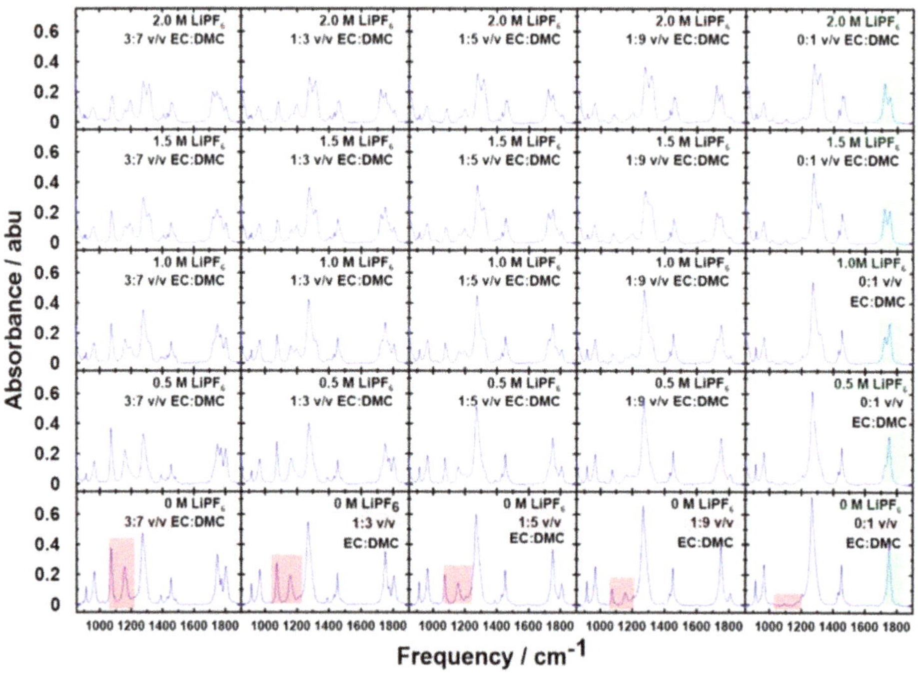

Figure 4.7. (A) ATR-FTIR spectra of electrolyte solutions with a range of $LiPF_6$, EC, and DMC concentrations. Green highlights the evolution of a spectral feature that is strongly correlated with [$LiPF_6$] in DMC solutions. Red highlights the evolution of a spectral feature that is correlated with [EC] in DMC solutions [25].

batteries. As Li^+ ions are extracted out of (and then reinserted back into) the cathode during the battery operation, the local arrangement of the atoms in the lattice shifts around to compensate for changes in the lithium occupancy. The local structure dynamics of the cathode during the battery operation are crucial to the overall cell performance and lifetime, and several studies have shown that FTIR can probe the local structural changes of the cathode as a function of lithium intercalation and de-intercalation [26–30]. For example, M Kunduraci *et al* found that in the high-voltage spinel cathode $LiMn_{1.5}Ni_{0.5}O_4$ in its ordered phase ($P4_332$), Ni-O bands blue shift to higher wavenumbers as lithium is removed from the lattice during charging, but the Mn–O bonds are mostly unaffected, as shown in figure 4.8 [26]. From this, it can be inferred that the nickel ions are the dominant redox species in the material.

4.2.1.2 *Raman*

Raman spectroscopy offers several advantages over non-vibrational surface-characterization techniques: its relatively high resolution and laser focused spot size enables individual particle sampling and can be used to identify spatial heterogeneity in a composite electrode, while its sensitivity to coordination environments in the electrode complements the structural information derived from x-ray diffraction techniques and further reveals unexplored details including local and electronic structure of the material. Confocal Raman is ideal for studying chemical and structural changes that occur both in the electrolyte and the electrode. W A Henderson *et al* utilized Raman spectroscopy to build Raman band peak position libraries of the various lithium (Li) salt anions (i.e., bis(trifluoromethanesulfonyl)imide ($TFSI^-$), hexafluorophosphate (PF_6^-), tetrafluoroborate (BF_4^-), difluoro(oxalato)borate ($DFOB^-$)) for the different crystalline solvates[18, 31–33] and to analyze the solution structure focusing on the degree of solvation and ionic association for the Li salt–solvent mixtures [34–49]. For example, in some cases, $LiBF_4$-based solvate crystal structures can be prepared and determined utilizing a single crystal x-ray powder diffractometer, which provide direct structural information about solvation and ionic association interactions [32]. Using this detailed structural information, the anion vibrational bands of crystalline mixtures can be analyzed to assign band positions to specific forms of anion coordination (figure 4.9(a)), such as solvent-separated ion pairs (SSIPs), contact ion pairs (CIPs), or aggregates (AGGs) if the anions are coordinated to zero, one, or more than one Li^+ cations, respectively [32, 40]. This information can then be extended to the liquid phase

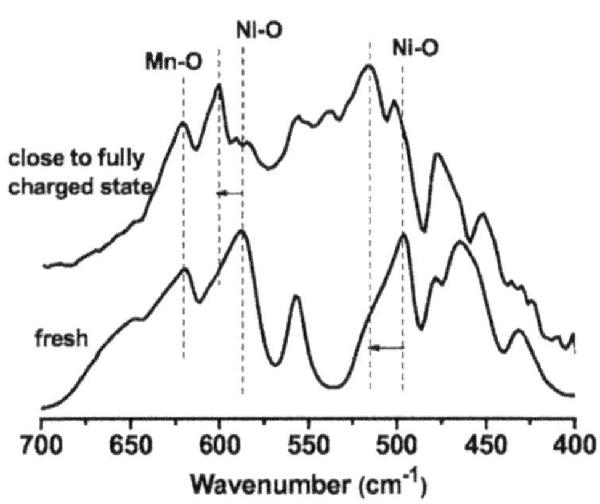

Figure 4.8. Infrared spectra of a cation ordered $P4_332$ spinel [$Li_xMn_{1.5}Ni_{0.5}O_4$] at fresh and almost fully charged states. The dashed lines show the shift of the Ni–O bands to higher wavenumbers upon charging [26].

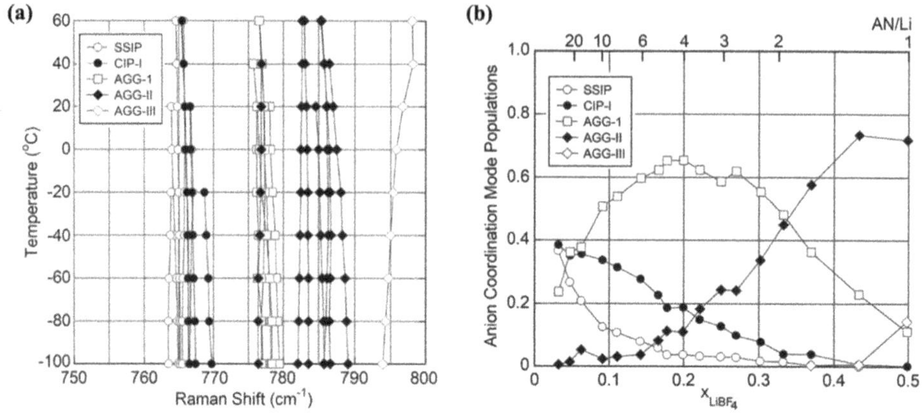

Figure 4.9. (a) Raman data assessment tool for the BF_4^- anion ν_1 vibrational band showing the Raman band peak positions for various crystalline solvates with the indicated anion coordination modes, [40] and (b) populations of anions with differing Li^+ cation coordination modes in the $(AN)_n$-$LiBF_4$ electrolytes at 60 °C [37].

in order to determine the fraction and identity of the solvate species present, as well as the amount of uncoordinated solvent in a bulk electrolyte (figure 4.9(b)) [32, 37, 40]. As the authors mention, however, interpreting the Raman data in terms of correlating the spectroscopic vibrational bands with specific solvate species requires extra caution to prevent erroneous conclusions with regard to the identity and population (distribution) of the SSIP, CIP and AGG solvate species [40].

The solution structure information in a bulk electrolyte is more meaningful when it is correlated directly with the physicochemical and electrochemical properties (e.g., density, viscosity, ionic conductivity, transference number, etc) of the electrolyte [34–49]. This research provides a detailed understanding of how solvent and anion structures determine the solvation state of Li^+ cations in solvent–Li salt mixtures and how this, in turn, dictates the electrolyte properties, which strongly influence battery performance. In addition, using the suggested electrolyte characterization framework, one can analyze the Li salts mixtures with newly developed promising solvents with or without additives to understand how their solution structures and properties can be changed to influence the surface chemistry and evolution of electrode–electrolyte interface.

On the other hand, confocal Raman spectroscopy can provide point scans—the confocal nature of the instrument allows for a spot size of approximately 1 μm—as well as lateral mapping of the electrode surface [50]. For instance, single particle Raman spectra can reveal the lithium transition metal oxides ($LiMO_2$, M = transition metals, such as Ni, Co, Mn) lattice structure, focusing on the M–O stretching and O–M–O bending, and the delithiation and lithiation behaviors of individual oxide particles in the composite cathode as a function of state of charge (SOC), depth of discharge (DOD) and cycle numbers. As described in the previous studies [51, 52], two Raman active modes for each transition metal in the $LiMO_2$ unit cell are the E_g mode (at lower wavenumber band) originating from an opposite motion of oxygens along adjacent O-layers (O–M–O bending) and the A_{1g} mode (at higher wavenumber band) where oxygens move symmetrically along the *c*-axis (M–O stretching) and is mostly stretching in character. E J Berg *et al* summarized the Raman spectrum of pristine $LiCoO_2$ and $LiNiO_2$ electrode powders, along with several substitutional variants, and provided detailed explanation for spectral variations (e.g., band positions, widths and intensities) mostly attributed to local environments resulting from complex interactions between homogeneously distributed transition metals and even different setups (e.g., oxide crystal size, orientation, purity, internal stresses, etc) and experimental conditions (e.g., temperature, incident light polarization, wavelength, exposure time, etc) [52].

Previous Raman studies demonstrate lithiation/delithiation behaviors and material stress of individual oxide particles in the composite cathodes—$LiCoO_2$ [53, 54] and lithium transition metal oxides [55–58]—as a function of SOC, DOD and cycle numbers. Compared to pristine electrode, the cycled ones show a negative Raman shift for the A_{1g} and E_g bands that correspond to retention of material stress and electrochemical degradation via capacity fading, respectively [53], or spectra evolution in a similar manner to the parent material phases with Raman shifts that correlate to the changes in the *c*-lattice parameter [54]. H A Gasteiger *et al* utilized Raman spectroscopy to study the formation of surface contamination on a single layered lithium nickel manganese cobalt oxide ($LiNi_xMn_yCo_zO_2$, generally referred to as NMC) particle in the NMC electrodes by measuring several spots and showed that extracted Ni from the layered NMC structure is part of the surface impurity decomposing the surface of the NMC [57].

Raman spectroscopy can resolve single particles and their aggregates in an electrode, which permits the identification of spatial heterogeneities within the composite electrodes and monitors inhomogeneous electrode reactions. Furthermore, Raman images of a cycled electrode compared with a pristine electrode can provide insight into the deterioration mechanisms and localized lithiation of the electrode with an inhomogeneous surface layer [59–62]. Raman images can be made by plotting the band position with the maximum intensity in the specific frequency range based on the material (i.e., silicon (Si) for 490 to 520 cm^{-1}). J Nanda *et al* demonstrate Raman maps identify electrochemically inactive regions of the silicon (Si)-graphite (Gr) composite anodes at different cycles by deconvoluting each spectrum into the sum of the four components (i.e., crystalline silicon, amorphous silicon, graphite and carbon black), and emphasize the improvement of electrode uniformity for the development of high-energy batteries with alloy anodes [59]. R M Kostecki *et al*'s research visualizes the highly non-uniform cathode surface even at deep discharge due to heterogeneous kinetic behavior of individual oxide particles that contributed to the degradation of electronic pathways within the composite cathodes and the subsequent electrochemical performance deterioration of composite electrodes [60]. Based on the reported previous studies [59–62] mapping of reaction inhomogeneities in the composite electrodes could be further developed to reveal the underlying causes including electrical disconnection, mechanical stress and cracking of electrode active materials under varying operating conditions (e.g., applied current density, temperatures, constant voltage step, etc).

The Raman activity of most electrode materials is inherently weak, so collecting high-quality spectra requires long exposure times that may reduce time-resolution.

However, higher laser power, longer exposure times and higher accumulation numbers can damage the samples, thus surface-enhanced Raman spectroscopic (SERS) [63–83] and tip-enhanced Raman scattering (TERS) [84–88] techniques receive increasing attention and provide unique insights into the local structure of the electrodes [89, 90]. SERS is a powerful surface sensitive technique that involves enhancement in Raman scattering (by a factor of ~10^6) for the species located at roughened metal—typically, gold (Au), silver (Ag) and copper (Cu)—surfaces or the sample with noble metal (e.g, Au or Ag) nanoparticles. The intensity of the Raman scattering from a specific species of interest can be enhanced by the presence of certain nanostructured metals possibly due to a polarizability enhancement induced by direct interaction between the metal surface and the adsorbed molecule and/or the interaction of laser light with the nanostructured metal surface, which excites an oscillation of the conduction electrons (localized surface plasmon resonance) [76, 89].

A Tornheim *et al* demonstrated that the intensity of the Raman scattering from the NCM523 phonons can be plasmon-enhanced by around two orders of magnitude when the electrode laminates are electroplated with sub-micron gold clusters [64]. G Yang *et al* utilized SERS to study the electrolyte solution structures (e.g., ion solvation and de-solvation) within the SEI based on a five orders of magnitude increase in Raman intensity for electrolyte components and their mixtures by employing a gold nanoparticle monolayer [65, 66]. M J Piernas-Muñoz *et al* investigated the SEI layer on a silicon anode with electrodeposited gold using SERS and determined the partial composition of SEI including organo-phosphate-derivatives [67]. As demonstrated in the above studies, the SERS technique can reduce the spectra collecting times from fractions of an hour to a few seconds, but the highly reactive gold clusters are likely unstable in the electrolyte at high anodic potentials and may react with other electrodes components. Some research groups, therefore, employed shell-isolated nanoparticle-enhanced Raman spectroscopy (SHINERS) by coating the chemically reactive surfaces of the plasmonic nanoparticles with a thin layer of an inert oxide [76, 78–83].

The SERS technique is limited to studying electrochemical reactions on roughened metal electrodes or noble metal nanoparticles electroplated electrodes, but TERS offers a promising solution of overcoming this constraint by employing a plasmonic responsive nanoscale metallic tip (generally made of Au or Ag) controlled via scanning probe microscopy [76, 89]. The nanometer-scale tip (50–100 nm) concentrates the incident light field at the apex and enhances the electromagnetic field locally confined immediately below the tip apex, greatly improving the Raman sensitivity (by a factor of 10^3–10^7). Recently, J Nanda *et al* employed TERS to study the SEI layer formation and evolution on model amorphous silicon (a-Si) thin film anodes over cycles for the first time, and proposed a nanomosaic-multilayer hybrid SEI model—a mosaic-type nano-island structure in the sample plane and a multi-layered structure through the sample plane—with chemical and topographical heterogeneity based on their observations [88]. The study shows that TERS has great potential to be a powerful tool for surface analysis, specifically sensitive to dynamical evolution of the SEI at a lateral resolution of <10 nm and a sampling depth of <5 nm compared with relatively bulk sensitive Raman and FTIR.

Ex situ FTIR and Raman analysis as a first quick approach can provide a variety of information and specific guidance for more detailed characterization, but there are several limitations in *ex situ* spectroscopic analysis: (i) limited or no information for dynamic behaviors of electrode components with respect to their natural occurrence in the battery system, (ii) potential contamination and surface chemistry changes during sample transferring and washing, (iii) discrepancy in the SOC and DOD states of electrodes due to self-discharge, lateral and axial (de)lithiation inhomogeneities and difficulty in precise electrochemical control of cycling cells, and (iv) differences (e.g., loading weight, morphology, surface chemistry, particle size, crystalline orientations) from sample to sample.

4.2.2 *In situ* FTIR and Raman

Many spectroscopic and analytical techniques have been developed to characterize the properties of battery components and understand cell capacity fade, but most often, these techniques are applied *ex situ* to a component of the battery that has been recovered from a disassembled cell after testing. Once the component is removed and exposed, physical and chemical changes may occur that obscure important characteristics of the sample. Furthermore, while *ex situ* measurements are informative, they cannot elucidate changes happening to the cell during testing, thus potentially failing to provide critical information about correlated processes underpinning the cell performance. Vibrational IR spectroscopy is amenable to *in situ* application because it is non-desctructive, relatively low energy, and compatible with most materials.

4.2.2.1 FTIR

While studying surface films on lithium metal anodes, Aurbach *et al* noted that both the lithium and the surface species formed in solution are very sensitive to impurities (i.e., O_2, H_2O, N_2, and CO_2) ever-present either in the glove box or the spectrometer atmospheres, and this caused some doubts about the accuracy of *ex situ* measurements. To avoid this, they developed a few *in situ* FTIR apparatus, where the active surface was prepared and measured in solutions while controlling the cell potential [91, 92]. Figure 4.10(a) illustrates one of their spectroelectrochemial cell designs in which the working electrode is a very thin film of sputtered platinum metal (200 Å) on an NaCl window [91]. At very low cell potentials (−0.2 to 0.0 V versus Li/Li^+), lithium metal is plated onto the semi-transparent Pt film and IR spectra are measured without the light passing through a significant amount of strongly-absorbing electrolyte solution which would obscure the absorptions of surface species of interest. Figure 4.10(b) shows an example of the *in situ* FTIR spectra obtained with their method [92]. In this study, they acquired two noteworthy results: 1. After *in situ* formation of lithium on Pt, there is electrochemical reduction of the propylene carbonate which results in observable FTIR signatures from lithium alkylcarbonate ($ROCO_2Li$) species. 2. In the presence of CO_2, new peaks around 1500 and 880 cm^{-1} are observed at low potentials, indicating significant lithium carbonate (Li_2CO_3) formation, and showing that additives to the electrolyte solution can affect the SEI composition.

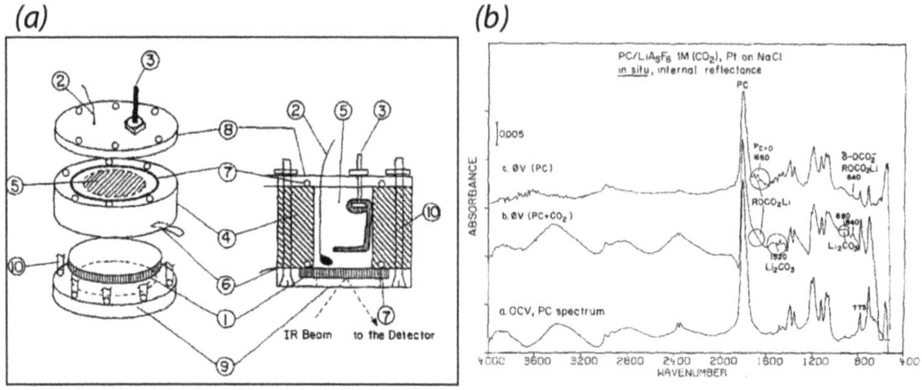

Figure 4.10. (a) Spectroelectrochemical cell for single internal reflectance spectroscopy. 1, Working electrode —a thin film of platinum (200 Å, deposited by sputtering) on an NaCl disk (3 mm thick, 3 cm diameter) which is also the optical window. 2, The reference electrode formed from a lithium piece adhered to a Pt wire. 3, The counterelectrode formed from lithium foil mounted on a brass or stainless steel bolt. 4, Polypropylene cell body cylinder shape. 5, Space for solution. 6, Electrical contact for the working electrode (nickel foil, attached to the edge of the Pt-covered NaCI window). 7, 'O' rings, 8, Upper cover (polypropylene disk). 9, A brass holder for the optical window (and the working electrode). 10, 6 bolts which hold the cell [91]; (b) FTIR spectra obtained using internal reflectance mode from Pt electrode (deposited on NaCl) in PC-LiAsFs 1M solution. (The cell is described in (a)). Each spectrum was obtained by subtraction of the spectrum of the empty cell from the measured spectrum, a, An OCV (3 V, Li/Li^+) spectrum, b, Solution was saturated with CO_2, the electrode was polarized to 0 V (Li/Li^+). c, CO_2-free solution, 0 V (Li/Li^+) [92].

Several other groups have developed custom spectroelectrochemical cells for *in situ* FTIR studies of electrodes beyond lithium metal anodes [93–98]. In most of these studies, the *in situ* FTIR spectra were dominated by the electrolyte signals. Solvation solution structure was found to be dynamic at the electrode interface as a function of cell potential. Although none of these studies were able to directly detect the formation and evolution of the SEI or changes in the electrode structure with lithiation/delithiation, they showed that understanding the complex changes in electrolyte solvent/ion solvation behavior can be correlated to the interfacial phenomena, particularly when coupled to complimentary characterization techniques such as *in situ* Raman microscopy [97].

Recently, S-D Han *et al* have developed an *in situ* FTIR technique applied to custom battery cells studying the interfacial properties of a high-Ni, low-Co cathode ($LiNi_{0.6}Mn_{0.2}Co_{0.2}O_2$, [NMC622]) during cell operation [99]. In this study, they were able to not only track electrolyte solution solvation structure changes with cell potential, but also directly measure vibrational absorption frequency and intensity changes stemming from the local order changes in the cathode lattice with varying lithium occupancy. A recreated illustration of their custom *in situ* FTIR spectroelectrochemical cell is shown in figure 4.11(a). Figure 4.11(b) is a series of *in situ* spectra collected during the first three charging and discharging cycles of one of their NMC622/Li half cells. The frequency range between 500 to 620 cm^{-1} contains vibrational absorptions primarily from the Ni–O octahedral stretching mode in the NMC622, although there is some overlapping absorption from the salt's PF_6^- anion

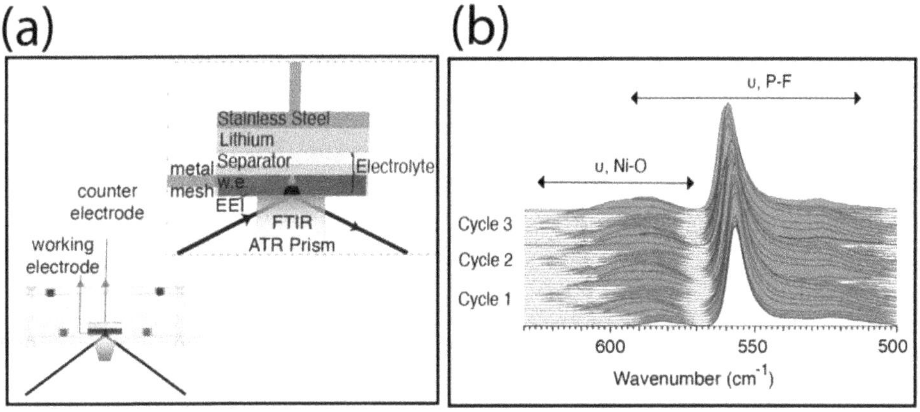

Figure 4.11. (a) Spectroelectrochemical cell for *in situ* ATR-FTIR measurement of mesh electrodes; (b) ATR-FTIR spectra for the first three cycles of an NMC-622/Li half cell.

stretching modes between 500 to 550 cm^{-1}. As the cell is charged and lithium is removed from the NMC lattice, the vibrational absorption band frequency blue shifts to higher energy as the Ni–O bond length shortens. Reversible behavior is observed upon discharging. In their paper, the change in bond length was confirmed with complimentary x-ray absorption measurements [99]. Importantly, their study showed that *in situ* ATR-FTIR can be used to investigate the cathode–electrolyte interfacial region during electrochemical cycling, and revealed three important aspects of the cell evolution. (1) Li^+ ion solvation by the electrolyte solvent molecules near the cathode interface is affected by the polarization of the electrode surface and the transfer of Li^+ ions out of (and into) the NMC622 cathode during charging (and discharging) of the battery. (2) Local structural changes within the NMC lattice, as affected by the transition-metal (predominantly Ni) ion valence state during redox, can be quantified directly by *in situ* FTIR measurements. (3) They found some FTIR evidence for the formation and evolution of a cathode –electrolyte interphase (CEI) containing carbonate species, thus elucidating the nature of the cathode surface passivation.

4.2.2.2 Raman

Raman spectroscopy is a non-destructive technique and uses a visible excitation source, allowing measurements through glass or other encapsulation methods that enable a spectro-electrochemical cell suitable for *in situ* analysis. Deleterious performance of a battery is, in part, attributed to the chemical composition of the electrode–electrolyte interphase (EEI) and the irreversible phase transformations of the electrode surface that occur under bias, so *in situ* confocal Raman is ideal for studying the chemical and structural changes occurring in the EEI during cycling. To obtain both optimum electrochemical performance and spectra quality, however, there are several experimental considerations including working distance between the light source and the probed sample, laser excitation wavelength, laser specific power, *in situ* cell sealing, electrode pressure, electrode size, electrolyte choice and reactivity of cell components [52]. A wide variety of studies have used *in situ* Raman spectroscopy, SERS and TERS to investigate the chemical composition and phase transformation of the electrode surface, the electrochemical reactions in an electrolyte and the EEI formation and evolution during cycling [52, 68–83, 100–112].

The mechanism of sulfur (S) reduction in Li–S battery slurry cathodes with 1.0 M lithium bis(trifluoromethane sulfonyl)imide in tetraethylene glycol dimethyl ether/1,3-dioxolane (1/1, v/v) electrolyte and the structure of an electrolyte solutions in the vicinity of a graphite anode were investigated using *in situ* Raman microscopy [101, 102]. The kinetic characteristics of Li^+ intercalation/deintercalation into/from individual $LiNi_{0.8}Co_{0.15}Al_{0.05}O_2$ particles in a composite cathode and the link between the local structure evolution of high voltage cathodes and its electrochemical response were studied during cell cycling [52, 103–105]. The detailed *in situ* Raman measurements allowed analysis of morphology changes of uncoated and carbon-coated Si nanowires as well as interphase formation from electrolyte components during the first lithiation and delithiation cycle, and to understand the lattice vibration mechanics and microstructure of carbon-based anode materials with chemical structure information at the molecular level for alkali metal-ion batteries [106–108].

P G Bruce *et al* utilized *in situ* SERS to study oxygen (O_2) reduction chemistry, focusing on lithium peroxide (Li_2O_2) and superoxide (O_2^-) formation, in non-aqueous electrolytes for the $Li{-}O_2$ battery [68–70]. In addition, SEI formation and evolution on a Li metal or Si-based anodes, Li deposition and dissolution processes on Cu, copper oxide (CuO) nanoparticle formation on Cu electrodes and redox behaviors of Si-based anodes were investigated using *in situ* SERS [71–77]. In some cases, *in situ* ShINERS technique was employed to analyze the formation and evolution of EEI species, monitor redox reactions on Li-rich or Ni-rich cathodes and study the mechanism of the O_2 reduction reaction on carbon substrates relevant to $Li{-}O_2$ cells by coating the chemically reactive surfaces of the plasmonic nanoparticles with a thin layer of an inert oxide (i.e., silica coated Au nanoparticles) [76, 78–83].

There are few studies demonstrating *in situ* Raman imaging and mapping of the electrode surface to investigate the local SOC distribution and the Li^+ ion transport in a cathode and to monitor the lithiation/de-lithiation behavior of a Si-based anode and electrical disconnection of the active material from the current collector during cycling [109, 110]. B Ren *et al* developed an electrochemical TERS technique by introduction of the light horizontally to the integrated electrochemical-scanning tunneling microscopy (STM) cell to keep the TERS measurement under a well-controlled condition by minimizing the optical distortion [111]. They could collect potential-dependent electrochemical-TERS signals from monolayer species on Au (111) single-crystal surfaces and observe a subtle change in the molecule configuration based on the protonation and deprotonation of the molecule as a function of potential. Furthermore, *in situ* electrochemical atomic force microscopy (AFM) TERS was developed to spatially resolve local heterogeneity at the nanoscale in redox behavior on an electrode surface [112]. The study demonstrates TERS intensity maps representing the spatially dependent redox behavior of Nile Blue molecules on a structurally well-defined Au(111) nanoplate on a polycrystalline indium tin oxide electrode at the nanoscale. Advanced *in situ* SERS and TERS characterization is experimentally challenging, but a further development of those techniques could be more readily applicable for electrochemical energy storage studies, and offer a highly dynamic and convenient tool to provide design guidelines for new electrolyte formulations and next-generation Li-ion battery electrodes.

4.3 Future perspective

Despite the unquestionable usefulness of IR vibrational spectroscopy for advancing rechargeable battery technology, it is important to keep in mind some of the key limitations to the FTIR and Raman techniques in this field. One of the most obvious ones is that alkali metal batteries such as lithium are reactive with oxygen and moisture. For safety reasons, these batteries must be studied in inert atmosphere conditions. Thus, any *in situ* spectroscopic techniques usually require specialized cell configurations that keep the sample protected from air while also allowing the light to penetrate the sample and then be collected. A plethora of cell designs for either *in situ* FTIR or *in situ* Raman can be found throughout the literature. When considering a custom spectroelectrochemical cell design, it is imperative that the cell design reproduces accurate electrochemical performance that is found in standard battery cell formats like a coin, pouch, or cylindrical cell. Subtle deficiencies in the *in situ* cell design, for example inadequate pressure, often lead to inferior performance and can be difficult to assess unless researchers explicitly show direct comparisons to a standard cell.

Another challenge, which is general to battery research, is that these systems are quite complicated and messy. It is often thought that the primary parts of a battery are just a negative electrode, electrolyte, separator, and a positive electrode. However, each of these parts is typically a multi-component system of its own. For example, a cathode electrode is often made up of a lithium transition metal oxide solid mixed with a small amount of conductive carbon to improve electrical connectivity and a polymeric binder to hold everything in place onto a current-collecting metal foil. In addition to the salt, nonaqueous electrolytes are rarely made of a single solvent and instead contain two or more co-solvents as well as small amounts of additives to help with stable SEI formation. The ramifications of the battery complexity to spectroscopic analysis are that collected spectra usually contain many signals and often the signals from different components overlap with each other. Thus, one of the primary challenges in spectroscopy of batteries is identifying and deconvolving complex spectra into more manageable pieces. Thankfully, several tried and true approaches can help a spectroscopist tackle these problems. The first is to build a reference library whenever possible. If individual components can be attained and analyzed (or a simplified model system can be used), then it will make assigning peaks in a complex multicomponent spectrum easier. The caveat to this is that sometimes strong interactions between individual components can result in new signals that exist in none of the individual spectra. For example, refer back to the sections of this chapter that discuss Li-ion solvation and new FTIR and Raman peaks that arise from solvent–ion interactions.

Another tool that can be indispensable to complex spectral analysis is computation and simulation. Well-developed approaches exist to predict electronic, structural, and chemical properties of many battery materials [113, 114]. For example, density functional theory (DFT) provides assignment of IR and Raman vibrational modes [22]. Additionally, molecular dynamics (MD) simulation has been used to better understand the dynamics of ion solvation in bulk electrolytes and across the electrolyte–electrode interface [19]. The predictions produced by these computational methods inform experimental validation.

Therefore, if enough information is known about an FTIR or Raman spectrum, it is useful to perform multi-peak fitting and spectral deconvolution. As an example, we show in figure 4.12 how each individual spectrum (bottom left) in an *in situ* FTIR measurement series (top left) can be fitted such that vibrational absorptions of the electrolyte (top right) and of the electrode (bottom right) can be separated from each other. In this case, we can separate the peaks from perchlorate ion (ClO_4^-) stretching modes in propylene carbonate (PC) from the cathode ($LiNi_{0.95}Mn_{0.05}Co0._{05}O_2$, [NMC90–5–5]). The result is that we can easily track how the Ni–O vibrational bond shifts in energy and intensity while charging and discharging the battery.

IR vibrational spectroscopy will continue to be used heavily in future energy storage research. Despite limitations, simple *ex situ* measurements still provide crucial chemical characterization of new materials and identification of chemical products that are important in developing next-generation technologies. In addition, *in situ* applications of Raman and FTIR will continue to evolve. Novel cell designs and advancements in instrumentation will enable experiments that could not be done before. For example, time-resolved multi-dimensional IR spectroscopy is an emerging method to attain dynamic and orientational characteristics of electrolyte

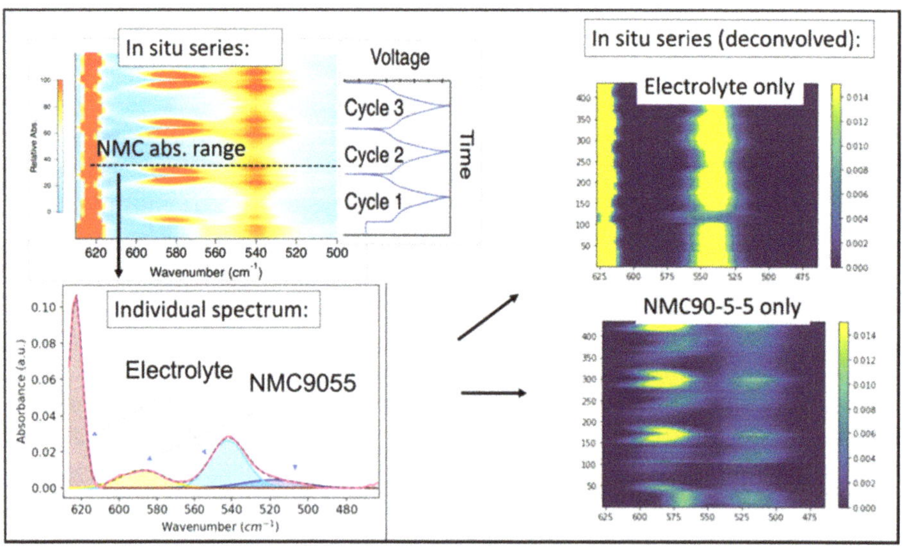

Figure 4.12. Fitting a series of *in situ* FTIR spectra to separate electrolyte from cathode absorption signals provides a clearer understanding of how the cathode local structure changes with state of charge.

solvation changes influenced by the kinetics of Li^+ ion interfacial transport [115, 116]. Furthermore, coupled *in situ* techniques continue to develop in ways that provide broader fundamental and mechanistic understanding of chemistry within the electrode–electrolyte interface, for example, *in situ* Raman microscopy in concert with scanning probe-based 3D potential may enable mapping on the electrical double layer (EDL) at the electrode/electrolyte interface. FTIR of the electrolyte combined with electron paramagnetic resonance may elucidate the formation of transition-metal organometallic complexes that originate from the dissolution of cathode transition-metal ions, i.e. Mn^{3+} ions, at high cell voltages. Additionally, FTIR electrode–electrolyte reduction products and formed gases may be analyzed using multi-modal approaches with mass-spectrometry techniques such as GC-MS. Generally, there is a push towards *in situ* and *operando* IR spectroscopy combined with various other advanced characterization techniques and it will be exciting to see how new frontiers emerge in metal-ion, solid-state, and other rechargeable battery systems.

References

[1] Griffiths P R and De Haseth J A 2006 *Fourier Transform Infrared Spectrometry* 2nd edn (New York: Wiley) pp 1–529

[2] Infrared spectroscopy absorption table 2020 https://chem.libretexts.org/@go/page/22645

[3] Larkin P J 2017 *Infrared and Raman Spectroscopy: Principles and Spectral Interpretation* (Amsterdam: Elsevier) pp 1–286

[4] Milosevic M 2013 On the nature of the Evanescent wave *Appl. Spectrosc.* **67** 126–31

[5] Skoog D A 1998 *Principles of Instrumental Analysis* ed F J Holler and T A Nieman 5th edn (Philadelphia, PA: Saunders College Pub.)

[6] Flores Cedeño E J 2018 Development of operando diagnostics for Li-ion cathodes by Raman spectroscopy *PhD Dissertation* ETH Zurich

[7] Baker M J, Hughes C S and Hollywood K A 2016 *Biophotonics: Vibrational Spectroscopic Diagnostics* (San Rafael, CA: Morgan & Claypool) pp 1–94

[8] Aurbach D, Daroux M L, Faguy P W and Yeager E 1987 Identification of surface films formed on lithium in propylene carbonate solutions *J. Electrochem. Soc.* **134** 1611–20

[9] Dey A N 1977 Lithium anode film and organic and inorganic electrolyte batteries *Thin Solid Films.* **43** 131–71

[10] Nazri G and Muller R H 1985 Composition of surface layers on Li electrodes in PC, $LiClO_4$ of very low water content *J. Electrochem. Soc.* **132** 2050–4

[11] Peled E 1983 Film forming reaction at the lithium/electrolyte interface *J. Power Sources* **9** 253–66

[12] Peled E 1979 The electrochemical behavior of alkali and alkaline earth metals in non-aqueous battery systems—the solid electrolyte interphase model *J. Electrochem. Soc.* **126** 2047–51

[13] Peled E and Menkin S 2017 Review—SEI: past, present and future *J. Electrochem. Soc.* **164** A1703–19

[14] Peled E, Golodnitsky D and Ardel G 1997 Advanced model for solid electrolyte interphase electrodes in liquid and polymer electrolytes *J. Electrochem. Soc.* **144** L208–10

[15] Gauthier M, Carney T J, Grimaud A, Giordano L, Pour N and Chang H-H *et al* 2015 Electrode–electrolyte interface in Li-ion batteries: current understanding and new insights *J. Phys. Chem. Lett.* **6** 4653–72
[16] Xu K 2004 Nonaqueous liquid electrolytes for lithium-based rechargeable batteries *Chem. Rev.* **104** 4303–418
[17] Xu K 2014 Electrolytes and interphases in Li-ion batteries and beyond *Chem. Rev.* **114** 11503–618
[18] Han S-D, Yun S H, Borodin O, Seo D M, Sommer R D and Young V G *et al* 2015 Solvate structures and computational/spectroscopic characterization of $LiPF_6$ electrolytes *J. Phys. Chem.* C **119** 8492–500
[19] Seo D M, Borodin O, Han S-D, Ly Q, Boyle P D and Henderson W A 2012 Electrolyte solvation and ionic association *J. Electrochem. Soc.* **159** A553
[20] Ein-Eli Y, McDevitt S F and Laura R 1998 The superiority of asymmetric alkyl methyl carbonates *J. Electrochem. Soc.* **145** L1–3
[21] Ein-Eli Y, Thomas S R, Koch V, Aurbach D, Markovsky B and Schechter A 1996 Ethylmethylcarbonate, a promising solvent for Li-ion rechargeable batteries *J. Electrochem. Soc.* **143** L273–7
[22] Chapman N, Borodin O, Yoon T, Nguyen C C and Lucht B L 2017 Spectroscopic and density functional theory characterization of common lithium salt solvates in carbonate electrolytes for lithium batteries *J. Phys. Chem. C.* **121** 2135–48
[23] Barthel J, Buchner R and Wismeth E 2000 FTIR spectroscopy of ion solvation of $LiClO_4$ and LiSCN in acetonitrile, benzonitrile, and propylene carbonate *J. Solution Chem.* **29** 937–54
[24] Seo D M, Reininger S, Kutcher M, Redmond K, Euler W B and Lucht B L 2015 Role of mixed solvation and ion pairing in the solution structure of lithium ion battery electrolytes *J. Phys. Chem. C.* **119** 14038–46
[25] Ellis L D, Buteau S, Hames S G, Thompson L M, Hall D S and Dahn J R 2018 A new method for determining the concentration of electrolyte components in lithium-ion cells, using fourier transform infrared spectroscopy and machine learning *J. Electrochem. Soc.* **165** A256–62
[26] Kunduraci M and Amatucci G G 2006 Synthesis and characterization of nanostructured 4.7V $Li_xMn_{1.5}Ni_{0.5}O_4$ spinels for high-power lithium-ion batteries *J. Electrochem. Soc.* **153** A1345
[27] Julien C and Massot M 2002 Spectroscopic studies of the local structure in positive electrodes for lithium batteries *Phys. Chem. Chem. Phys.* **4** 4226–35
[28] Ben-Kamel K, Amdouni N, Mauger A and Julien C M 2012 Study of the local structure of $LiNi_{0.33+\delta}Mn_{0.33+\delta}Co_{0.33-2\delta}O_2$ ($0.025 \leqslant \delta \leqslant 0.075$) oxides *J. Alloys Compd.* **528** 91–8
[29] Julien C M, Zaghib K, Mauger A, Massot M, Ait-Salah A and Selmane M *et al* 2006 Characterization of the carbon coating onto $LiFePO_4$ particles used in lithium batteries *J. Appl. Phys.* **100** 063511
[30] Kunduraci M, Al-Sharab J F and Amatucci G G 2006 High-power nanostructured $LiMn_{2-x}Ni_xO_4$ high-voltage lithium-ion battery electrode materials: electrochemical impact of electronic conductivity and morphology *Chem. Mater.* **18** 3585–92
[31] Han S-D, Allen J L, Jónsson E, Johansson P, McOwen D W and Boyle P D *et al* 2013 Solvate structures and computational/spectroscopic characterization of lithium difluoro (oxalato)borate (LiDFOB) electrolytes *J. Phys. Chem.* C **117** 5521–31
[32] Seo D M, Boyle P D, Allen J L, Han S D, Jónsson E and Johansson P *et al* 2014 Solvate structures and computational/spectroscopic characterization of $LiBF_4$ electrolytes *J. Phys. Chem.* C **118** 18377–86

[33] Seo D M, Boyle P D, Sommer R D, Daubert J S, Borodin O and Henderson W A 2014 Solvate structures and spectroscopic characterization of LiTFSI electrolytes *J. Phys. Chem.* B **118** 13601–8

[34] Han S D, Borodin O, Allen J L, Seo D M, McOwen D W and Yun S H *et al* 2013 Electrolyte solvation and ionic association: IV. Acetonitrile-lithium difluoro(oxalato)borate (LiDFOB) mixtures *J. Electrochem. Soc.* **160** A2100–10

[35] Han S D, Borodin O, Seo D M, Zhou Z B and Henderson W A 2014 Electrolyte solvation and ionic association: V. Acetonitrile-lithium Bis(fluorosulfonyl)imide (LiFSI) mixtures *J. Electrochem. Soc.* **161** A2042–53

[36] Seo D M, Borodin O, Balogh D, O'Connell M, Ly Q and Han S D *et al* 2013 Electrolyte solvation and ionic association III. Acetonitrile-lithium salt mixtures-transport properties *J. Electrochem. Soc.* **160** A1061–70

[37] Seo D M, Borodin O, Han S-D, Ly Q, Boyle P D and Henderson W A 2012 Electrolyte solvation and ionic association: I. Acetonitrile-lithium salt mixtures: Intermediate and highly associated salts *J. Electrochem. Soc.* **159** A553–65

[38] Seo D M, Borodin O, Han S D, Boyle P D and Henderson W A 2012 Electrolyte solvation and ionic association II. Acetonitrile-lithium salt mixtures: highly dissociated salts *J. Electrochem. Soc.* **159** A1489–500

[39] Borodin O, Han S D, Daubert J S, Seo D M, Yun S H and Henderson W A 2015 Electrolyte solvation and ionic association: VI. Acetonitrile-lithium salt mixtures: highly associated salts revisited *J. Electrochem. Soc.* **162** A501–10

[40] Henderson W A, Seo D M, Han S-D and Borodin O 2020 Electrolyte solvation and ionic association. VII. Correlating Raman spectroscopic data with solvate species *J. Electrochem. Soc.* **167** 110551

[41] McOwen D W, Delp S A, Paillard E, Herriot C, Han S D and Boyle P D *et al* 2014 Anion coordination interactions in solvates with the lithium salts LiDCTA and LiTDI *J. Phys. Chem.* C **118** 7781–7

[42] Seo D, Han S D, Ly Q and Henderson W A 2011 A 'looking glass' into electrolyte properties: acetonitrile-lithium salt (AN-LiX) mixtures *ECS Trans.* **33** 71

[43] Allen J L, Seo D M, Ly Q D, Boyle P D and Henderson W A 2019 Solvent-$LiBF_4$ phase diagrams, ionic association and solubility - cyclic carbonates and lactones *ECS Trans.* **41** 41–5

[44] Allen J L, Seo D M, McOwen D W, Han S D, Knight B A and Boyle P D *et al* 2012 Thermal phase behavior and electrochemical/physicochemical properties of carbonate and ester electrolytes with $LiBF_4$, LiDFOB and LiBOB *ECS Trans.* **50** 381–7

[45] Foley M P, Seo D M, Boyle P D, Henderson W A, De Long H C and Trulove P C 2019 Phase behavior and solvation of lithium triflate γ-butyrlactone *ECS Trans.* **35** 3–8

[46] Foley M P, Seo D M, Worosz C J, Boyle P D, Henderson W A and De Long H C *et al* 2019 Phase behavior and solvation of lithium triflate in ethylene carbonate *ECS Trans.* **41** 99–105

[47] Han S-D, Allen J L, Boyle P D and Henderson W A 2019 Delving into the properties and solution structure of nitrile-lithium difluoro(oxalato)borate (LiDFOB) electrolytes for Li-ion batteries *ECS Trans.* **41** 47–51

[48] Seo D M, Afroz T, Ly Q D, O'Connell M, Boyle P D and Henderson W A 2019 A 'Looking Glass' into electrolyte properties: cyclic carbonate and ester-$LiClO_4$ mixtures *ECS Trans.* **41** 11–5

[49] Seo D M, Allen J L, Gardner L A, Han S D, Boyle P D and Henderson W A 2013 Electrolyte solvation and ionic association: cyclic carbonate and ester-LiTFSI and -$LiPF_6$ mixtures *ECS Trans.* **50** 375–80

[50] Baddour-Hadjean R and Pereira-Ramos J-P 2010 Raman microspectrometry applied to the study of electrode materials for lithium batteries *Chem. Rev.* **110** 1278–319

[51] Julien C 2000 Local cationic environment in lithium nickel–cobalt oxides used as cathode materials for lithium batteries *Solid State Ionics.* **136–137** 887–96

[52] Flores E, Novák P and Berg E J 2018 *In situ* and operando Raman spectroscopy of layered transition metal oxides for Li-ion battery cathodes *Front. Energy Res.* **6** 82

[53] Inaba M, Iriyama Y, Ogumi Z, Todzuka Y and Tasaka A 1997 Raman study of layered rock-salt $LiCoO_2$ and its electrochemical lithium deintercalation *J. Raman Spectrosc.* **28** 613–7

[54] Adams R A, Li B, Kazmi J, Adams T E, Tomar V and Pol V G 2018 Dynamic impact of $LiCoO_2$ electrodes for Li-ion battery aging evaluation *Electrochim. Acta* **292** 586–93

[55] Ruther R E, Callender A F, Zhou H, Martha S K and Nanda J 2014 Raman microscopy of lithium-manganese-rich transition metal oxide cathodes *J. Electrochem. Soc.* **162** A98–102

[56] Ghanty C, Markovsky B, Erickson E M, Talianker M, Haik O and Tal-Yossef Y *et al* 2015 Li^+-Ion extraction/insertion of Ni-Rich $Li_{1+x}(Ni_yCo_zMn_z)_wO_2$ ($0.005 < x < 0.03$; $y{:}z = 8{:}1$, $w \approx 1$) electrodes: in situ XRD and Raman Spectroscopy study *Chemelectrochem.* **2** 1479–86

[57] Otoyama M, Ito Y, Hayashi A and Tatsumisago M 2016 Raman spectroscopy for $LiNi_{1/3}Mn_{1/3}Co_{1/3}O_2$ composite positive electrodes in all-solid-state lithium batteries *Electrochemistry.* **84** 812–4

[58] Jung R, Morasch R, Karayaylali P, Phillips K, Maglia F and Stinner C *et al* 2018 Effect of ambient storage on the degradation of Ni-rich positive electrode materials (NMC811) for Li-ion batteries *J. Electrochem. Soc.* **165** A132–41

[59] Ruther R E, Hays K A, An S J, Li J, Wood D L and Nanda J 2018 Chemical evolution in silicon–graphite composite anodes investigated by vibrational spectroscopy *ACS Appl. Mater, Interfaces* **10** 18641–9

[60] Kerlau M, Marcinek M, Srinivasan V and Kostecki R M 2007 Studies of local degradation phenomena in composite cathodes for lithium-ion batteries *Electrochim. Acta* **52** 5422–9

[61] Kostecki R and McLarnon F 2004 Local-probe studies of degradation of composite $LiNi_{0.8}Co_{0.15}A_{0.05}O_2$ cathodes in high-power lithium-ion cells *Electrochem. Solid-State Lett.* **7** A380

[62] Nanda J, Remillard J, O'Neill A, Bernardi D, Ro T and Nietering K E *et al* 2011 Local state-of-charge mapping of lithium-ion battery electrodes *Adv. Funct. Mater.* **21** 3282–90

[63] Liu K, Yu Z, Zhu X, Zhang S, Zou F and Zhu Y 2016 A universal surface enhanced Raman spectroscopy (SERS)-active graphene cathode for lithium–air batteries *RSC Adv.* **6** 102272–9

[64] Tornheim A, Maroni V A, He M, Gosztola D J and Zhang Z 2017 Enhanced Raman scattering from NCM523 cathodes coated with electrochemically deposited gold *J. Electrochem. Soc.* **164** A3000–5

[65] Yang G, Ivanov I N, Ruther R E, Sacci R L, Subjakova V and Hallinan D T *et al* 2018 Electrolyte solvation structure at solid-liquid interface probed by nanogap surface-enhanced Raman spectroscopy *ACS Nano.* **12** 10159–70

[66] Yang G, Sacci R L, Ivanov I N, Ruther R E, Hays K A and Zhang Y *et al* 2019 Probing electrolyte solvents at solid/liquid interface using gap-mode surface-enhanced raman spectroscopy *J. Electrochem. Soc.* **166** A178–87

[67] Piernas-Muñoz M J, Tornheim A, Trask S, Zhang Z and Bloom I 2021 Surface-enhanced Raman spectroscopy (SERS): a powerful technique to study the SEI layer in batteries *Chem. Commun.* **57** 2253–6

[68] Freunberger S A, Chen Y, Peng Z, Griffin J M, Hardwick L J and Bardé F *et al* 2011 Reactions in the rechargeable lithium–O_2 battery with alkyl carbonate electrolytes *J. Am. Chem. Soc.* **133** 8040–7

[69] Peng Z, Freunberger S A, Hardwick L J, Chen Y, Giordani V and Bardé F *et al* 2011 Oxygen reactions in a non-aqueous Li^+ electrolyte *Angew. Chem. Int. Ed.* **50** 6351–5

[70] Peng Z, Freunberger S A, Chen Y and Bruce P G 2012 A reversible and higher-rate Li–O_2 battery *Science* **337** 563–6

[71] Schmitz R, Ansgar Müller R, Wilhelm Schmitz R, Schreiner C, Kunze M and Lex-Balducci A *et al* 2013 SEI investigations on copper electrodes after lithium plating with Raman spectroscopy and mass spectrometry *J. Power Sources* **233** 110–4

[72] Tang S, Gu Y, Yi J, Zeng Z, Ding S-Y and Yan J-W *et al* 2016 An electrochemical surface-enhanced Raman spectroscopic study on nanorod-structured lithium prepared by electrodeposition *J. Raman Spectrosc.* **47** 1017–23

[73] Miroshnikov Y and Zitoun D 2017 Operando plasmon-enhanced Raman spectroscopy in silicon anodes for Li-ion battery *J. Nanopart. Res.* **19** 372

[74] Sardari B and Özcan M 2017 Real-time and tunable substrate for surface enhanced Raman spectroscopy by synthesis of copper oxide nanoparticles via electrolysis *Sci. Rep.* **7** 7730

[75] Miroshnikov Y, Yang J, Shokhen V, Alesker M, Gershinsky G and Kraytsberg A *et al* 2018 Operando micro-Raman study revealing enhanced connectivity of plasmonic metals decorated silicon anodes for lithium-ion batteries *ACS Appl. Energy Mater.* **1** 1096–105

[76] Cowan A J and Hardwick L J 2019 Advanced spectroelectrochemical techniques to study electrode interfaces within lithium-ion and lithium-oxygen batteries *Ann. Rev. Anal. Chem.* **12** 323–46

[77] Ha Y, Tremolet de Villers B J, Li Z, Xu Y, Stradins P and Zakutayev A *et al* 2020 Probing the evolution of surface chemistry at the silicon-electrolyte interphase via *in situ* surface-enhanced Raman spectroscopy *J. Phys. Chem. Lett.* **11** 286–91

[78] Li J F, Huang Y F, Ding Y, Yang Z L, Li S B and Zhou X S *et al* 2010 Shell-isolated nanoparticle-enhanced Raman spectroscopy *Nature* **464** 392–5

[79] Hy S, Felix, Chen Y-H, Liu J-Y, Rick J and Hwang B-J 2014 *In situ* surface enhanced Raman spectroscopic studies of solid electrolyte interphase formation in lithium ion battery electrodes *J. Power Sources* **256** 324–8

[80] Hy S, Felix F, Rick J, Su W-N and Hwang B J 2014 Direct *In situ* observation of Li_2O evolution on Li-rich high-capacity cathode material, $Li[Ni_xLi_{(1-2x)/3}Mn_{(2-x)/3}]O_2$ ($0 \leqslant x \leqslant 0.5$) *J. Am. Chem. Soc.* **136** 999–1007

[81] Galloway T A, Cabo-Fernandez L, Aldous I M, Braga F and Hardwick L J 2017 Shell isolated nanoparticles for enhanced Raman spectroscopy studies in lithium–oxygen cells *Faraday Discuss.* **205** 469–90

[82] Li J-F, Zhang Y-J, Ding S-Y, Panneerselvam R and Tian Z-Q 2017 Core–shell nanoparticle-enhanced Raman spectroscopy *Chem. Rev.* **117** 5002–69

[83] Li C-Y, Yu Y, Wang C, Zhang Y, Zheng S and Li J-F *et al* 2020 Surface changes of $LiNi_xMn_yCo_{1-x-y}O_2$ in Li-ion batteries using *In Situ* surface-enhanced Raman spectroscopy. *J. Phys. Chem. C.* **124** 4024–31

[84] Stöckle R M, Suh Y D, Deckert V and Zenobi R 2000 Nanoscale chemical analysis by tip-enhanced Raman spectroscopy *Chem. Phys. Lett.* **318** 131–6

[85] Mauser N and Hartschuh A 2014 Tip-enhanced near-field optical microscopy *Chem. Soc. Rev.* **43** 1248–62

[86] Liao M, Jiang S, Hu C, Zhang R, Kuang Y and Zhu J *et al* 2016 Tip-enhanced Raman spectroscopic imaging of individual carbon nanotubes with subnanometer resolution *Nano Lett.* **16** 4040–6

[87] Wang X, Huang S-C, Huang T-X, Su H-S, Zhong J-H and Zeng Z-C *et al* 2017 Tip-enhanced Raman spectroscopy for surfaces and interfaces *Chem. Soc. Rev.* **46** 4020–41

[88] Nanda J, Yang G, Hou T, Voylov D N, Li X and Ruther R E *et al* 2019 Unraveling the nanoscale heterogeneity of solid electrolyte interphase using tip-enhanced Raman spectroscopy *Joule.* **3** 2001–19

[89] Kim H, Kosuda K M, Van Duyne R P and Stair P C 2010 Resonance Raman and surface- and tip-enhanced Raman spectroscopy methods to study solid catalysts and heterogeneous catalytic reactions *Chem. Soc. Rev.* **39** 4820–44

[90] Kawata S, Ichimura T, Taguchi A and Kumamoto Y 2017 Nano-Raman scattering microscopy: resolution and enhancement *Chem. Rev.* **117** 4983–5001

[91] Goren E, Chusid O and Aurbach D 1991 The application of *In Situ* FTIR spectroscopy to the study of surface films formed on lithium and noble metals at low potentials in Li battery electrolytes *J. Electrochem. Soc.* **138** L6–9

[92] Aurbach D 1993 *In situ* FTIR spectroelectrochemical studies of surface films formed on Li and nonactive electrodes at low potentials in Li salt solutions containing CO_2 *J. Electrochem. Soc.* **140** L155

[93] Santner H J, Korepp C, Winter M, Besenhard J O and M ller K C 2004 In-situ FTIR investigations on the reduction of vinylene electrolyte additives suitable for use in lithium-ion batteries *Anal. Bioanal.Chem.* **379** 266–71

[94] Li J-T, Chen S-R, Ke F-S, Wei G-Z, Huang L and Sun S-G 2010 *In situ* microscope FTIR spectroscopic studies of interfacial reactions of Sn–Co alloy film anode of lithium ion battery *J. Electroanal. Chem.* **649** 171–6

[95] Ikezawa Y and Ariga T 2007 *In situ* FTIR spectra at the Cu electrode/propylene carbonate solution interface *Electrochim. Acta* **52** 2710–5

[96] Ikezawa Y and Nishi H 2008 *In situ* FTIR study of the Cu electrode/ethylene carbonate +dimethyl carbonate solution interface *Electrochim. Acta* **53** 3663–9

[97] Pérez-Villar S, Lanz P, Schneider H and Novák P 2013 Characterization of a model solid electrolyte interphase/carbon interface by combined *in situ* Raman/Fourier transform infrared microscopy *Electrochim. Acta* **106** 506–15

[98] Marino C, Boulaoued A, Fullenwarth J, Maurin D, Louvain N and Bantignies J-L *et al* 2017 Solvation and dynamics of lithium ions in carbonate-based electrolytes during cycling followed by operando infrared spectroscopy: the example of NiSb2, a typical negative conversion-type electrode material for lithium batteries *J. Phys. Chem.* C **121** 26598–606

[99] Tremolet de Villers B J, Bak S M, Yang J and Han S D 2021 *In Situ* ATR-FTIR study of the cathode–electrolyte interphase: electrolyte solution structure, transition metal redox, and surface layer evolution *Batteries Supercaps.* **4** 778–84

[100] Julien C M and Mauger A 2018 *In situ* Raman analyses of electrode materials for Li-ion batteries *AIMS Mater. Sci.* **5** 650–98

[101] Wu H-L, Huff L A and Gewirth A A 2015 *In Situ* Raman spectroscopy of sulfur speciation in lithium–sulfur batteries *ACS Appl. Mater. Interfaces* **7** 1709–19
[102] Song H-Y, Fukutsuka T, Miyazaki K and Abe T 2016 *In situ* Raman investigation of electrolyte solutions in the vicinity of graphite negative electrodes *Phys. Chem. Chem. Phys.* **18** 27486–92
[103] Lei J, McLarnon F and Kostecki R 2005 *In Situ* Raman microscopy of individual $LiNi_{0.8}Co_{0.15}Al_{0.05}O_2$ particles in a Li-ion battery composite cathode *J. Phys. Chem.* B **109** 952–7
[104] Flores E, Vonrüti N, Novák P, Aschauer U and Berg E J 2018 Elucidation of $LixNi_{0.8}Co_{0.15}Al_{0.05}O_2$ redox chemistry by operando Raman spectroscopy *Chem. Mater.* **30** 4694–703
[105] Flores E, Novák P, Aschauer U and Berg E J 2019 Cation ordering and redox chemistry of layered Ni-rich $Li_xNi_{1-2y}Co_yMn_yO_2$: an operando Raman spectroscopy Study *Chem. Mater.* **32** 186–94
[106] Krause A, Tkacheva O, Omar A, Langklotz U, Giebeler L and Dörfler S *et al* 2019 *In Situ* Raman spectroscopy on silicon nanowire anodes integrated in lithium ion batteries *J. Electrochem. Soc.* **166** A5378–85
[107] Cheng X-Q, Li H-J, Zhao Z-X, Wang Y-Z and Wang X-M 2021 The use of *in situ* Raman spectroscopy in investigating carbon materials as anodes of alkali metal-ion batteries *New Carbon Mater.* **36** 93–105
[108] Zheng Y, Deng T, Yue N, Zhang W, Zhu X and Yang H *et al* 2021 Raman spectroscopy and correlative-Raman technology excel as an optimal stage for carbon-based electrode materials in electrochemical energy storage *J Raman Spectros.* 1–12
[109] Nishi T, Nakai H and Kita A 2013 Visualization of the state-of-charge distribution in a $LiCoO_2$ cathode by *In Situ* Raman imaging *J. Electrochem. Soc.* **160** A1785–8
[110] Yang J, Kraytsberg A and Ein-Eli Y 2015 In-situ Raman spectroscopy mapping of Si based anode material lithiation *J. Power Sources* **282** 294–8
[111] Zeng Z-C, Huang S-C, Wu D-Y, Meng L-Y, Li M-H and Huang T-X *et al* 2015 Electrochemical tip-enhanced Raman spectroscopy *J. Am. Chem. Soc.* **137** 11928–31
[112] Kang G, Yang M, Mattei M S, Schatz G C and Van Duyne R P 2019 *In Situ* nanoscale redox mapping using tip-enhanced Raman spectroscopy *Nano Lett.* **19** 2106–13
[113] Urban A, Seo D-H and Ceder G 2016 Computational understanding of Li-ion batteries *NPJ Comput. Mater.* **2** 16002
[114] Jain A, Shin Y and Persson K A 2016 Computational predictions of energy materials using density functional theory *Nat. Rev. Mater.* **1** 15004
[115] Lim J, Lee K-K, Liang C, Park K-H, Kim M and Kwak K *et al* 2019 Two-dimensional infrared spectroscopy and molecular dynamics simulation studies of nonaqueous lithium-ion battery electrolytes *J. Phys. Chem. B.* **123** 6651–63
[116] Fulfer K D and Kuroda D G 2017 A comparison of the solvation structure and dynamics of the lithium ion in linear organic carbonates with different alkyl chain lengths *Phys. Chem. Chem. Phys.* **19** 25140–50

Chapter 5

Differential electrochemical mass spectrometry (DEMS) for batteries

Boyu Shi, Kewei Liu, Eungje Lee and Chen Liao

5.1 Introduction

The evolution of gas species is common in electrochemical reactions. The immediate detection and quantitative analysis of gas products can help electrochemists to investigate the reaction mechanism by figuring out the detailed steps of a series of reactions. It is especially important for applications to catalysts, fuel cells, and battery systems. However, real-time detection can be very challenging because of the small amounts of gas products involved [1]. In 1970, Buckenstein first used a mass spectrometer (MS) to detect volatile gas products evolved from electrochemical reactions. The electrochemical cell, where the reactions occur, was connected to the MS by a non-wetting, porous polytetrafluoroethylene (PTFE) membrane. However, real-time detection was not achieved [2, 3]. In 1984, Wolter and Heitbaum significantly improved on the vacuum system of Buckenstein's setup while still using a PTFE membrane interface. A differential pumped vacuum system was employed for Wolter and Heitbaum's setup. The new pump system dramatically shortened the response time and thus enabled real-time detection of volatile species. To distinguish its near-instantaneous nature from other integration methods, their method was named 'differential electrochemical mass spectrometry' (DEMS) [3, 4].

DEMS has been widely explored and applied for gas analysis since then, and several modified DEMS setups have been reported. The improvements were primarily related to modifications of the electrochemical cell [3–6]. However, later researchers found that the application of conventional DEMS to lithium-ion or lithium–oxygen battery gas studies was limited. This is because the amount of gas evolution in those systems is even smaller. In 2007, Petr Novák's group reported a new DEMS setup. They redesigned the conventional DEMS electrochemical cell and established a headspace-based gas analysis principle. They named the system 'online electrochemical mass spectrometry' (OEMS). OEMS is capable of analyzing

doi:10.1088/978-0-7503-2682-7ch5

the gas evolution of lithium battery systems because it uses a carrier gas to make up for the insufficient flow rate of gas products [7]. However, the time resolution of OEMS is typically longer than for DEMS. This is the reason for using 'online' instead of 'differential' in the name. In 2011, McCloskey's group reported an 'intermittent DEMS' setup. This system allowed an intermittent test protocol: gas products were accumulated in the headspace of the electrochemical cell during the testing interval and expanded into the MS at the testing point. The system can analyze the gas products of battery systems because of the high flow rate at the testing point [8]. In 2013, Gasteiger's group introduced a novel OEMS setup. They claimed that their system could achieve a continuous test for several without using carrier gas. The absence of carrier gas can reduce the possibility of contamination [9].

DEMS is an analytical technique that allows for an instantaneous, quantitative measurement of volatile-species reactants and reaction products. It combines electrochemical experiments with mass spectrometry [10]. By collecting mass signals of different volatile species from the MS, electrochemists can determine what kind of reactants, intermediates, by-products, and products are involved in the reactions of interest. The relative intensity of each volatile species can also be calculated to correlate the evolution of gaseous products with tested electrochemical processes. DEMS is a very valuable measurement method for elucidating electrochemical reaction mechanisms and processes in bulk materials and at interfaces. DEMS is usually combined with electrochemical measurements such as galvanostatic cycling; the combined technique has been named mass spectroscopic cyclic voltammetry. Other electrochemical test techniques such as potential step and galvanic step were also employed [1].

5.2 General principles of DEMS

The general principles of DEMS are laid out as follows: an introduction to the classification of DEMS techniques is given in section 5.2.1, followed by an introduction to mass spectrometry in section 5.2.2. The configurations of different types of DEMS will be discussed in section 5.2.3. In section 5.2.4, instrument design, including electrochemical cell and vacuum system configurations, will be addressed. In section 5.2.5, the quantification process for different DEMS setups will be shown. Finally, the advantages and limitations of different DEMS systems will be discussed in section 5.2.6.

5.2.1 Classification of DEMS

Since the launch of the first DEMS system in 1984 [11], various kinds of DEMS systems have been built. Three categories are defined, based on the design of the electrochemical cells, gas inlet systems, and vacuum systems, and the applications: conventional DEMS, OEMS, and intermittent DEMS, as listed in table 5.1.

Conventional DEMS was initially used for studying the oxidation mechanism in fuel cells [1]. It usually uses a membrane-based electrochemical cell and a differential

Table 5.1. Comparison of different DEMS systems.

	Electro-chemical cell	Gas inlet system	Vacuum system	System seal	Response time	Application
Conventional DEMS (C-DEMS) [1, 5, 12, 13]	Membrane cell	Membrane	Differential pump	Sealed system	<2 s	Fuel cell/lithium-ion battery
Sealed OEMS (S-OEMS) [9]	Headspace-analysis-based cell	Capillary	Single pump	Sealed system	1 s	Lithium–oxygen battery/lithium-ion battery
Continuous OEMS (C-OEMS) [14]	Headspace-analysis- based cell	Capillary	Differential pump/single pump	Open system	30 s	Lithium–oxygen battery/lithium-ion battery
Intermittent DEMS (I-DEMS) [8]	Headspace-analysis- based cell	Eight-valve inlet	Differential pump	Half-sealed system	>15 min	Lithium–oxygen battery/lithium-ion battery

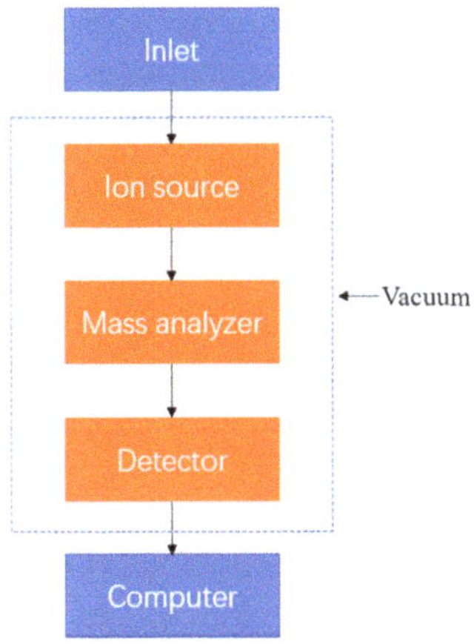

Figure 5.1. A diagram showing the elements of a mass spectrometer 15.

pumped vacuum system. The gas inlet system is also membrane-based, matching the electrochemical cell. The conventional DEMS is a sealed system.

OEMS is designed for investigating electrochemical reactions of lithium–oxygen and lithium-ion battery systems, owing to its ability to detect very small amounts of gas species. A headspace-analysis-based cell is used for OEMS, and a capillary is connected to the electrochemical cell for transferring gas from the cell to the MS. For OEMS, the system can be either sealed (sealed OEMS) or unsealed (continuous OEMS), depending on the usage of an external gas source. A single pump (part of the MS) or differential pump is used to evacuate the system, depending on the gas flow rate.

Intermittent DEMS is capable of testing gas products intermittently, thanks to a specially designed eight-valve gas inlet system [8]. Intermittent DEMS is also designed for analyzing gas behavior of lithium–oxygen and lithium-ion battery systems; since it allows gas products to accumulate during the testing interval, a high signal intensity can be obtained. A headspace-analysis-based cell is also used for intermittent DEMS as a container where gas species accumulate. A differential pumped vacuum system is used for intermittent DEMS. since the gas flow rate at the testing point is relatively high and the dual pumps can help to maintain the high-vacuum environment ($<10^{-6}$ mbar) that MS requires. Intermittent DEMS can be regarded as a half-sealed system, since the cell is hermetically sealed during the testing interval and open at the testing point.

5.2.2 Mass spectrometry

Mass spectrometry is a widely used analytical technique for identifying chemical compositions of pure samples as well as complex mixtures. It is based on the detection of the characteristic mass-to-charge ratio (m/z) of ions. The results are usually presented as a plot of relative abundance of the ions versus m/z. This plot is also called a mass spectrum. A typical MS contains the following elements (figure 5.1): Sample inlet, ion source, mass analyzers, detectors, and computers [15].

The ion source produces gas-phase ions from samples by multiple types of ionization methods, including electron, chemical, and field ionization. During the ionization process, the atoms or molecules from the sample lose one or more electrons to become positively charged fragments. Most atoms or molecules only lose one electron after ionization, but large molecules can lose multiple electrons [15].

The mass analyzers measure the *m*/*z* of gas-phase ions and separate them accordingly. Several types of mass analyzers have been developed, including quadrupole (Q) analyzers, ion trap (IT) analyzers, time-of-flight (TOF) analyzers, and magnetic and electromagnetic (B) analyzers. Although each mass analyzer has its own advantages and shortcomings, they can be separated into two categories according to their working mechanism. The first type of mass analyzer allows for the transmission of gas-phase ions with a certain *m*/*z* at a given time. This type includes Q analyzers and magnetic sector analyzers. The second type of mass analyzer allows simultaneous transmission of all gas-phase ions; these include IT analyzers, magnetic analyzers and TOF analyzers. For the DEMS system, a Q analyzer is usually employed. A Q analyzer consists of four parallel cylindrical rods, as shown in figure 5.2. The Q analyzer is capable of selecting ions with different *m*/*z* ratios. The gas-phase ions are separated on the basis of the stability of their trajectories in an oscillating electric field [15].

The selected gas-phase ion beam going through mass analyzers will be detected by a detector. The detector contains a Faraday cup and electron multipliers. The Faraday cup is a metal box with a small opening. Only selected gas-phase ions can go through the opening. The ions that enter will hit the wall of the metal box. As a consequence, their charge will be neutralized by the electrons transferred from the metal. The flow of electrons leads to an electric current. The current is then amplified by electron multipliers. The current density depends on the ion abundance. The more ions hit the metal box, the higher the current density. A computer can record the current density, plot it according to the *m*/*z*, and present the result as a mass spectrum [15].

It should be emphasized that the MS requires high vacuum, typically below 10^{-5} mbar, to avoid potential collisions of gas-phase ions with air molecules.

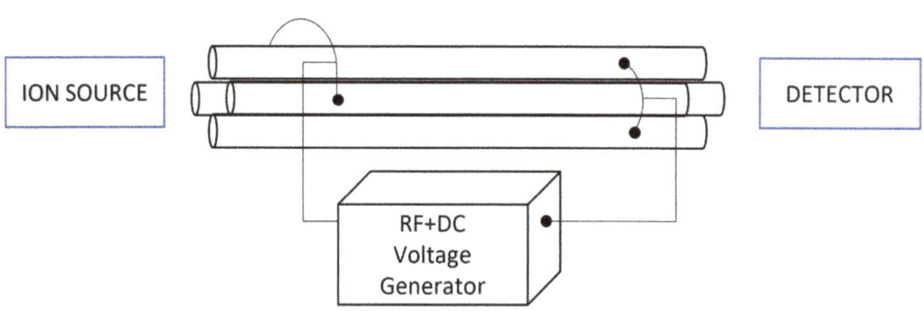

Figure 5.2. A schematic drawing of a quadrupole mass spectrometer. Reprinted with permission from [16]. Copyright 1986 American Chemical Society.

5.2.3 DEMS setup

5.2.3.1 Conventional DEMS setup

The first DEMS system was introduced by O Wolter and J Heitbaum in 1984 [11]. Since it is the first generation of DEMS systems, and to distinguish it from the other types of DEMS systems, here we classify it as conventional DEMS, denoted as C-DEMS, for simplification in later discussion. C-DEMS was initially invented to elucidate fuel-cell-related reactions such as methanol oxidation, by virtue of its ability to quantitatively determine gas evolution during the oxidation process. Gas products such as CO_2 ($m/z = 44$) or $C_2H_4O_2$ methylformate ($m/z = 60$) can act as very useful indicators for some products and intermediates. For example, the presence of methylformate proves the formation of formic acid, which is an intermediate during oxidation [17]. Later, C-DEMS was also used to determine hydrogen origin during formaldehyde oxidation, which is of great importance in studying electroless metal deposition [18]. C-DEMS was also used to study surface organic adsorbates and desorption products [1].

A C-DEMS system typically contains four parts: electrochemical cell, membrane gas inlet system, vacuum system, and MS. A schematic drawing of a C-DEMS system is shown in figure 5.3. A membrane-based electrochemical cell is used in C-DEMS. The cell is connected to a vacuum system by a hydrophobic membrane. The electrochemical cell is designed for holding a working electrode (WE), counter electrode (CE), reference electrode (RE), and electrolyte.

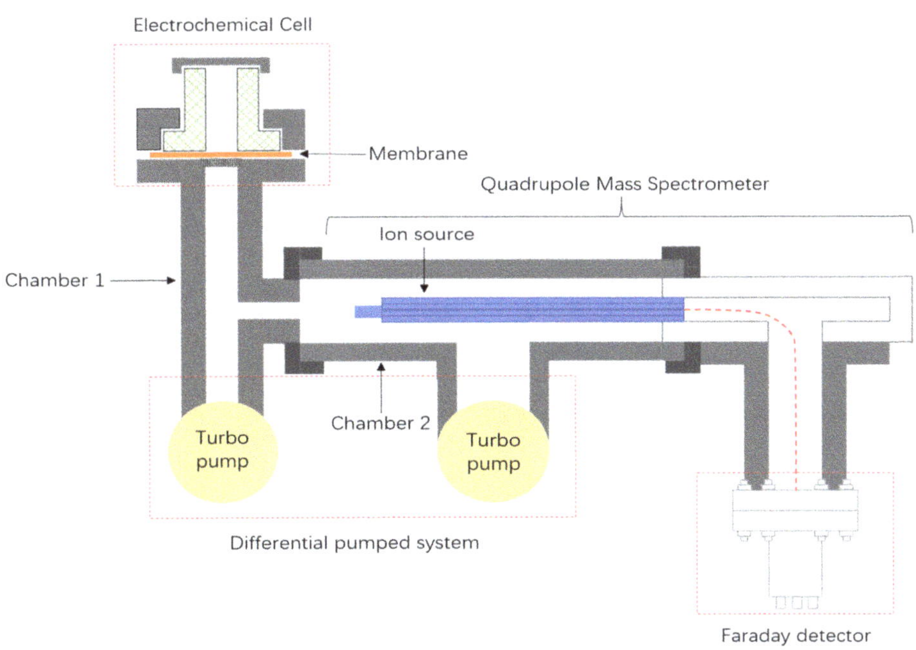

Figure 5.3. A schematic drawing of a C-DEMS system.

Although various kinds of membrane-based cells have been designed to study different reactions, the mechanism remains the same. We will discuss different types of electrochemical cell designs in section 2.2.1.

The membrane gas inlet system typically includes a microporous PTFE membrane as part of the membrane-based electrochemical cell, as shown in figure 5.4. PTFE is hydrophobic and can act as a barrier that allows the transmission of volatile species from the aqueous electrolyte into the vacuum chamber while inhibiting the aqueous electrolyte itself from penetrating the membrane pores. Figure 5.4 shows a traditional electrolyte–membrane interface [1]. The driving force for gaseous transportation is the pressure drop between the electrochemical cell chamber and the MS. The thickness and pore size of the PTFE membrane should be carefully selected in order to prevent aqueous electrolyte from passing through the PTFE membrane while retaining sufficiently high permeability for gaseous and volatile species. More details about the membrane inlet system, like pore size and thickness, will be discussed in section 5.2.4.1.1.1.

The C-DEMS system has a high time resolution (<2 s) as a result of the short transfer distance of the membrane inlet system, which allows a rapid transfer of volatile species. However, this short passage distance also causes more electrolyte solvent molecules to enter the MS, bringing in more contamination [19]. In addition, the high signal intensity of the electrolyte can undermine the resolution of the DEMS system, making it hard to analyze gaseous products that only occupy a small fraction of the total [14].

The term 'differential pumped vacuum system' refers to a two-stage pressure reduction system, where the gas stream is first expanded into a relatively low-pressure (0.1 to 1-mbar) state and then expanded into high vacuum (10^{-6} mbar) [9]. Since the gas flow rate for C-DEMS is relatively high (1 ml min^{-1}) with a high possibility of gaseous fractionation and the MS requires high-vacuum (typically $<10^{-6}$ mbar) working conditions to avoid ion interactions [19], an additional turbo pump is necessary to establish a gas bypass to pump out the large amount of residual

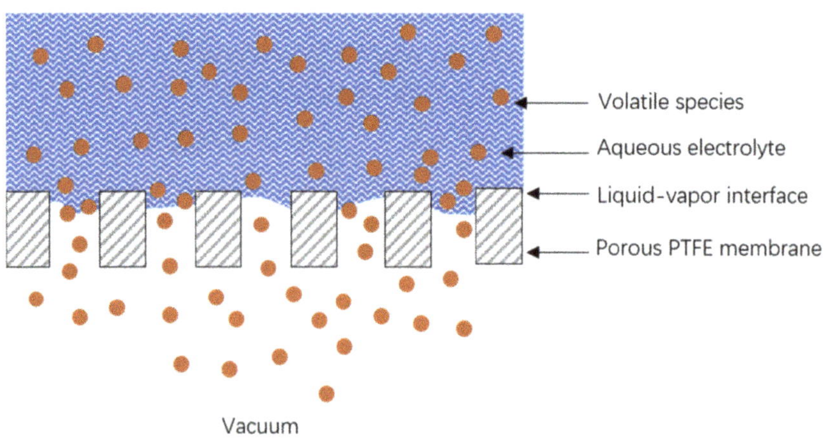

Figure 5.4. Liquid–vapor interface in a C-DEMS system.

gases and solvent. This dual-pump system, or differential pumped vacuum system [10, 20], is depicted in figure 5.3. The first-stage scroll pump is connected to the electrochemical cell via Chamber 1, in which the pressure is on the order of 0.1–1 mbar; the second-stage scroll pump on the right is attached to Chamber 2, where the pressure is reduced to 10^{-6} mbar. The advantage of the differential pumped vacuum system is that gas fractionation can be avoided, simplifying the quantification process that follows. The limitation of the differential pumped vacuum system is its requirement for a high gas-stream flow rate (around 1 ml min^{-1}). In the C-DEMS system, the differential pumped vacuum system plays a critical role in both the gaseous transport and the detection processes. The utilization of the differential pumped vacuum system make it possible to quantitatively determine the gas evolution rates in alignment with faradic currents [20].

In summary, the operation principles of the C-DEMS system can be described as follows: (1) volatile species including reactants, intermediates and products will evaporate directly into pump chamber 1 because of the pressure difference between the cell and chamber 1. (2) Most of the gas species will be pumped out by scroll pump 1; only a small amount of gas products (e.g. 25 μl min^{-1} out of 0.8 ml min^{-1}) will be transferred into the QMS ionized chamber. (3) The gas products will be analyzed instantaneously by the QMS. The mass spectrum (a plot of intensity versus *m*/*z*) gives the identities of gas species, and a plot of signal intensity as a function of time or voltage reveals the gaseous products' correlation with electrochemical response.

C-DEMS is a primitive version of the DEMS system. It is widely used in measuring the gas behavior of fuel cells and gas-related electrochemical reactions because of its advantages of short response time (<2 s) and simple quantification method. However, the potential for contamination of C-DEMS systems by electrolyte and their low detection limit cannot be overlooked.

As noted above, C-DEMS was used to study lithium-ion battery solid–electrolyte interphase (SEI) formation, where the major gas products are identified as hydrogen, ethylene and propylene [21]. A characteristic of these gas products is that they are produced in relatively high quantities. Meanwhile C-DEMS is vulnerable to electrolyte contamination and is incapable of testing small-fraction gas products. Analysis of gas products such as CO_2 and O_2, with a gas generation rate on the level of μmol min^{-1} m^{-2} [14] is very challenging for C-DEMS. In order to further study the gas behavior of lithium-ion and lithium–oxygen batteries, the improved OEMS method was introduced.

5.2.3.2 Online electrochemical mass spectrometry (OEMS)

One key drawback of C-DEMS is its low testing resolution, caused by the strong electrolyte background. As shown in figure 5.5(a), although the semi-permeable PTFE porous membrane used by C-DEMS is hydrophobic, small numbers of electrolyte solvent molecules can still go through the membrane and cause an increased electrolyte background. In order to solve this problem, a strategy combining a newly designed capillary gas inlet system with headspace analysis based on the electrochemical cell was introduced. Figure 5.5(b) is a schematic

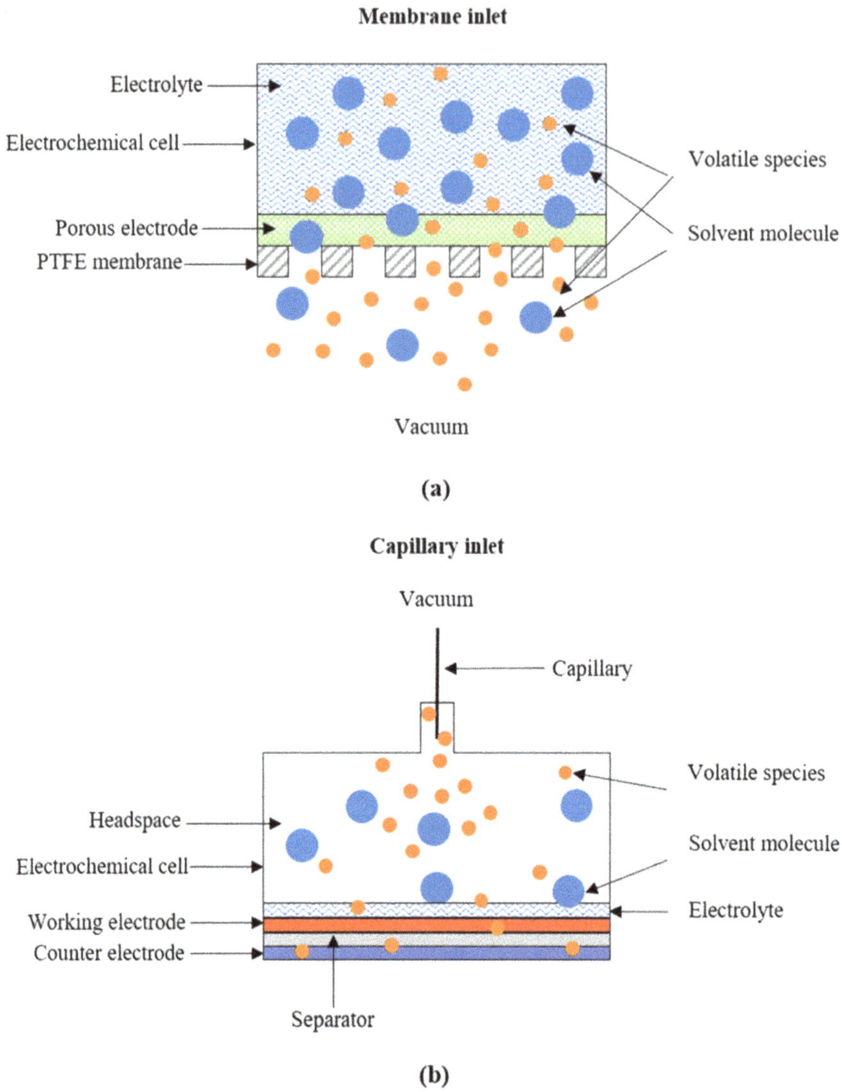

Figure 5.5. Comparison of membrane inlet system (a) and capillary inlet system (b) 19.

diagram of the capillary gas inlet system, showing the creation of a headspace in the electrochemical cell where volatile species can accumulate; hence the term 'head-space-analysis-based electrochemical cell' [19]. A key feature of the headspace-analysis-based cell is that the ratio of active material to electrolyte is significantly (almost 100 times) higher than for the membrane-based cell, making it most suitable for suppressing the electrolyte's strong signal intensity [7]. Furthermore, the capillary can also trap volatile electrolyte components, protecting the MS from electrolyte contamination and increasing its sensitivity [22]. More details about the headspace-analysis-based cell will be discussed in section 5.2.4.1.

Another problem in applying C-DEMS to battery systems is the low gas evolution rate. The differential pumped vacuum system requires a minimum flow rate of approximately 1 ml min^{-1}. As the gas volume is limited in a closed battery system, a higher flow rate is desired to enable fast analysis. There are two approaches to resolving this problem: (1) continuously purging carrier gas into the cell to achieve a sufficiently high flow rate; or (2) intermittently purging headspace volume into the differential pumped vacuum system [23]. The intermittent testing pattern allows volatile species to accumulate at the top of the headspace-analysis-based cell during the testing interval, so that more gas species can be transferred to the MS at the testing point, making use of the differential pumped vacuum system possible.

OEMS was initially designed to investigate the gas behavior of battery systems such as lithium-oxygen batteries because of its ability to identify much smaller quantities of gas species [14]. OEMS systems can be classified into two categories: C-OEMS and S-OEMS. The word 'online' rather than 'differential' highlights the fact that the response time of OEMS is usually longer than that of C-DEMS [24]. The capital 'C' in C-OEMS stands for 'continuous,' which denotes the continuous operation mode for OEMS. The 'S' in S-OEMS represents 'sealed,' which means that S-OEMS is a sealed system, unlike C-OEMS.

The first C-OEMS system using a headspace-analysis-based cell and capillary inlet system was introduced by P Novák's group in 2007 [7]. Figure 5.6 is a schematic drawing of Novák's C-OEMS system. The system typically consists of four parts plus external gas source consisting of carrier gas and calibration gas. Ar or He are usually used as carrier gas. Calibration gas consists of a series of known concentration gas sources. The components are a headspace-analysis-based electrochemical cell, capillary gas inlet system, vacuum system, and MS. The calibration gas and carrier gas cylinders are connected to the system with flow controllers 1, 2 and 3. The electrochemical cell is connected to the MS via a capillary inlet and vacuum system. For C-OEMS, a differential pumped vacuum system is used to maintain the high vacuum required by the MS.

Whereas the C-DEMS setup has a PTFE membrane inlet system, the C-OEMS setup features a capillary inlet system. The gaseous and volatile species are first evaporated into the headspace of the electrochemical cell and then transferred to the

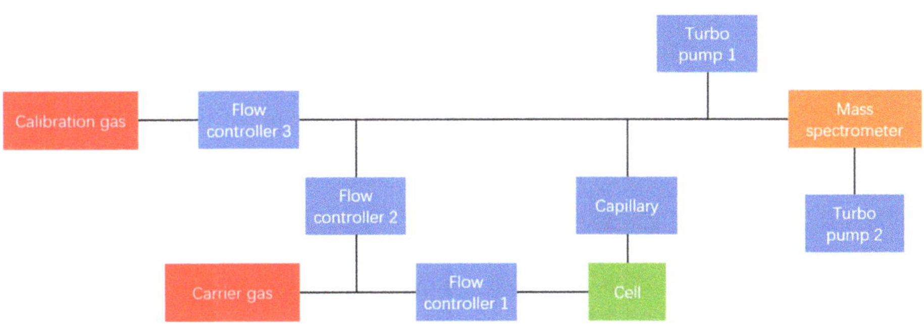

Figure 5.6. Schematic drawing of C-OEMS 19.

MS through a capillary. To achieve this goal, the electrochemical cell is designed to create a headspace volume for gas accumulation as shown in figure 5.5(b). Details about the headspace-analysis-based cell will be discussed in section 5.2.4.1.1.4.

Like the C-DEMS system, the C-OEMS system relies on a differential pumped vacuum system. To achieve the gas flow rate (~1 ml min^{-1}) required by the differential pumped vacuum system, carrier gas such as Ar or He needs to be continuously flowed into the system. As illustrated in figure 5.6, three flow controllers are connected to the carrier gas cylinders to regulate the carrier gas flow rate. An optimized gas flow rate of 0.85 ml min^{-1} is selected for Novák's C-OEMS system: the Ar gas flow rate into the electrochemical cell is set to 0.05 ml min^{-1} (flow controller 1), while the other gas flow rate is set to 0.8 ml min^{-1} (flow controller 2). The majority of the gas is pumped out by the differential pumped vacuum system. Only a small amount of gas mixture (25 μl min^{-1}) is transferred into the MS [25].

The C-OEMS system is capable of testing volatile organic species quantitatively and continuously. In order to complete a quantification test, a standard calibration process is required. To calibrate signal intensities for each gas species, standard calibration gases are flowed into the MS before the electrochemical test. Different concentrations of calibration gas can be obtained by adjusting the set points of flow controllers 2 and 3 to mix Ar at a particular flow rate with pure targeted gases. During an electrochemical test, a constant Ar stream with a fixed flow rate is continuously flowed into the electrochemical cell and introduces only a small amount of volatile products to the MS. Then, the ion intensities of each gas product are recorded by the MS. The quantitative measurement can establish the correlations between gas evolution and electrochemical characteristics.

The response time of C-OEMS is around 30 s, which is longer than C-DEMS (<2 s) because of the longer gas transport distance caused by the capillary gas inlet system, as illustrated in figure 5.5. The delay time can bring a certain degree of voltage uncertainty to the experiment, depending on the current density and scan rate of the tests [26–28]. This uncertainty can impair the analysis of the onset voltage of certain gas products. However, this uncertainty can be neglected for most battery studies that cycle for several hours or even days [22]. Another shortcoming of C-OEMS is that the system is a dynamic open system, making C-OEMS systems more easily contaminated, especially considering the relatively high gas flow rate (1 ml min^{-1}) compared to the small electrolyte volume of batteries (10~100 μl) [9]. Futhermore, the capillary is more prone to clogging because of the continuous depleting of volatile electrolyte components [19].

Since the high carrier gas flow rate can impair the sensitivity of C-OEMS, a modified C-OEMS system has been designed to reduce the carrier gas flow rate. The configuration of the modified C-OEMS is shown in figure 5.7. A micro-flow-rate capillary (Hiden Inc.) is selected as the gas inlet. The gas depletion rate of the micro-flow-rate capillary is 12 or 25 μl min^{-1} under atmospheric pressure, enabling a much smaller carrier gas stream compared with traditional C-OEMS systems (~1 ml min^{-1}). Carrier gas and calibration gas cylinders are connected to the system separately by two flow controllers. The functions of these flow controllers are

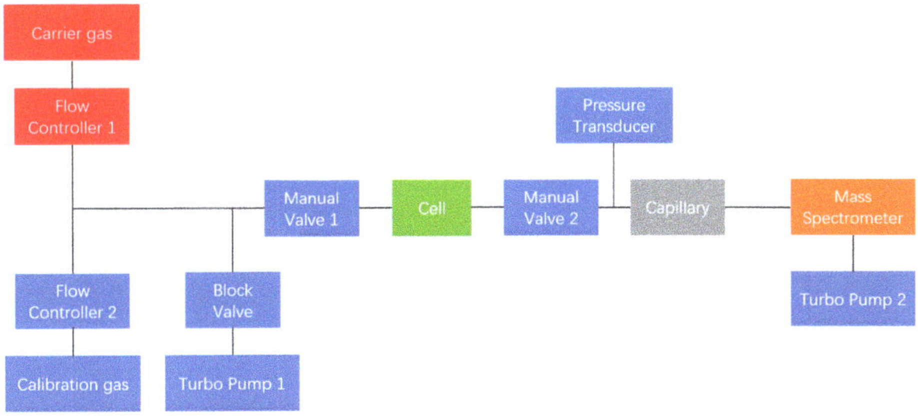

Figure 5.7. Schematic drawing of modified C-OEMS.

identical to those of traditional C-OEMS flow controllers. Two turbo pumps are included in the modified C-OEMS system. Turbo pump 1 is connected to the system via a blockvalve. It is used to exhaust the residual gas before testing. Turbo pump 2 is used to create the high vacuum needed by the MS. For the modified C-OEMS system, a single-stage vacuum system is sufficient, thanks to the micro-flow-rate capillary. Since no gas bypass is needed for modified C-OEMS, all the gas products expand into the MS, possibly increasing the background of targeted gas products. The operation principles are almost the same for the modified and the traditional C-OEMS systems, except that flow controller 1 will be set at around 11 μl min^{-1} for the modified system.

Unlike the traditional C-OEMS system, the modified C-OEMS system only requires an 11-μl min^{-1} carrier gas flow rate, and all gas products can be analyzed by the MS. Considering these two improvements, it is expected that modified C-OEMS should provide better resolution than traditional C-OEMS. Besides, the response time for modified C-OEMS is determined to be 16 s under atmospheric pressure, which is superior to traditional C-OEMS (~30 s). However, the time it took for the gas signal drop back to background is around 1–2 h which is longer than C-OEMS (<1 h) because of the lower carrier gas flow rate. Lastly, traditional C-OEMS usually suffers from capillary clogging because of the continuous removal of electrolyte components. However, this problem is greatly mitigated for modified C-OEMS, thanks to the 2 micron screen micro flow filter disk that comes with the micro-flow-rate flow capillary. The filter is fitted to the sampling end of capillary to protect the capillary from particulates.

A novel sealed OEMS system was also developed by Gasteiger's group; [9] We refer to it here as S-OEMS, where 'S' represents 'sealed.' figure 5.8 is a schematic diagram of Gasteiger's S-OEMS setup. The S-OEMS setup is very similar to that of traditional OEMS, except that S-OEMS uses no external gas source during the test period, thus achieving a sealed system. As shown in figure 5.8, the S-OEMS system is composed of four parts: electrochemical cell, crimped-leak gas inlet, vacuum system,

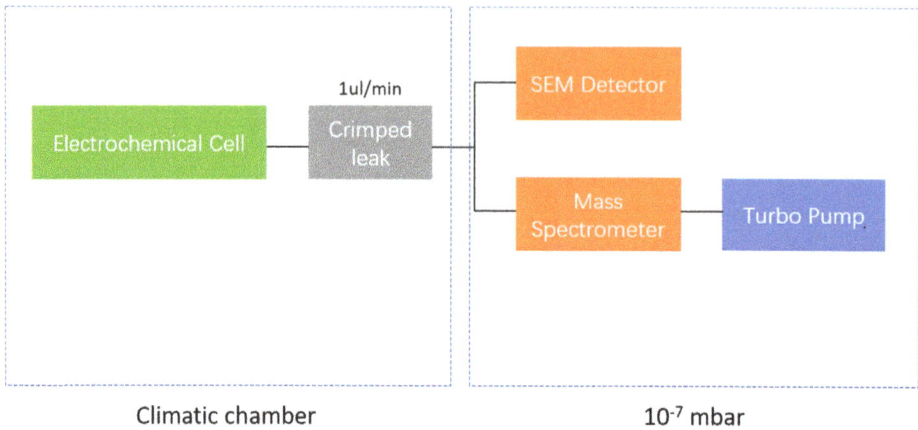

Figure 5.8. Schematic drawing of S-OEMS system.

and MS. Note that although an SEM detector is also included in Gasteiger's setup. it is not a must-have component of S-OEMS systems. For Gasteiger's specific configuration, a 1-μl min^{-1} calibrated capillary leak was used. The S-OEMS system was capable of measuring lithium–oxygen battery gas evolution continuously for up to 10 h with only minor changes of cell gas pressure. This result is a consequence of the balance between the gas evolution and depletion rates. For Gasteiger's setup, the electrochemical cell is assembled in an Ar glove box. Prior to the measurement, the MS needs to be calibrated with calibration gas. Pure Ar is flowed into the system to replace the calibration gas after calibration. As is required with a sealed system, no external carrier gas goes into the cell during the testing period. The MS signal can be quantified and converted to molarity by venting calibration gas [9]. More details about quantification methods can be found in section 5.2.5.

The essential part of the S-OEMS system is the 1-μl min^{-1} crimped leak, which makes it possible to test gas products continuously without the presence of carrier gas. This arrangement greatly reduces the chance of contamination compared to an open system like C-OEMS [9]. The response time for S-OEMS is only 1 s, which is remarkably smaller than for other OEMS systems. However, the crimped leak is easily clogged compared to the micro-flow-rate capillary mentioned above, limiting its application for long-term tests.

5.2.3.3 Intermittent DEMS

The intermittent DEMS system was first introduced by B D McCloskey in 2011 [8]. Figure 5.9 is a schematic drawing of the system established by McCloskey. The reason for using the term 'intermittent' is that the system samples gas species intermittently. Here, we simplify the name to 'I-DEMS.' The I-DEMS system can be used for testing the gas behavior of both lithium–air and lithium-ion battery systems. The I-DEMS system is made of five parts, namely, an external gas source (Ar/O_2), 8-way valve gas inlet system, differential pumped vacuum system, electrochemical cell, and MS. The cell configuration resembles the C-OEMS cell, since both of them

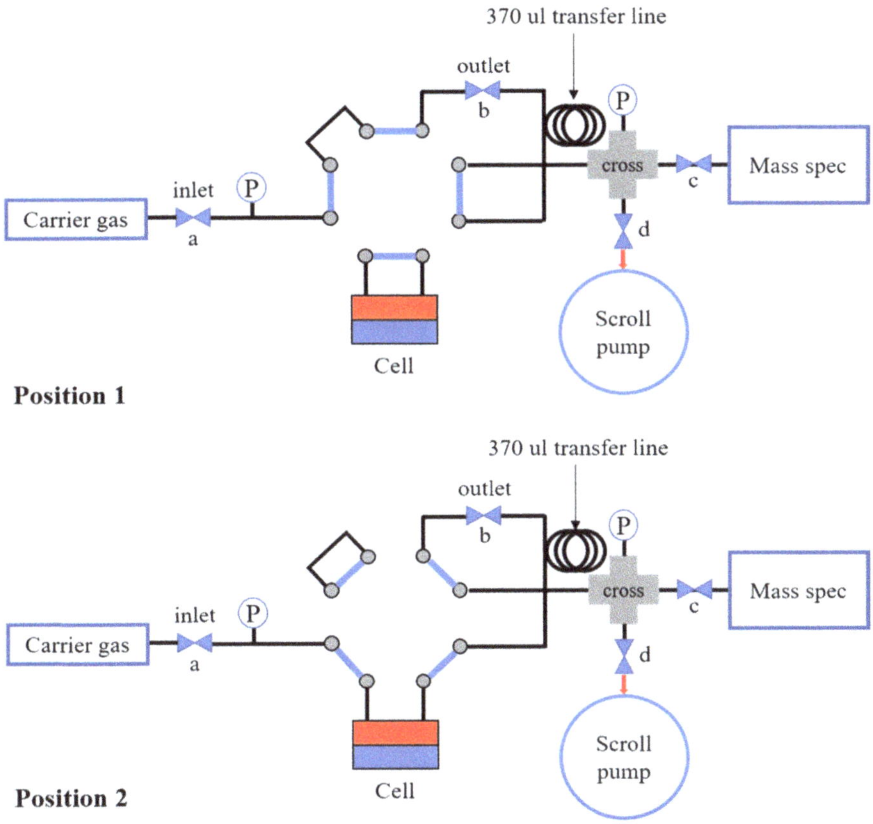

Figure 5.9. Schematic drawing of I-DEMS system 8.

are headspace-analysis-based cells. The electrochemical cell is connected to an 8-way valve that has two working positions, as shown in figure 5.9. In position 1, the electrochemical cell is isolated from the system, allowing an accumulation of gas products in the cell. In position 2, the cell gas inlet is connected to the Ar/O_2 cylinder with valve a and the gas outlet is connected to a cross tube with valve b. The cross tube, together with a dry pump and mass spec pump, make up a differential pumped vacuum system. Two pressure transducers are added to monitor the pressure changes in the electrochemical cell and cross tube separately. Blocks c and d are used for controlling the dry pump and the MS.

To quantitatively analyze the result, the headspace volume needs to be determined; the ideal gas law equation is used for this purpose. First, the lithium–oxygen cell is replaced by a standard-volume tube. The system is set at position 2 with valve a opened and valve b closed. Valve d is opened to evacuate the cross tube and part of the transfer line. Then a known amount of Ar is flowed into the standard-volume tube and the steady-state gas pressure is recorded. When valve b is opened, the pressure difference will cause the gas in the standard-volume tube to be transferred into the cross tube. Valve b is closed and the steady-state pressure value is read.

According to the ideal gas law, $P_1V_1 = P_2V_2$, the volume of the cross tube can be related to the standard tube volume. By using the same method, the headspace volume and transfer line volume can be determined.

The testing procedure for a lithium–oxygen battery with the I-DEMS system can be described as follows: first, the lithium–oxygen cell is filled with O_2 at a certain pressure with valve a opened and valve b closed. During the discharge process, the I-DEMS stays in position 2 with both valve a and valve b closed. The gas pressure drop inside the lithium–oxygen cell is recorded by a pressure transducer. Since the headspace volume has been calibrated, the oxygen consumption rate can be monitored continuously. Before the charging process begins, the residual O_2 is flushed by pure Ar with valves b, c, and d opened. The MS works to confirm that the O_2 signal is minimized and all the O_2 has been purged. The system is switched to position 1 during the charging process, allowing gas products to accumulate in the cell headspace. Block valves b, d and c are opened at the same time to evacuate the transfer line and cross tube. The time interval is usually 15 min After this period, the I-DEMS system is switched to position 2, with valve a opened and valve b closed. The gas products in the lithium–oxygen cell are swept into the transfer line by pure Ar because of the pressure difference. Then the system is switched back to position 1, transferring the gas products into the cross tube, and the lithium–oxygen cell is isolated again, waiting for the next testing point. The differential pumped vacuum system is monitored so that only a tiny portion of the gas products is sampled by the MS. I-DEMS can also be applied to a lithium-ion battery system with only minor changes [29].

5.2.4 Instrument design

In section 5.2.1, the general principles of different types of DEMS systems were introduced. The designs of specific components will be introduced in section 5.2.4. In section 5.2.4.1, several kinds of electrochemical cells will be discussed. In section 5.2.4.2, vacuum system design will be discussed.

5.2.4.1 Electrochemical cell design

Electrochemical cells can be classified into two categories by their operation mechanisms, as shown in figure 5.10. The first type is the membrane-based electrochemical cell. It includes the C-DEMS cell, thin-layer cell, and dual thin-layer cells. Although various configurations of membrane-based electrochemical cells exist, the working mechanism remains the same: they all use a membrane-based inlet. The second type of cell is the headspace-analysis-based cell. Two generations of this cell type have been developed. Their common feature is that a headspace is created inside the cell for accumulation of gas products. The second generation uses much less electrolyte compared to the first generation.

5.2.4.1.1 Membrane-based electrochemical cell

5.2.4.1.1.1 Conventional DEMS cell

The C-DEMS cell is also known as a membrane-based cell, owing to the usage of a hydrophobic porous membrane. It was first introduced by O Wolter and J Heitbaum

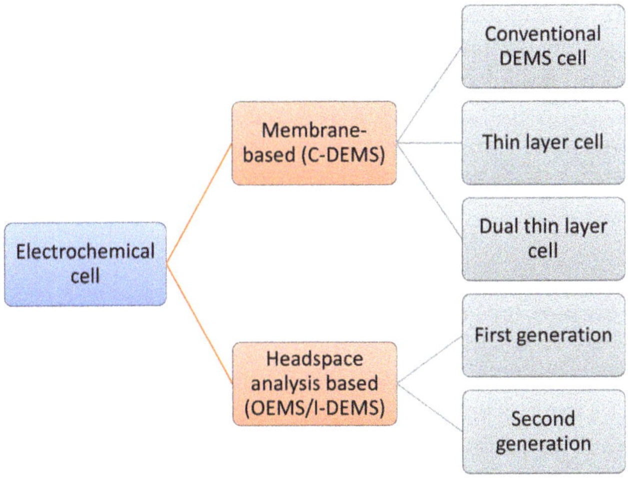

Figure 5.10. Classifications of electrochemical cells.

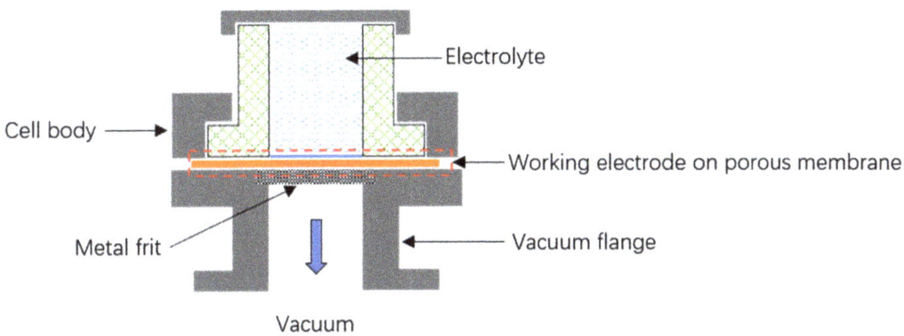

WE are usually placed in the same compartment.

Figure 5.11. A schematic draw of conventional DEMS cell 11.

in 1984 [11]. A conventional design of this kind of cell is shown in figure 5.11 [1, 30]. The C-DEMS cell includes a modified vacuum flange with a metal frit placed at the middle. The frit acts as a support for the membrane and WE. The porous membrane, sputter deposited with catalyst or cathode material, is attached with a PTFE sealing ring. The cell is hermetically sealed by compressing the cell body onto the vacuum flange. The cell body can be made of different kinds of materials, such as glass, polychlorotrifluoroethylene (Kel F), or polypropylene [11, 12, 21]. PTFE may also be used, but it can adsorb some reaction species owing to its porosity [1, 21]. The rest of the cell is made of stainless steel. A metal wire (Pt or Cu) contacts the WE to achieve a closed circuit (not shown in the figure). The cell chamber can be filled with several milliliters of electrolyte, as indicated in figure 5.11 [12, 21]. Note that, for the C-DEMS cell, the CE, RE, and WE are usually placed in the same compartment.

One of the most important parameters of porous membranes is the pore size, which depends on the contact angle between liquid and membrane, and on surface tension. The pores should be able to create a liquid-to-vapor interface that only allows penetration of volatile species; that is, a contact angle of >90° with the solvent should be realized. The pore size can be determined from equation (5.1) [11]:

$$r < -\frac{2\sigma\cos\theta}{p}\# \tag{5.1}$$

where σ is the surface tension (N cm^{-1}), θ is the contact angle, and p is applied pressure.

Appling equation (5.1) to water, a critical pore size $r < 0.8$ μm is obtained. For a given liquid, if its critical pressure is higher than that of water, narrower pores are required to prevent wettability. However, a narrow pore can hinder gas transport and eventually decrease time resolution. Considering this fact, membranes that feature high porosities are preferred [31]. For a C-DEMS membrane of PTFE, a pore size of 0.02 μm with a porosity of 50% and a thickness of 75 μm is usually chosen.

Although a PTFE membrane is adequate for some *in situ* gas analysis, it may not be suitable for battery studies owing to the unique properties of battery electrolyte, since most of the organic electrolytes exhibit a low vapor pressure. This property can result in a high electrolyte evaporation rate, eventually leading to contamination of the MS. To solve this problem, some researchers switched to a headspace-analysis-based cell, which we will discuss in section 5.2.4.1.1.4. Other researchers switched to investigating new membranes. Jusys's group introduced a non-porous fluoropolymer membrane in 2019. The electrochemical cell is similar in design to a C-DEMS cell, but owing to the application of a non-porous membrane, the electrolyte contamination can be significantly reduced. However, this membrane also suffers from a long response time (about 100 s) compared with that of the PTFE porous membrane, which is typically less than 0.1 s [1, 24].

For the C-DEMS cell, the WE is usually homogeneously sputter-deposited on the porous membrane to minimize the transport distance [32]. With a transport distance on the order of microns between the electrode surface and vacuum, a rapid mass transport rate is achieved: the response time is typically less than 0.1 s, as mentioned above. However, this proximity can cause an electrolyte depletion problem. When reactants with high volatility are used, a reactant depletion region will form near the cathode surface as a result of the high evaporation rate of the volatile reactant species [13]. Also, the competitive reactant evaporation and adsorption processes near the membrane and vacuum surface can lead to a shift in the equilibrium electrochemical reaction when gas species are involved. To summarize, the proximity of electrode surface and vacuum is beneficial for obtaining a high mass transport rate. This proximity, however, leads to an overlap of the reaction surface and diffusion layer. Thus, the reaction process can be deeply affected by the diffusion layer [1].

The conventional DEMS cell can be customized to meet the demands of various kinds of electrochemical reactions, and could be optimized to alleviate the reactant depletion problem mentioned above. For instance, a forced continuous electrolyte circulation can be achieved by simply adding PTFE capillaries to the cell. If the capillaries are placed close to the electrode surface, the convection at the surface can mitigate the depletion problem to a certain extent. To supplement electrolyte convection, Masamichi and Toshimitsu [33] added a rotating electrode inlet between the MS and the electrochemical cell to obtain steady transport properties near the electrode. This steadiness is crucial for the quantification process discussed in section 5.2.5.2. A schematic drawing of this setup is shown in figure 5.12.

The major problem faced by the C-DEMS cell is that the electrode materials can only be deposited on a porous membrane. This restriction largely limited its use with smooth-surface electrodes, which is attracting more interest for electrochemical studies nowadays [10]. A newly designed thin-layer cell was created to solve this problem, and will be discussed in the next section.

5.2.4.1.1.2 Thin-layer cell

The thin-layer cell was introduced in 1989 by Hartung and Baltruschat [4]. Its most important advantage is that it allows the use of a massive electrode, like a single-crystal electrode, rather than a porous electrode. The use of a smooth electrode is of great significance for electrochemical studies, since electrode surface roughness can greatly affect cell performance. A schematic drawing of a thin-layer cell is shown in figure 5.13. Unlike in the C-DEMS cell, the massive WE of the thin-layer cell is separated from the PTFE membrane by a thin layer (50 to 100 μm) of electrolyte. The PTFE membrane is supported by a metal frit. A spacer is positioned between the WE and PTFE membrane to ensure a fixed WE-to-membrane distance. Electrolyte inlet and outlet capillaries are attached to the cell for continuous electrolyte flow. These capillaries also serve as connections to the RE and CE, as shown in figure 5.13. Titanium is used for the cell body because of its high mechanical stability and chemical resistance [4].

The thin-layer cell can be operated under both static electrolyte and continuous-flow electrolyte conditions. During a static-electrolyte electrochemical test (desorption experiment), almost all volatile products or reactants will diffuse from the massive electrode to the MS through the PTFE spacer and membrane. The

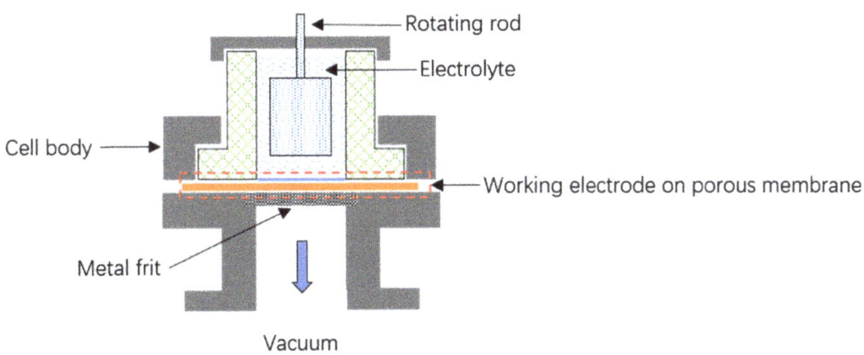

Figure 5.12. A schematic drawing of a rotating-electrode inlet 13.

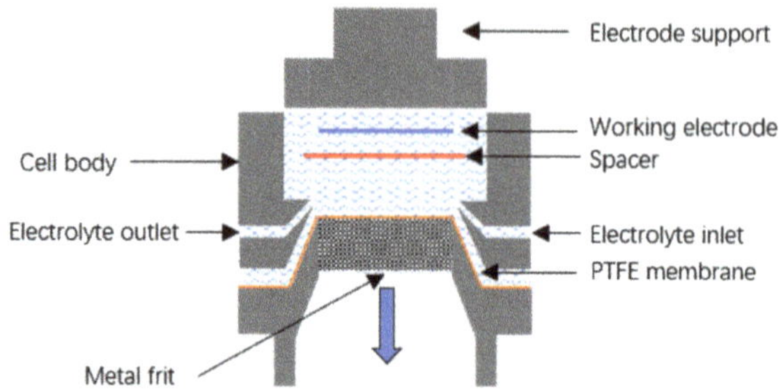

Figure 5.13. Schematic drawing of thin-layer cell 4.

collection efficiency for the static-electrolyte condition is usually in the range of 0.9 to 0.95. For the continuous-flow condition, however, the collection efficiency drops greatly, to around 0.2 for 1-μl s^{-1} flow rates. The reason is that most gas products will be exhausted by the flowing electrolyte and will not reach the MS. Thus, the thin-layer cell is not suitable for continuous-electrolyte-flow research [33].

The response time for the thin-layer cell is longer than for the C-DEMS cell (typically around 2 s), because the increased transport distance between membrane and massive electrode decreases the time resolution. The thin-layer cell, like the C-DEMS cell, also suffers from the problem of depletion of gaseous reactants [10].

In summary, the thin-layer cell is more suitable for static-electrolyte electrochemical experiments. The usage of massive electrodes makes the cell a great improvement compared to the C-DEMS cell. However, the relatively low collection efficiency and reactant-depletion problem limit its application in continuous-flow electrolyte and reactant tests. In order to meet the requirements for such experiments, a dual thin-layer cell was invented.

5.2.4.1.1.3 Dual thin-layer cell

The dual thin-layer cell is more suitable for conditions of continuous electrolyte flow, and it is widely used for faradic reactions. A schematic drawing of a dual thin-layer cell is shown in figure 5.14 [3]. The dual thin-layer cell contains two compartments; the WE is placed in the upper compartment and the PTFE porous membrane is placed in the lower compartment. The electrolyte flows into the upper compartment first for faradic reactions and will be transferred into the lower compartment through a capillary, accompanied by gas products. The gas products reaching the PTFE membrane will be tested by the MS [1].

Two major factors make the dual thin-layer cell suitable for continuous-electrolyte-flow conditions. First, the separation of the WE and the PTFE membrane can help improve the collection efficiency (from 0.2 to 0.4) under these conditions: the relatively longer residence time of electrolyte near the PTFE membrane surface allows enough time for the diffusion of gas products. Second, the gas reactant depletion problem is ameliorated by the two-compartment configuration [10]. One of the shortcomings of the dual thin-layer cell is that the response time (~2 s) is longer than that of the C-DEMS cell, owing to the separation of the WE and the

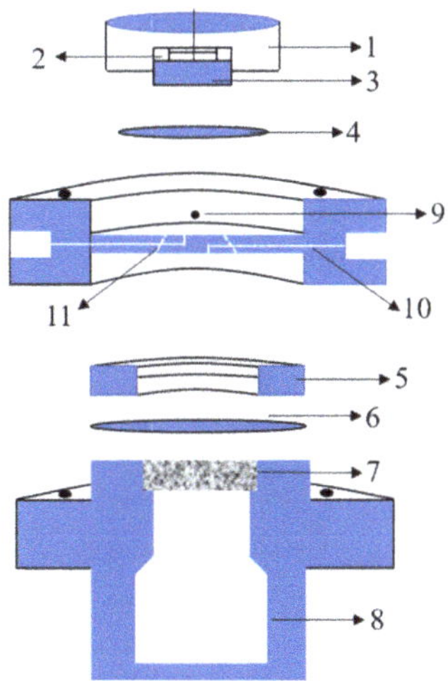

Figure 5.14. A schematic drawing of dual thin layer cell. DEMS/EQCM assembly: (1) Kel-F support; (2) Kalrex gasket; (3) quartz crystal; (4, 5) Teflon gaskets; (6) porous Teflon membrane; (7) stainless steel frit; (8) stainless steel connection to MS; (9) capillaries for contact wires; (10) inlet-outlet capillaries; (11) connecting capillaries 3.

PTFE membrane into two compartments. Another shortcoming is that the usage of capillaries may increase the internal resistance and result in a voltage drop of the electrochemical cell [10].

In summary, the dual thin-layer cell is more suitable for faradic reactions compared to the thin-layer cell and tradition DEMS cell because of its higher collection efficiency and mitigated reactant-depletion problem.

5.2.4.1.1.4 Headspace-analysis-based cell

With the conventional membrane-based DEMS cell, gas analysis of various kinds of reactions can be performed. However, for some reactions, like lithium-ion battery cathode reactions, which produce very small amounts of gas products, the C-DEMS cell is no longer suitable because of the high depletion rate of both electrolyte and gas products. The headspace-analysis-based cell was designed specifically for analyzing reactions in which only very small amounts of gas products are evolved. The choice of name for this kind of cell was meant to differentiate its mechanism from that of a conventional membrane-based cell, where gas products diffuse through a porous membrane to the MS. In a headspace-analysis-based cell, a headspace is created in the cell where gas species accumulate, and these species are continuously transferred to the MS by a capillary during the test. One of the advantages of this configuration is that the ratio of active material to electrolyte can be significantly increased and thus a higher product signal can be obtained. The term OEMS is usually used for a headspace-analysis-based cell, as mentioned in section 5.2.3.2.

The first-generation headspace-analysis-based cell was introduced by P Novák in 2005 [14]. A schematic drawing of Novák's cell is shown in figure 5.15. The WE and

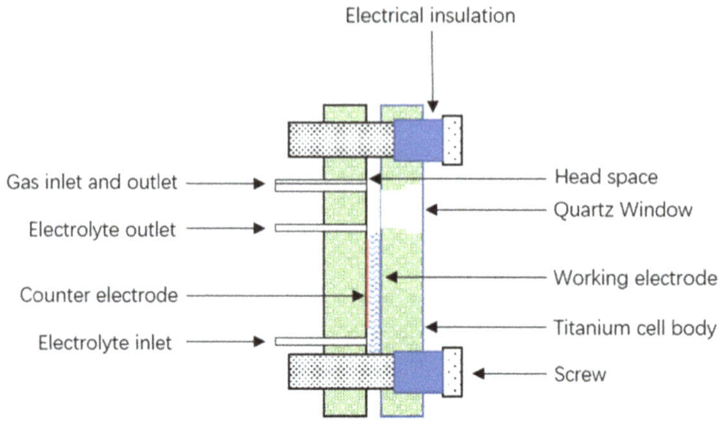

Figure 5.15. A schematic drawing of a headspace-analysis-based cell 14.

CE are vertically placed and parallel, isolated from each other by a 1-μm electrolyte layer without a separator. Both electrodes are coated directly onto the inner cell wall. The evolved gas products are continuously expelled from the headspace to the MS via a capillary. Ar is used as carrier gas and is continuously flowed into the system during the test. The cell is hermetically sealed by two screws. A quartz window is also created for observation. The cell body is made of titanium metal, chosen for its high chemical resistance and excellent mechanical performance. Compared to a C-DEMS cell, a headspace-analysis-based cell requires a much smaller amount of electrolyte (~4 ml), largely avoiding the strong background signals from the electrolyte. This design is compatible with various kinds of CEs. However, the headspace and electrolyte volumes for the first-generation headspace-analysis-based cell are still relatively high, resulting in the dissolution of more gas species into the organic electrolyte.

To further minimize headspace volume and inhibit electrolyte background, a second-generation headspace-analysis-based cell was invented. The ECC-DEMS cell is an example of this classification of cells. Figure 5.16 shows a schematic drawing of the ECC-DEMS cell configuration. The CE, a glass fiber separator, and the WE are vertically stacked at the bottom of the cell. The WE can be either a standard lithium-ion electrode or a lithium–oxygen gas diffusion electrode. Lithium metal can be used for the CE or RE. The amount of electrolyte needed for the ECC-DEMS cell is similar to that of a coin cell and is much smaller than in the first-generation headspace-analysis-based cell [34]. During the charge and discharge process, a controlled stream of carrier gas is flowed into the cell through the gas inlet. The gas outlet is connected to a MS for gas composition analysis. Thus, unlike the first-generation headspace-analysis-based cell, the second-generation cell can achieve a cell configuration resembling coin cells. This configuration helps avoid the deviations in electrochemical performance described for other cell setups. Also, the ECC-DEMS cell uses a polyethylene seal, which can provide excellent seal performance [28].

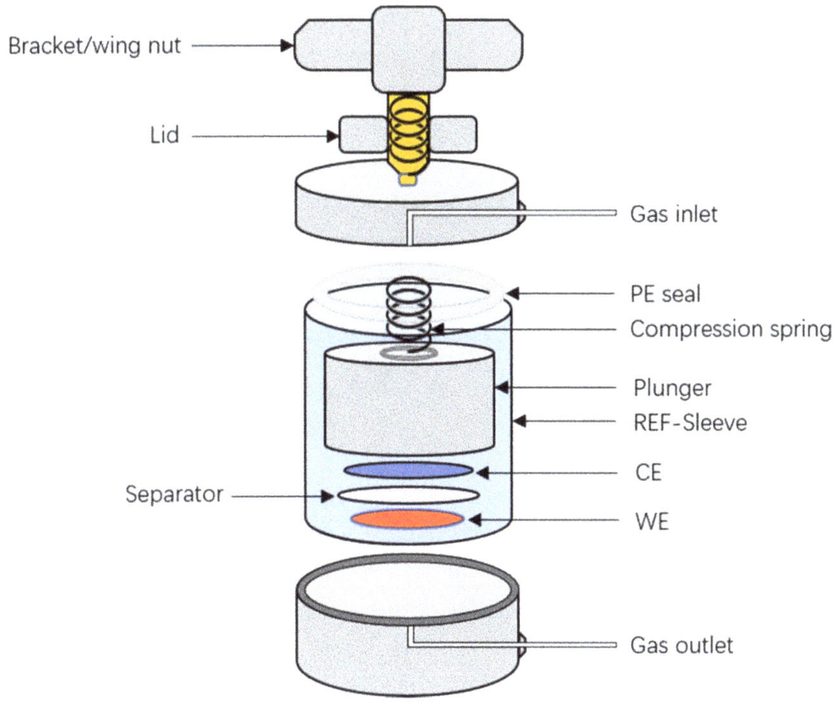

Figure 5.16. A schematic drawing of the ECC-DEMS cell.

5.2.4.2 Vacuum system

The vacuum system is an essential part of the DEMS setup. In 1984, Wolter *et al* established the first C-DEMS system by significantly improving the vacuum system relative to membrane inlet mass spectrometry (MIMS) [11]. The name 'differential pumped system' was applied to the two-stage pumping system used by Wolter's group. It typically contains two pumps which are connected by an aperture or skimmer. As described in the literature, various types of pumps have been employed for the first pump, including rotary vane pump, diffusion pump, and turbo pump. But for the second pump, which is connected to the MS, all setups use a turbo pump [10].

The necessity for using a differential pumped vacuum system arises from two issues. First, although the porous PTFE membrane can prevent liquid electrolyte from going into the MS, a large quantity of electrolyte vapor still passes through the membrane. Second, the required operation pressure for the MS is typically smaller than 10^{-5} mbar. To maintain this high vacuum under the high stream of vapor gas, the pumping rate of the system should be able to reach 0.09 mbar l s^{-1} cm^{-2} [1]. The unit mbar l s^{-1} cm^{-2} is a leak rate unit which means the amount of gas necessary to be removed from a 1 L container in 1 s per centimeter square area to reduce the pressure by 1 mbar. However, it has been calculated that the membrane interface area is confined to 0.033 cm^2 if only a single pump is used. Such a small interface area is impractical for electrode applications [10]. Thus, the differential pumped vacuum system is essential for DEMS. A typical setup for a differential pumped

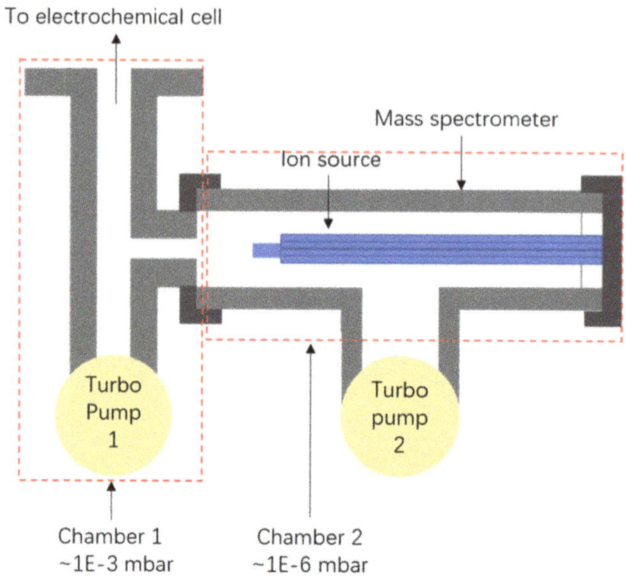

Figure 5.17. A typical setup for a differential pumped vacuum system.

vacuum system is shown in figure 5.17. The pressure in chamber 1 (first stage) is roughly 10^{-3} mbar and the pressure in chamber 2 (second stage) is around 10^{-5} mbar to achieve the required operating pressure for the MS. With this setup, if turbo pumps with a pumping speed of 90 l s^{-1} (pump 1) and 200 l s^{-1} (pump 2) are employed, a ~1 cm^2 membrane area can be achieved.

As mentioned in section 5.2.3.2, modified C-OEMS and S-OEMS systems are equipped with a single-pump system rather than a differential pumped vacuum system because both of these systems use a micro-flow capillary or crimped leak. The gas stream that flows into the MS will be reduced to the order of microliters. In this situation, a single pump is enough to maintain the vacuum required by the MS. It should be noted that the single-pump system is associated with a gas fractionation problem, and a specific quantification method is needed for modified C-OEMS and S-OEMS systems.

5.2.5 Quantification methods

Three types of DEMS setups were introduced in section 5.2.3, that is, C-DEMS, OEMS and I-DEMS systems. The objective for these DEMS systems is to analyze gas products quantitatively. To achieve this goal, it is very important to figure out a way to quantitatively process the gas signal. However, because of the different operation mechanisms as well as the distinct application fields of these DEMS setups, quantification methods need to be specifically designed for every DEMS setup.

5.2.5.1 Quantification method for C-DEMS

C-DEMS is widely used in gas analysis for various kinds of catalysts and electrochemical reactions. For quantification purposes, the faradic current should be able to be quantitatively correlated with the ion current of a specific gas product. The collection efficiency is a key value needed for the quantification process. Collection efficiency can be defined as the amount of gas products transferred into the MS divided by the total amount of gas evolved. Collection efficiency is determined by the transport properties of the catalyst layer and the diffusion layer between the catalyst and electrolyte. So for a given C-DEMS setup, the collection efficiency needs to be specifically calibrated. Some early-stage C-DEMS setups may not have a stable collection efficiency because of the thickness changes of the diffusion layer. For those setups, the gas products can only be qualitatively determined [6].

Equations (5.2)–(5.4) show a typical quantification method for membrane-based DEMS [6]. The ion current I_i of a species is proportional to the incoming flow rate J_E; K is the MS instrument parameter and can be determined by flowing a known amount of standard gas into the MS vacuum chamber.

$$I_i = K_* J_E \# \tag{5.2}$$

For electrochemically formed gas species, the incoming flow of the species is correlated to the faradic current by equation (5.3):

$$J_E = \frac{1}{n_* F} {}_* N_* A_* I_F, \ \# \tag{5.3}$$

where I_F is the current at the electrode, A is the current efficiency (typically, its value is 1), n is the molar number of electrons transferred in the formation of one molar gas product, F is the Faraday constant, and N is the collection efficiency. Combining equations (5.2) and (5.3), the ion current is given by

$$I_i = \frac{K_* N}{F} {}_* A_* \frac{1}{n} {}_* I_F \# \tag{5.4}$$

By solving equation (5.4), collection efficiency N can be calculated. If collection efficiency remains unchanged, we can correlate faradic current with ion current of a gas species.

5.2.5.2 Quantification method for OEMS and I-DEMS

OEMS and I-DEMS are predominantly used for lithium-ion or lithium–oxygen battery gas analysis. The quantification of gas products for OEMS and I-DEMS is based on headspace-based analysis [19]. Headspace-based analysis uses the ideal gas law to convert partial pressure to molarity, following the steps shown below. The cell pressure is recorded continuously during OEMS tests and intermittently during I-DEMS tests. The headspace volume can be obtained from the cell manufacturer. Note that, since OEMS is an open system, the molar gas flow rate is used in the ideal gas law calculation instead of headspace volume. The difference between these two methods is shown in step 5 below.

1. Perform calibration before the test begins. The system's machine-specific sensitivity factor $S_{m/z}$ can be determined by flowing a series of calibration gases with step increases of concentration.

2. Record background intensity, $B_{m/z}$. The background intensity consists of machine background and electrolyte background.
3. Calculate relative signal intensity x_i (concentration) by using equation (5.5):

$$x_i = \frac{I_{m/z*}S_{m/z} - B_{m/z}}{I_{\text{total}}}, \# \qquad (5.5)$$

where I_{total} is the total signal intensity and $I_{m/z}$ is the signal of a specific gas species.
4. Use cell pressure data and concentration to obtain the partial pressure of each gas species:

$$P_i = P_{\text{total}*}x_i, \# \qquad (5.6)$$

where P_{total} is the pressure read from the pressure transducer.
5. For I-DEMS: Convert partial pressure to molarity by using the ideal gas law:

$$n_i = \frac{P_{i*}V}{\text{RT}}, \# \qquad (5.7)$$

where n_i is the molarity of the specific gas species and V is headspace volume.

For OEMS:
Use volume flow rate to convert partial pressure to molarity.

$$r_{\text{gas}} = \frac{P_{i*}r_{\text{flow}}}{\text{RT}} \# \qquad (5.8)$$

where r_{gas} is the molar flow rate of gas species (mol min^{-1}), r_{flow} is the volume flow rate of total gas flow into the cell, R is ideal gas constant, T is temperature.

5.2.6 System comparison

In the previous sections, we have introduced several DEMS and OEMS setups. All have their advantages and limitations. In general, C-DEMS is most suitable for fuel cell and catalyst studies, while OEMS is widely used in gas analysis for battery systems. This is because C-DEMS typically requires a large amount of gas evolution. For systems in which only tiny amounts of gas products are involved, like lithium-ion and lithium–oxygen batteries, C-DEMS application is limited. Instead, OEMS is predominantly used for battery systems because of its ability to detect and analyze tiny amounts of gas products. This difference can be attributed to the designs of the gas inlet system and vacuum system of C-DEMS and OEMS. For C-DEMS, a membrane-based gas inlet system is used, which enables a fast gas species transfer rate. However, a considerable amount of gaseous electroyte can penetrate the porous membrane and significantly increase the background, making C-DEMS unsuitable for testing small amounts of gas products. In contrast, OEMS relies on headspace-based gas analysis: the gas species initially accumulate in the headspace of the electrochemical cell and are then transferred into the MS via a capillary. This design can help to reduce the strong background from the electrolyte. Another difference between C-DEMS and OEMS is that carrier gas is used for OEMS (except for S-OEMS). The purpose of carrier gas is to compensate for the insufficient gas stream from battery systems and meet the flow-rate requirement of the differential pumped vacuum system.

C-DEMS usually has a higher time resolution than OEMS. The response time for C-DEMS can be as short as 0.1 s, compared with 30 s for OEMS [10]. This short response time reflects the microscopic transport distance of the membrane gas inlet system. For the capillary inlet system, gas products have to evaporate into the headspace first. This process inevitably increases the response time.

OEMS can be classified into two types: C-OEMS and S-OEMS. While both C-OEMS and S-OEMS use a headspace-analysis-based cell, C-OEMS differed from S-OEMS in the presence of carrier gas. The advantage of using carrier gas is that more rapid gas convection can be achieved. This means that a shorter time is needed for the gas signals to return to background. However, carrier gas can make the system more vulnerable to contamination, especially considering the limited gas volume of battery systems. For S-OEMS, carrier gas is not needed because a crimped leak is employed for reducing the gas flow rate to 1 ul min^{-1}. S-OEMS uses a sealed system, and the low background can help to detect even smaller amounts of gas products. However, crimped leaks are prone to clogging, making the system less reliable. Another difference between C-OEMS and S-OEMS is that C-OEMS uses a differential pumped vacuum system, while S-OEMS uses a single-pump system. A differential pumped vacuum system cannot be applied to S-OEMS because of the absence of carrier gas. The single-pump system usually leads to a mass fractionation problem and thus makes the quantification process more complicated [9]. Lastly, the gas depletion rate of S-OEMS is very low (1 μl min^{-1}) because of the usage of a crimped leak. So gas signals obtained from S-OEMS reflect an integration of the total gas evolved. For C-OEMS, owing to the relatively rapid gas convection, the signal can drop to background in about 15 min

The most significant difference between I-DEMS and OEMS is the testing mode; i.e., I-DEMS tests gas products intermittently while OEMS tests gas products continuously. For I-DEMS, the gas products accumulate in the headspace during the test interval in order to increase the amount of gas products being transferred into the MS. This makes the I-DEMS system a half-sealed system. The advantage of this approach is that the clogging problem can be alleviated and more gas products can be analyzed at the testing point. However, the time resolution for I-DEMS is limited, since it tests gas species intermittently. The testing interval for I-DEMS is usually more than 15 min This duration may also lead to another problem: gas products such as O_2 and CO_2 may react with electroyte during the testing interval. The detection and analysis of those active gas products could be difficult with I-DEMS [8].

5.3 Research applications

We will show examples of the application of different DEMS in energy storage systems, namely lithium–oxygen batteries, and lithium-ion batteries with application

Table 5.2. Gas evolution in Li–oxygen batteries.

Reaction	Gas product
$2\ H_2O + 2\ e^- \rightarrow 2\ OH^- + H_2 \uparrow$	H_2
$2\ H\bullet \rightarrow H2 \uparrow$	H_2
$LiPF_6 + H_2O \rightarrow LiF + 2\ HF + POF_3\uparrow$	POF_3
OH^- + ethylene carbonate → $-OCH_2CH_2OH + CO_2\uparrow$	CO_2
POF_3 + carbonates → organofluoro compounds + $CO_2 \uparrow$	CO_2
2 li+ + ethylene carbonate → $Li_2CO_3 \downarrow/(CH2CH2OLi)_2 \downarrow + C_2H_4$	C_2H_4
2 li+ + ethylene carbonate → $LiOCH_2CH_2OLi + CO\uparrow$	CO

of studies in electrolytes, solid–electrolyte interface, and cathode stability studies. The pros of DEMS include the fast response time, high sensitivity, and quantitative feature. The technique can be used towards all types of energy storage systems as long as there is gas generation, and its mass spectrometric analysis mode allows the identification of the exact species to provide mechanistic insight. The online coupling with a potentiostat also makes the technique very versatile, and can be used together with different electrochemical techniques, including cyclic voltammetry (for fast analysis), galvanostatic cycling (for *operando* measurement), potentiostatic hold (for detection of concomitant gaseous generation together with electrochemical decomposition/passivation), and many more. Table 5.2 listed some potential reactions that occur on the interfaces inside the battery. As demonstrated in table 5.2, H_2 can be a product resulting from both oxidation and reduction, depending on the species origins and reactions.

5.3.1 Lithium–oxygen batteries

The Li–oxygen battery, with its high theoretical capacity, has attracted considerable research interest. The battery chemistry is based on the oxidation of the Li at the anode and reduction of oxygen at the cathode. Under practical test conditions, the degradation of the carbonaceous cathode together with the adjacent aprotic electrolyte in the highly oxidative environment creates a very complex process [35]. DEMS is a powerful tool to investigate the gaseous product in Li–oxygen batteries. The gas evolution in Li–oxygen batteries is summarized in table 5.2.

Kang *et al* measured the stability and performance of tetraethylene glycol dimethyl ether (TEGDME) electrolyte compared with N-dimethylacetamide (DMA) electrolyte by DEMS [36]. In the gas product analysis during charging, they observed that DMA is more stable, exhibiting 19% higher oxygen efficiency and 5.1% less CO_2 evolution compared with TEGDME. These results reveal that DMA is a more desirable electrolyte for Li–oxygen battery systems.

Using DEMS, Kim *et al* sought to determine an optimum electrolyte solvent for use with lithium salt (1 M LiTFSI) to achieve high performance of Li–oxygen batteries [37]. Figure 5.18 illustrates their *in situ* DEMS setup. According to the DEMS analysis, dimethyl sulfoxide (DMSO) demonstrated CO_2 evolution when a potential of 4.4 V was

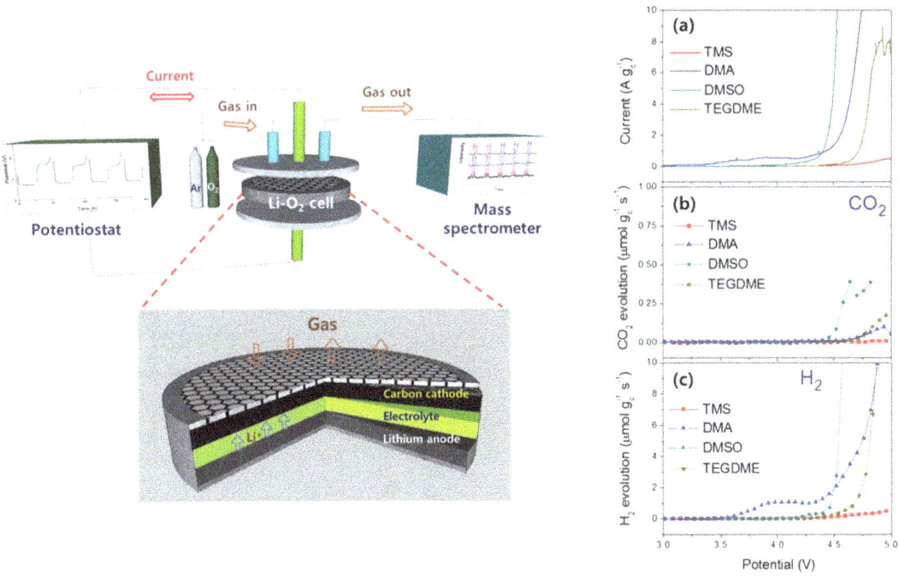

Figure 5.18. Schematic illustration of the DEMS setup for studying Li–oxygen batteries. Gas evolution was measured with DEMS during a linear oxidative scan from open-circuit voltage to 5 V at a scan rate of 0.1 mV s^{-1} with 1 M LiTFSI in TMS, DMA, DMSO, and TEGDME. Adapted from [37].

reached. DMA and TEGDME showed a higher oxidation-onset potential of 4.7 V. For H_2 gas evolution, DMA showed the lowest onset potential, at 3.5 V, compared with DMSO and TEGDME at 4.5 V and 4.7 V, respectively. The authors claimed that the H_2 evolution was most likely related to the reductive decomposition of the solvent with Li metal. These results suggest that tetramethylene sulfone (TMS) exhibited better oxidative stability and was more stable to lithium metal than other typical lithium–oxygen electrolyte solvents, including DMA, DMSO, and TEGDME.

On the other hand, low energy efficiency is a critical issue for Li–oxygen batteries. High overpotentials are usually required to oxidize lithium peroxide back to oxygen. Noble metal-based catalysts are regarded as promising materials to promote lithium peroxide oxidation. Wang *et al* reported *in situ* DEMS characterizations of Li–oxygen cells containing carbon nanotube (CNT) and Ru@CNT cathodes [38]. A mixed gas of O_2/Ar (80% O_2) was introduced during the discharge processes, and O_2 consumption was observed at both cathodes as shown in figure 5.19(a) and (b). In the oxygen evolution reaction process, the gas evolution during the charging process was also analyzed. It can be observed from figure 5.19(c), for the Li–oxygen cell with the CNT cathode, O_2 was released immediately at a voltage of about 3.3 V. A steady O_2 evolution was then observed at 3.7 V. As the cell voltage reached 4.0 V, the oxygen evolution gradually diminished and an obvious burst of CO_2 evolution was observed, indicating the decomposition of carbonates formed in the previous discharge process or from electrolyte decomposition [39]. In the cells with the Ru@CNT cathode, O_2 evolution was observed at the beginning of charge at a low

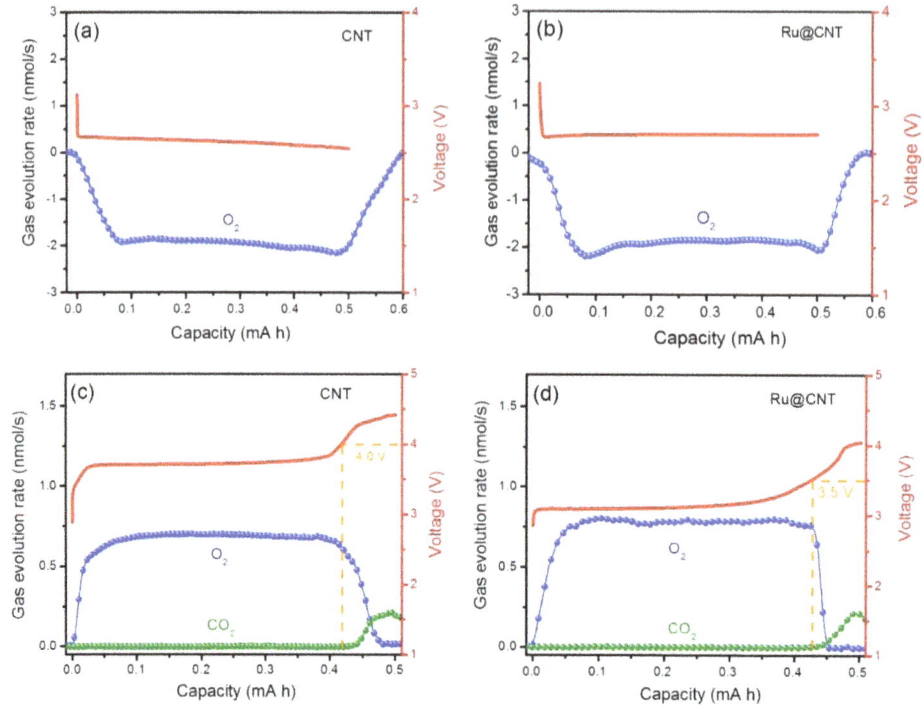

Figure 5.19. Gas evolution with (a) CNT and (b) Ru@CNT cathodes during charge, and (c) CNT and (d) Ru@CNT cathodes during discharge. Reprinted from [38], copyright (2016) with permission from Elsevier.

voltage of 3.1 V, as shown in figure 5.19(d). The potential then increased up to 3.5 V, accompanied by CO_2 evolution. Considering the much lower potential compared with that of the cells with the CNT cathode, this observation indicates that Ru may catalyze the decomposition of the by-product.

Recently, CO_2 was also demonstrated to be an energy carrier in a system of Li–CO_2 batteries. Zhou *et al* reported a CO_2 fixation strategy based on a rechargeable/irreversible oxidation process, in which Li_2CO_3 is decomposed and the carbon obtained remains fixed [40]. When a relatively lower charge rate (500 mA g^{-1}) is applied, only the evolution of CO_2 can be detected and no obvious oxygen evolution is observed, as shown in figure 5.20(D). The related charge-to-mass ratio of CO_2 (nearly 3e/2CO_2) indicates that the solo decomposition of Li_2CO_3 dominates in the charging process. However, at a higher current rate (2000 mA g^{-1}, figure 5.20(E)), both CO_2 and O_2 are released. This finding reveals that the Li_2CO_3 decomposition reaction can be defined as $2Li_2CO_3 \rightarrow 2CO_2 + O_2 + 4Li^+ + 4e^-$. After extending the discharge capacity to the Li_2O-related plateau (1.0 mAh), the evolution of O_2 can be observed at the lower potential at the initial stage of subsequent charging, as shown in figure 5.20(F). Furthermore, based on the 4e-/O_2 charge-to-mass ratio obtained at the corresponding stage, the solo decomposition pathway of the Li_2O obtained ($2Li_2O \rightarrow O_2 + 4Li^+ + 4e^-$) can be further confirmed.

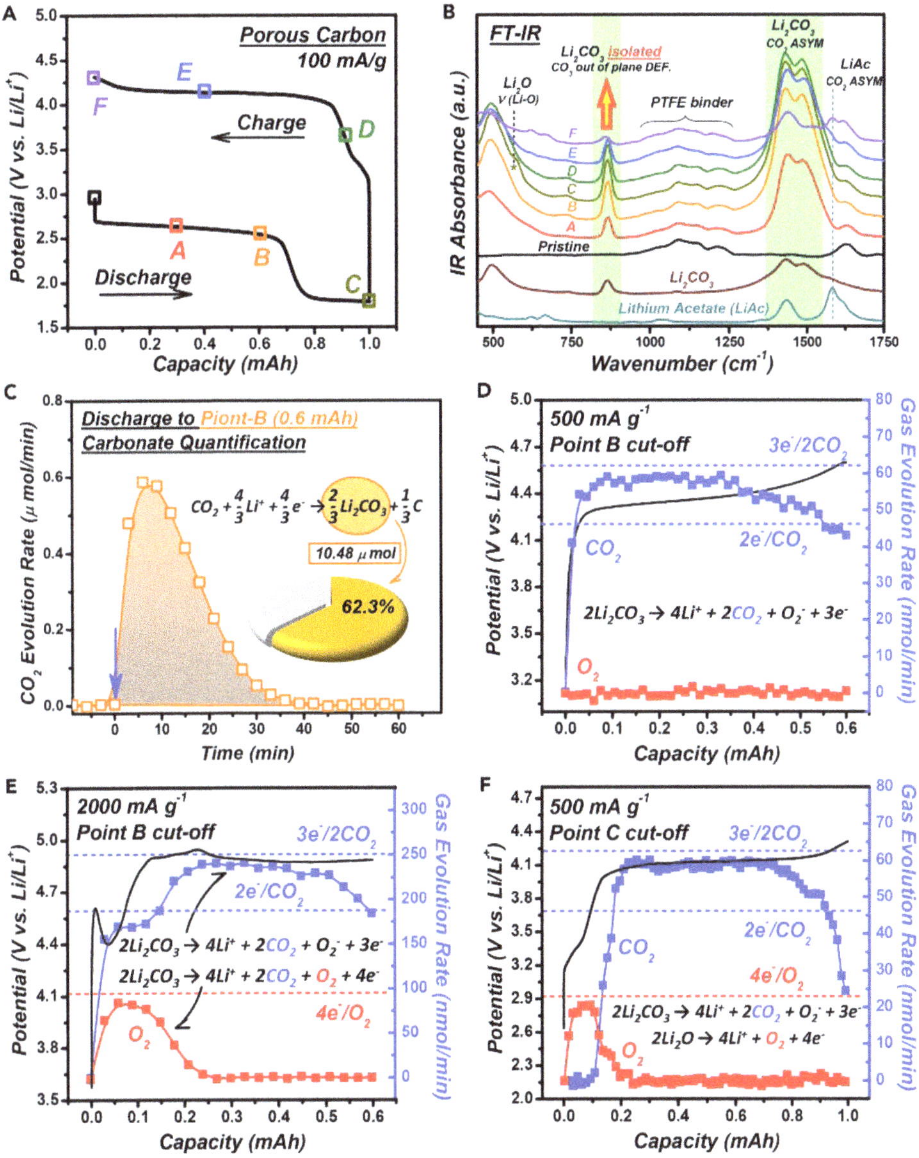

Figure 5.20. (A) Voltage profile of a Li–CO_2 cell with a discharge capacity of 1.0 mAh and charge cutoff voltage of 4.5 V. (B) *Ex situ* FTIR spectra performed on the cathode collected at different discharge/charge stages marked by varying colors for clarity. (C) CO_2 gas evolution rate during carbonate quantification conducted on a cathode discharged at 0.6 mAh (point B). Blue arrow corresponds to the injection of H_2SO_4 to decompose Li_2CO_3. The yield shown on the inset pie chart indicates the percentage of fixed Li_2CO_3 species (obtained by the area integral of the CO_2 evolution rate) versus the quantity of reduced CO_2 (calculated based on the discharge capacity). CO_2 and O_2 evolution during cell charging: at (D) 500 mA g^{-1} and (E) 2000 mA g^{-1} after discharge to point-B; (F) 500 mA g^{-1} after discharge to point-C. Reprinted from [40], copyright (2017) with permission from Elsevier.

5.3.2 Lithium-ion battery

5.3.2.1 Electroyte

The electrolyte conducting the lithium in lithium-ion batteries is an important component, considering its insufficient thermodynamic stability under battery testing conditions.

Gasteiger and coworkers used an OEMS system to study the oxidative stability of vinylene carbonate (VC) and its effect on the interfacial resistances in $LiNi_{0.5}Mn_{1.5}O_4$ (LNMO)/graphite cells [41]. The authors demonstrated that the oxidation of VC occurs at an onset potential of ~4.3 V versus Li/Li+, leading to a significant impedance buildup on the LNMO cathode in graphite/LNMO cells, as shown in figure 5.21.

Novak and coworkers proposed an OEMS system based on an X-shaped cell with the conventional WE, CE, and RE, and an additional connector linked to a modified gas cell-MS system [42]. A potentiostatic aging experiment was carried out on the $LiNi_{1/3}Co_{1/3}Mn_{1/3}O_2$ (NCM)/Li half cell with 1 M $LiPF_6$ in ethylene carbonate (EC): ethyl methyl carbonate (EMC) (1:1, by wt.) as the electrolyte. CO_2 was regarded as the main product of the electrolyte decomposition, which was clearly observed at the onset voltage of 4.6 V versus Li/Li^+. As the potential increased to 5.4 V versus Li/Li^+, silicon tetrafluoride (SiF_4) was generated, and was attributed to a reaction between HF and the glass fiber separator that was used. Furthermore, the evolution of triethyl phosphate was observed at 5.5 V versus Li/Li^+, and was related to the $LiPF_6$ decomposition. This finding is also consistent with the cell thermal aging results reported by Weber *et al* [43]. However, oxygen evolution was not detected at even 5.5 V versus Li/Li^+, possibly because the concentration of the target analyte was below the detection limit of the system.

Baltruschat and coworkers used DEMS to investigate the oxygen reduction and evolution in a Li-containing electrolyte [44]. CO_2 evolution was observed along with oxygen evolution during charging. Also, the quantity of the evolved CO_2 was linearly dependent on the oxygen that was reduced in the previous discharge, which indicates that the DMSO decomposition accompanied with oxygen reduction, considering DMSO as the only carbon source in the system when gold was used as the cathode. This work has led to new insights into by-product formation in lithium–oxygen batteries.

Behm *et al* reported an approach to studying the decomposition of two ionic liquids (ILs): 1-butyl-1-methylpyrrolidinium bis(trifluoromethanesulfonyl)amide ([BMP][TFSA]) and 1-ethyl-3-methylimida-zolium bis(trifluoromethanesulfonyl) amide ([EMIM][TFSA]) on a gold electrode by DEMS [45]. Both ILs begin decomposition from the cations, at a potential of −1.0 V for the $[BMP]^+$ and −1.75 V for the $[EMIM]^+$ cation from the cathodic scan. The anion decomposition begins at potentials more negative than −2.6 V in [BMP][TFSA] and −2.3 V in [EMIM][TFSA]. At anodic potentials, the $[TFSA]^-$ anion decomposes at 1.8 V for [EMIM][TFSA] and at 2.2 V for [BMP][TFSA]. Behm *et al* also investigated the Li^+ and Mg^{2+} functions in the decomposition of 1-butyl-1-methylpyrrolidinium bis (trifluoromethanesulfonyl)imide (BMP-TFSI)-based electrolytes [46]. According to

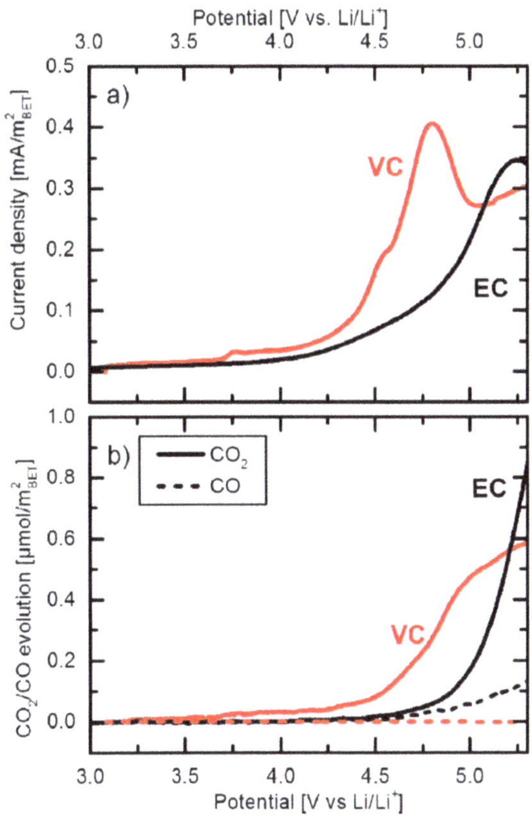

Figure 5.21. Evolution of $^{12}CO_2$ and ^{12}CO from the 1 M $LiPF_6$ in EC or VC electrolyte during a linear scan from open-circuit voltage to 5.3 V vs. Li/Li^+ (0.1 mV s^{-1}) on a ^{13}C carbon electrode. Adapted from [41].

the DEMS data, CF_3 is the main decomposition product of the $TFSI^-$ anion; the BMP^+ undergoes a cation ring-opening and scission of either the methyl or the butyl side chain. The addition of Li^+ greatly suppresses the decomposition of volatile BMP-TFSI. In comparison, Mg^{2+} acts to enhance the decomposition of $TFSI^-$ at the expense of the BMP^+ decomposition.

Zeis and coworkers reported a DEMS study on a carbon felt electrode for vanadium redox flow batteries [47]. Increases in the evolution of CO_2 and H_2 were observed at high and low potentials in the vanadium environment. At oxidative potentials, the presence of vanadium ions in the system resulted in 50% higher CO_2 evolution with a lower onset potential. This result indicates that the oxidized V^{5+} behaved as an oxidizing agent towards the carbon electrode.

5.3.2.2 Solid–electrolyte interphase formation

The concept of the SEI was introduced by Peled in 1979 [48]. This passivation layer originates from the decomposition of electrolytes on the electrode surface. The SEI layer has an ionic conductive nature and prevents further electrolyte decomposition, which allows stable cycling and thus prevents undesired gas evolution (table 5.3) [49].

Table 5.3. Gas evolution from SEI formation.

Reaction	Gas product
Reductive reactions	C_2H_4, CO, CO_2, C_3H_6, H_2
Oxidative reactions	O_2, CO_2

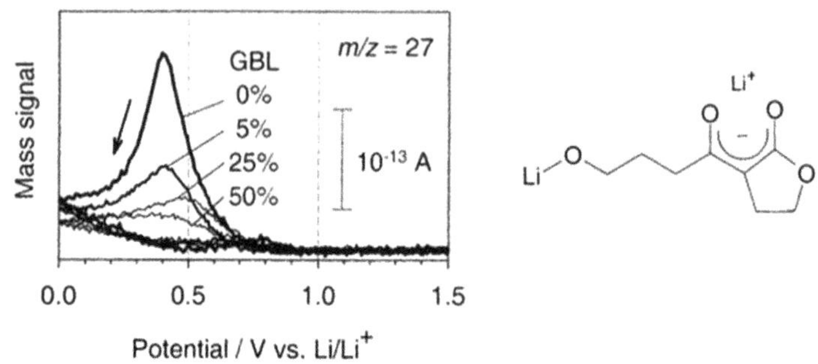

Figure 5.22. The gas evolved from graphite SFG 15 in 1 M $LiBF_4$ electrolyte in EC/DMC/GBL mixed solvents. Adapted from [50], copyright (2001) with permission from Elsevier.

Novák *et al* studied gas evolution on thick graphite electrodes in the EC/Dimethyl carbonate (DMC) electrolyte [50]. The results are shown in figure 5.22. Hydrogen and ethylene were observed on the graphite anode when the cell voltage was below 1.0 V versus Li/Li^+. Gamma butyrolactone (GBL) as a co-solvent was found to reduce the gas evolution. Ethylene evolution on an SFG 15 graphite electrode with 1 M $LiBF_4$ in EC/DMC/GBL mixed solvent was demonstrated. As the ratio of GBL increased to 50% (DMC was replaced with GBL), the ethylene evolution was decreased to its lowest level. This finding could be attributed to the GBL decomposition having a role in SEI formation by forming a cyclic α-alkoxy-β-keto ester Li salt.

Janek *et al* investigated the continuous destruction of the SEI on graphite electrodes in the LNMO/graphite full cell [51]. Significant H_2 evolution was observed during the charge and discharge. The effect of state of discharge (SOD) on H_2 evolution was determined: when SOD was below 80%, H_2 gas evolution was markedly suppressed. The gas evolution was observed to be extensive as the anode potential increased above 0.9 V during discharge. These observations suggest three measures for improving the cycle of LNMO/graphite cells: (1) Use graphite with preformed SEI. (2) Use sacrificial lithium source additives. (3) Use cathode materials with a higher irreversible capacity loss than graphite in the first cycle.

Cui *et al* reported DEMS characterization of the impact of hybrid functional additives on a SiO_x-C/LNMO lithium-ion battery system [52]. The gas evolution in the first discharge curves of the SiO_x-C/Li half cell was studied. The result was shown in figure 5.23. Compared to the basic electrolyte, an obvious reduction of C_2H_4 and CO_2 evolution was observed when a hybrid functional additive (FEC +

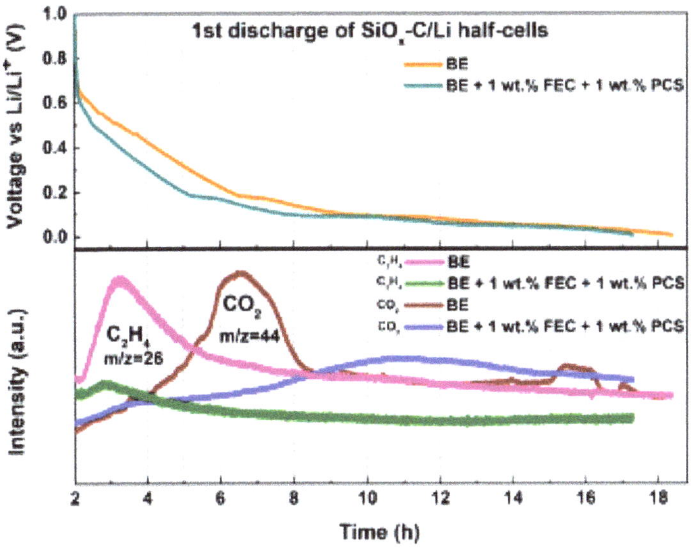

Figure 5.23. First discharge curve of SiO_x-C/Li half-cells (adopting a commercial *in situ* cell EL-CELL) with and without hybrid functional additives (1 wt % FEC + 1 wt % PCS) and the corresponding DEMS signals for $m/z = 26$ (C_2H_4) and $m/z = 44$ (CO_2). Adapted with permission from [52], copyright (2018) American Chemical Society.

1,3-propanediolcyclic sulfate) was introduced into the electrolyte. This finding reveals the more stable SEI formation assisted by the additive.

The same group also utilized DEMS to explore a sulfolane-based electrolyte for 5-V lithium-ion batteries. LiDFOB/Sulfolane (SL) in SF with or without TMSP was used as the electrolyte, and the gas evolution of MCMB/Li half cells during the first discharge was examined [53]. No obvious evolution of SO_2 ($m/z = 64$), CO ($m/z = 28$), CO_2 ($m/z = 44$), or C_2H_4 ($m/z = 28/27$) gas was observed with either 1 M LiDFOB/SL or 1 M LiDFOB/SL + 5 wt% TMSP (figure 5.24), indicating that unique reactions assisted by the additives occurred to form a stable SEI on the anode, increasing the resistance to gas evolution.

5.3.2.3 Cathode stability

High-voltage cathode material (such as low-cobalt and cobalt-free cathodes) has attracted great research interest because of its high voltage plateau and stable processes [54, 55]. However, this material raises safety concerns. Gas evolution in lithium-ion batteries is considered to be a major risk factor for ignition and explosion [56]. The high-voltage testing conditions for state-of-the-art electrolytes are harsh. The oxidation of the solvent and salt molecules is critical to battery performance. Abruna *et al* described a DEMS cell design with a three-electrode configuration [57]. The gas evolution was examined in the cell with propylene carbonate (PC), EC/DMC, and EC/diethyl carbonate (DEC) electrolytes, with $LiCoO_2$ (LCO) as the WE as shown in figure 5.25(a), (b) and (c), respectively. CO_2 evolution was detected in the PC and EC/DMC solutions at potentials between

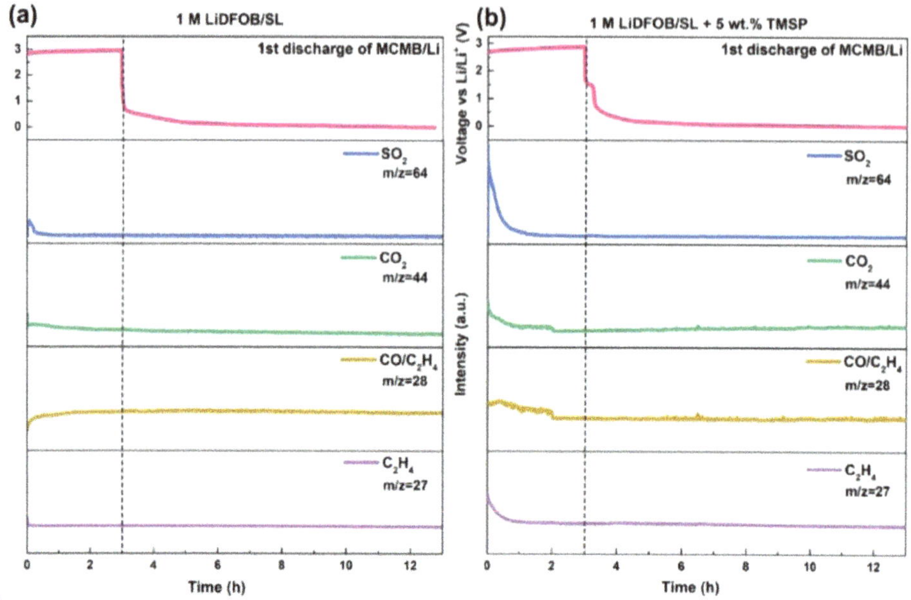

Figure 5.24. *In situ* DEMS signals from MCMB/Li half-cells equipped with (a) 1 m LiDFOB/SL and (b) 1 m LiDFOB/SL + 5 wt% TMSP. Adapted with permission from [53], copyright (2019) John Wiley & S0ns Ltd.

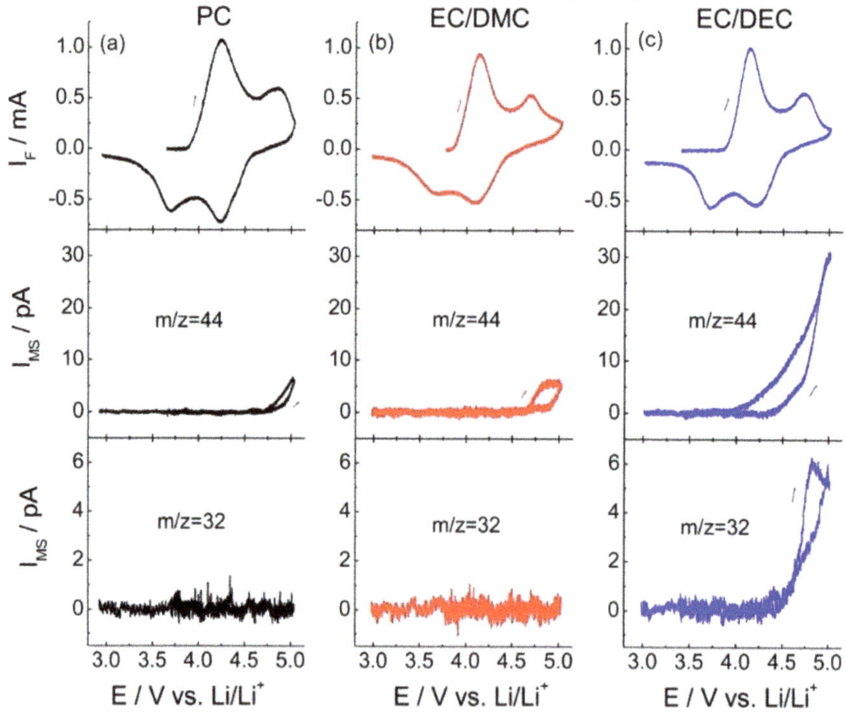

Figure 5.25. DEMS data for $LiCoO_2$ electrode in (a) 1 M $LiPF_6$ + PC, (b) 1 M $LiPF_6$ + EC/DMC (1:1 wt %), and (c) 1 M $LiPF_6$ + EC/DEC (1:1 wt %), respectively. Reprinted with permission from [57], copyright (2014) American Chemical Society.

4.7 and 5.0 V. Compared with the limiting amount of CO_2 evolved at 4.8 V, PC solutions kept decomposing even beyond 5.0 V. Meanwhile, no O_2 evolution was observed. In the EC/DEC electrolyte, apart from the CO_2 evolution beginning at 4.0 V, obvious O_2 evolution on the over-lithiated LCO was observed when the voltage was above 4.5 V.

When $LiMn_2O_4$ was used as the cathode, CO_2 evolution began at potentials higher than 4.7 V in the PC and EC/DEC electrolytes as shown in figure 5.26(a) and (b); in the EC/DMC electrolyte (figure 5.26(c)), CO_2 was observed at a higher potential of 4.9 V. It is worth mentioning that more CO_2 was evolved on LMNO compared to $LiMn_2O_4$ in the EC/DEC electrolyte. This observation indicates the catalytic effect of LMNO on EC/DEC electrolytes. However, less CO_2 was observed with LMNO when PC and EC/DMC electrolytes were used.

Berkes *et al* monitored the potential-resolved evolution of gases including H_2, C_2H_4, CO_2, and CO, in $LiNi_{0.5}Mn_{0.3}Co_{0.2}O_2$ (NMC532)/graphite full cells from cycle to cycle [58]. H_2, C_2H_2, and CO immediately reached their maximum values with a constant current applied at charging. In comparison, only a small amount of CO_2 gas was evolved during the early cycles. The high concentration of H_2 at the beginning of the first charge was attributed to water contamination. Ethylene is regarded as a characteristic peak for SEI formation. In the subsequent cycles, the low ethylene evolution indicates that relatively stable SEI formation occurred in the first cycle. Considering that the potential of the cathode is still not very high, the initial gas evolution was most likely contributed by the SEI formation. Also, temperature-dependent concentrations of CO_2 and H_2 were observed at open-circuit voltage, indicating that chemical reactions were occurring between the electrolyte and electrode. In the subsequent cycles, when the charge cutoff potentials were

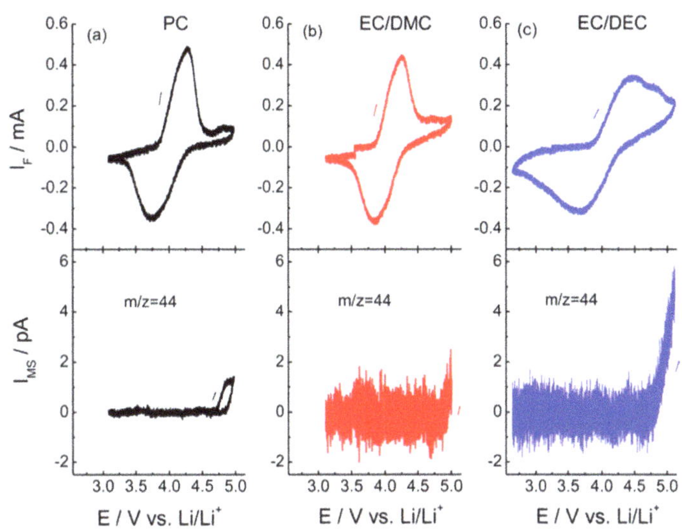

Figure 5.26. DEMS data for $LiMn_2O_4$ electrode in (a) 1 M $LiPF_6$ + PC, (b) 1 M $LiPF_6$ + EC/DMC (1:1 wt %), and (c) 1 M $LiPF_6$ + EC/DEC (1:1 wt %), respectively. Reprinted with permission from [57], copyright (2014) American Chemical Society.

increased from 4.3 to 4.4 and 4.5 V, a higher reaction rate was observed, according to the peak intensity increase from the MS. The authors stated that both CO_2 and CO were regarded as the products of electrolyte decomposition reactions at high potentials. No oxygen release was reported in this work.

Janek's group reported the gas evolution analysis of LNMO/graphite cells with a 'standard' carbonate-based electrolyte (LP57: 1 M $LiPF_6$ in EC/EMC) as shown in figure 5.27 [59]. On the basis of the DEMS results, the main oxidation and reduction products were determined to be CO_2, H_2, and C_2H_4. To differentiate between cathode and anode as the gas evolution source, graphite was replaced with delithiated $LiFePO_4$ (LFP) in different sets of experiments. In the LNMO/graphite cells, the first and fourth cycles were selected for DEMS signal analysis. The oxidation of Ni^{2+} to Ni^{3+} and Ni^{3+} to Ni^{4+}, respectively, are observed at charge plateaus at 4.57 and 4.64 V. During discharge, the plateaus at 4.58 and 4.53 V are attributed to the respective nickel reductions [60]. The other plateaus are associated with graphite. H_2, C_2H_4, and CO_2 evolution is observed immediately at the beginning of the first charge. The carbon monoxide (CO; m/z = 28) is not readily distinguished from ethylene or a fragment of CO_2 by DEMS. The evolution of C_2H_4 is related to SEI formation on graphite, as a result of the reductive decomposition of ethylene carbonate. The CO_2 evolution reaches a peak at the beginning of the charge cycle, which is related to the electrolyte decomposition. Then the signal decreases and eventually increases again at the onset of the second nickel oxidation plateau at 4.64 V. A sharp peak of CO_2 was also observed at the end of the charge. In the

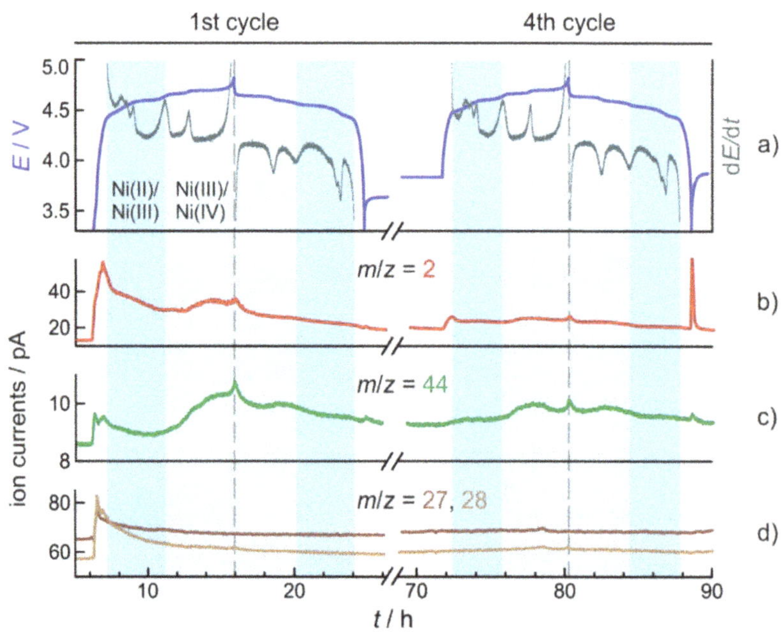

Figure 5.27. (a) Voltage profiles and corresponding derivative curves and DEMS signals for (b) m/z = 2, (c) m/z = 44, and (d) m/z = 27, 28 of an LNMO/graphite cell at C/10 rate. Reprinted with permission from [59], copyright (2016) American Chemical Society.

discharge, a peak is detected around 4.58 V ($Ni^{4+} + e^- \rightarrow Ni^{3+}$). Similar CO_2 evolution with Ni oxidation and reduction was also observed in the fourth cycle.

To further separate gas evolution on the cathode from the anode, an LNMO/LFP cell was utilized for the DEMS analysis, as shown in figure 5.28. Unlike with the LNMO/graphite system, CO_2 was the main gaseous product evolved during the cycle. This indicates that CO_2 evolution is mostly contributed by oxidative electrolyte decomposition on the cathode. Meanwhile, this CO_2 evolution also appeared along with the Ni^{3+}/Ni^{4+} redox couple. The authors claimed that the intermediate charge state of Ni plays a key role in the gas evolution of LNMO cells.

Yang's group reported a newly designed cation-disordered $Li_{1.25}Nb_{0.25}Fe_{0.50}O_2$/C cathode for high-voltage lithium-ion batteries [61]. Evolution of O_2 was observed from 4.30 V to 4.80 V in the first charge; the CO_2 gas evolution was attributed to the decomposition of electrolyte attacked by oxygen evolved from the cathode, and electrolyte decomposition at a high voltage. In the subsequent cycle, O_2 and CO_2 were still detected at a high potential but in much smaller quantities. The authors explained that this observation could be due to the formation of a passivation layer in the previous cycle and the electrode polarization. The CO_2 evolution at 3.0 V was

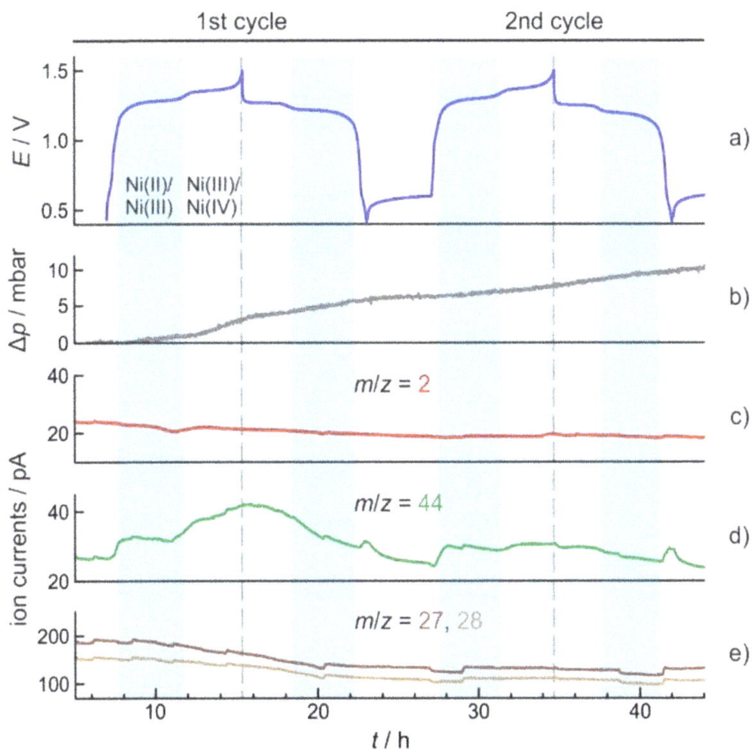

Figure 5.28. (a) Voltage profiles and corresponding derivative curves and DEMS signals for (b) $m/z = 2$, (c) $m/z = 44$, and (d) $m/z = 27, 28$ of an LNMO/LFP cell at C/10 rate. Reprinted with permission from [59], copyright (2016) American Chemical Society.

associated with the decomposition of Li_2CO_3 formed in the previous cycle, when CO_2 could react with the surface peroxide or superoxide species [62].

5.4 Summary and conclusion

This chapter aims to give a comprehensive understanding of the operation and design principles of DEMS instruments as well as demonstrate their research applications. To achieve this goal, the classification of DEMS instruments into C-DEMS, OEMS and I-DEMS systems was introduced in section 5.2.1. The operation principles of mass spectrometry were discussed in section 5.2.2. In section 5.2.3, the operation principles of C-DEMS, OEMS and I-DEMS were explained. In section 5.2.4, we focused on instrument design, including electrochemical cells and vacuum systems. In section 5.2.5, quantification methods were introduced for DEMS and OEMS separately. In section 5.2.6, we gave a brief comparison of different DEMS systems. Research applications for DEMS in the area of lithium–oxygen batteries were presented in section 5.3.1. In section 5.3.2, applications to lithium-ion batteries were demonstrated.

In general, each of the three DEMS systems has its own advantages and limitations. It is unfair to say which one is superior, since they were designed specifically for different research applications. For example, C-DEMS is more suitable for gas analysis of catalysts and fuel cells, while OEMS and I-DEMS are more frequently applied to energy storage systems such as batteries.

To conclude, the DEMS technique is a very powerful analytical characterization method for both online detecting and quantitively analyzing gas products from electrochemical reactions. The high sensitivity and time resolution make it superior to other gas analysis methods such as chromatography. Furthermore, DEMS is able to quantitatively determine gas evolution, making it a powerful method to investigate the mechanisms behind electrochemical reactions.

Abbreviations

[BMP][TFSA]	1-butyl-1-methylpyrrolidinium bis(trifluoromethanesulfonyl)amide
[EMIM][TFSA]	1-ethyl-3-methylimida-zolium bis(trifluoromethanesulfonyl)amide
A	Current efficiency
B	Magnetic electromagnetic
Bm/z	Background intensity
BMP-TFSI	1-butyl-1-methylpyrrolidinium bis(trifluoromethanesulfonyl)imide
$C_2H_4O_2$	Methylformate
C-DEMS	Conventional differential electrochemical mass spectrometry
CE	Counter electrode
CNT	Carbon nanotube
CO_2	Carbon dioxide
C-OEMS	Continuous online electrochemical mass spectrometry
DEC	Diethyl carbonate
DEMS	Differential electrochemical mass spectrometry

(Continued)

DMA	N-dimethylacetamide
DMC	Dimethyl carbonate
DMSO	Dimethyl sulfoxide
EC	Ethylene carbonate
EMC	Ethyl methyl carbonate
F	Faraday constant
FEC	Fluoroethylene carbonate
FTIR	Fourier-transform infrared spectroscopy
GBL	Gamma butyrolactone
I-DEMS	Intermittent differential electrochemical mass spectrometry
IF	Current at the electrode
I_i	The ion current of a species
ILs	Ionic liquids
$I_{m/z}$	Signal intensity of a specific gas species
IT	Ion trap
I_{total}	Total gas signal intensity
J_E	Incoming flow rate
K	MS instrument parameter
Kel F	Polychlorotrifluoroethylene
LCO	$LiCoO_2$
LFP	$LiFePO_4$
LiDFOB	$LiF_2BC_2O_4$
LNMO	$LiNi_{0.5}Mn_{1.5}O_4$
LP57	Carbonate-based electrolyte (1 M $LiPF_6$ in EC/EMC)
m/z	Mass-to-charge ratio
MIMS	Membrane inlet mass spectrometry
MS	Mass spectrometry
N	Collection efficiency
n	Molar number
NCM	$LiNi_{1/3}Co_{1/3}Mn_{1/3}O_2$
NMC532	$LiNi_{0.5}Mn_{0.3}Co_{0.2}O_2$
OEMS	Online electrochemical mass spectrometry
PC	Propylene carbonate
P_i	Partial pressure of gas species
PTFE	Polytetrafluoroethylene
P_{total}	Cell pressure
Q	Quadrupole
QMS	Quadrupole mass spectrometry
R	Ideal gas constant
RE	Reference electrode
r_{flow}	Volume flow rate of gas products
r_{gas}	Molar flow rate of gas products
SEI	Solid electrolyte interphase
SiF_4	Silicon tetrafluoride
SL	Sulfolane

$S_{m/z}$	Machine-specific sensitivity factor
SOD	State of discharge
S-OEMS	Sealed online electrochemical mass spectrometry
T	Temperature
TEGDME	Tetraethylene glycol dimethyl ether
TMS	Tetramethylene sulfone
TMSP	Tris(trimethylsilyl) phosphite
TOF	Time-of-flight
V	Headspace volume
VC	Vinylene carbonate
WE	Working electrode

Acknowledgements

This work was partially supported by the Joint Center for Energy Storage Research, an Energy Innovation Hub funded by the U.S. Department of Energy and also partially supported the Vehicle Technologies Office (VTO), David Howell (Manager), Battery R&D, Peter Faguy (Technology Manager), at the U.S. Department of Energy, Office of Energy Efficiency and Renewable Energy. The submitted manuscript has been created by UChicago Argonne, LLC, Operator of Argonne National Laboratory ('Argonne'). Argonne, a U.S. Department of Energy Office of Science laboratory, is operated under Contract No. DE-AC02–06CH11357.

References

[1] Baltruschat H 2004 Differential electrochemical mass spectrometry *J. Am. Soc. Mass. Spectrom.* **15** 1693–706

[2] Bruckenstein S and Gadde R R 1971 Use of a porous electrode for *in situ* mass spectrometric determination of volatile electrode reaction products *JACS* **93** 793–4

[3] Jusys Z, Massong H, Baltruschat H and New A 1999, Approach for simultaneous DEMS and EQCM: electro-oxidation of adsorbed CO on Pt and Pt-Ru *J. Electrochem. Soc.* **146** 1093

[4] Hartung T and Baltruschat H 1990 Differential electrochemical mass spectrometry using smooth electrodes: adsorption and hydrogen/deuterium exchange reactions of benzene on platinum *Langmuir* **6** 953–7

[5] Hartung T, Schmiemann U, Kamphausen I and Baltruschat H 1991 Electrodesorption from single-crystal electrodes: analysis by differential electrochemical mass spectrometry *Anal. Chem.* **63** 44–8

[6] Tegtmeyer D, Heitbaum J and Heindrichs A 1989 Electrochemical on line mass spectrometry on a rotating electrode inlet system *Ber. Bunsenges. Phys. Chem.* **93** 201–6

[7] Holzapfel M, Würsig A, Scheifele W, Vetter J and Novák P 2007 Oxygen, hydrogen, ethylene and CO_2 development in lithium-ion batteries *J. Power Sources* **174** 1156–60

[8] McCloskey B D, Bethune D S, Shelby R M, Girishkumar G and Luntz A C 2011 Solvents' critical role in nonaqueous lithium–oxygen battery electrochemistry *J. Phys. Chem. Lett.* **2** 1161–6

[9] Tsiouvaras N, Meini S, Buchberger I, Gasteiger H A and Novel A 2013 On-line mass spectrometer design for the study of multiple charging cycles of a Li-O_2 battery *J. Electrochem. Soc.* **160** A471–7
[10] Ashton S J 2012 Differential electrochemical mass spectrometry *Design, Construction and Research Application of a Differential Electrochemical Mass Spectrometer (DEMS)* (Berlin: Springer) pp 9–27
[11] Wolrer O and Heitbaum J 1984 Differential electrochemical mass spectroscopy (dems)- a new method for the study of electrode processes *Berichte der Bunsengesellschaft für phys. Chem.* **88** 2–6
[12] Eggert G and Heitbaum J 1986 Electrochemical reactions of propylenecarbonate and electrolytes solved therein—a dems study *Electrochim. Acta* **31** 1443–8
[13] Fujihira M and Noguchi T 1992 A highly sensitive analysis of electrochemical reduction products of CO_2 on gold by new differential electrochemical mass spectroscopy (dems) *Chem. Lett.* **21** 2043–6
[14] Novák P, Goers D, Hardwick L, Holzapfel M, Scheifele W, Ufheil J and Würsig A 2005 Advanced *in situ* characterization methods applied to carbonaceous materials *J. Power Sources* **146** 15–20
[15] De Hoffmann E 2000 Mass spectrometry *Kirk-Othmer Encyclopedia of Chemical Technology* (New York: Wiley)
[16] Miller P E and Denton M B 1986 The quadrupole mass filter: basic operating concepts *J. Chem. Educ.* **63** 617
[17] Wang H and Baltruschat H 2007 DEMS study on methanol oxidation at poly-and monocrystalline platinum electrodes: the effect of anion, temperature, surface structure, Ru adatom, and potential *J. Phys. Chem.* C **111** 7038–48
[18] Jusys Z and Vaškelis A 1997 The kinetic HD isotope effect in electroless copper plating. A DEMS study *Electrochim. Acta* **42** 449–54
[19] He M 2016 Elucidating interface reactions in Li-ion batteries and supercapacitors by *in situ* gas analysis *ETH Zurich.*
[20] Abd-El-Aziz A, Xu J, Bogolowski N, Königshoven P and Baltruschat H 2012 New cell for DEMS applicable to different electrode sizes *Electrocatalysis* **3** 39–47
[21] Imhof R and Novák P 1998 *In situ* investigation of the electrochemical reduction of carbonate electrolyte solutions at graphite electrodes *J. Electrochem. Soc.* **145** 1081
[22] Berkes B B, Jozwiuk A, Vračar M, Sommer H, Brezesinski T and Janek J 2015 Online continuous flow differential electrochemical mass spectrometry with a realistic battery setup for high-precision, long-term cycling tests *Anal. Chem.* **87** 5878–83
[23] Castel E, Berg E J, El Kazzi M, Novák P and Villevieille C 2014 Differential electrochemical mass spectrometry study of the interface of $x\mathrm{Li_2MnO_3}\cdot(1-x)\mathrm{LiMO_2}$ (M = Ni, Co, and Mn) material as a positive electrode in Li-ion batteries *Chem. Mater.* **26** 5051–7
[24] Jusys Z, Binder M, Schnaidt J and Behm R J 2019 A novel DEMS approach for studying gas evolution at battery-type electrode|electrolyte interfaces: high-voltage $\mathrm{LiNi_{0.5}Mn_{1.5}O_4}$ cathode in ethylene and dimethyl carbonate electrolytes *Electrochim. Acta* **314** 188–201
[25] Castel E, Berg E J, El Kazzi M, Novák P and Villevieille C 2014 Differential electrochemical mass spectrometry study of the interface of $x\mathrm{Li_2MnO_3}\cdot(1-x)$ $\mathrm{LiMO_2}$ (M= Ni, Co, and Mn) material as a positive electrode in Li-ion batteries *Chem. Mater.* **26** 5051–7
[26] Jung R, Metzger M, Maglia F, Stinner C and Gasteiger H A 2017 Oxygen release and its effect on the cycling stability of $\mathrm{LiNi_xMn_yCo_zO_2}$(NMC) cathode materials for Li-ion batteries *J. Electrochem. Soc.* **164** A1361–77

[27] Strehle B, Kleiner K, Jung R, Chesneau F, Mendez M, Gasteiger H A and Piana M 2017 The role of oxygen release from Li- and Mn-rich layered oxides during the first cycles investigated by on-line electrochemical mass spectrometry *J. Electrochem. Soc.* **164** A400–6

[28] He M, Castel E, Laumann A, Nuspl G, Novák P and Berg E J 2015 *In Situ* gas analysis of $Li_4Ti_5O_{12}$ based electrodes at elevated temperatures *J. Electrochem. Soc.* **162** A870–6

[29] Renfrew S 2019 Tracing interfacial reactivity of lithium transition metal oxides through outgassing *PhD Thesis* UC Berkeley

[30] Novák P, Panitz J-C, Joho F, Lanz M, Imhof R and Coluccia M 2000 Advanced *in situ* methods for the characterization of practical electrodes in lithium-ion batteries *J. Power Sources* **90** 52–8

[31] Drioli E, Calabro V and Wu Y 1986 Microporous membranes in membrane distillation *Pure Appl. Chem.* **58** 1657–62

[32] Willsau J and Heitbaum J 1984 The influence of Pt-activation on the corrosion of carbon in gas diffusion electrodes—a dems study *J. Electroanal. Chem. Interfacial Electrochem.* **161** 93–101

[33] Jusys Z and Behm R 2001 Methanol oxidation on a carbon-supported Pt fuel cell catalyst a kinetic and mechanistic study by differential electrochemical mass spectrometry *J. Phys. Chem.* B **105** 10874–83

[34] Peng Z, Freunberger S A, Chen Y and Bruce P G 2012 A reversible and higher-rate Li-O_2 battery *Science* **337** 563–6

[35] Lu J, Li L, Park J-B, Sun Y-K, Wu F and Amine K 2014 Aprotic and aqueous Li–O_2 batteries *Chem. Rev.* **114** 5611–40

[36] Kim D W, Ahn S M, Kang J, Suk J, Kim H K and Kang Y 2016 *In situ* real-time and quantitative investigation on the stability of non-aqueous lithium oxygen battery electrolytes *J. Mater. Chem.* A **4** 6332–41

[37] Ahn S M, Suk J, Kim D Y, Kang Y, Kim H K and Kim D W 2017 High-performance lithium-oxygen battery electrolyte derived from optimum combination of solvent and lithium salt *Adv. Sci.* **4** 1700235

[38] Sun B, Guo L, Ju Y, Munroe P, Wang E, Peng Z and Wang G 2016 Unraveling the catalytic activities of ruthenium nanocrystals in high performance aprotic Li–O_2 batteries *Nano Energy* **28** 486–94

[39] Itkis D M *et al* 2013 Reactivity of carbon in lithium–oxygen battery positive electrodes *Nano Lett.* **13** 4697–701

[40] Qiao Y, Yi J, Wu S, Liu Y, Yang S, He P and Zhou H 2017 Li-CO_2 Electrochemistry: a new strategy for CO_2 fixation and energy storage *Joule* **1** 359–70

[41] Pritzl D, Solchenbach S, Wetjen M and Gasteiger H A 2017 Analysis of vinylene carbonate (VC) as additive in graphite/$LiNi_{0.5}Mn_{1.5}O_4$ cells *J. Electrochem. Soc.* **164** A2625–35

[42] Hahn H, Wagner R, Schappacher F, Winter M and Nowak S 2016 In operando X-shaped cell online electrochemical mass spectrometry (OEMS): new online analysis enables insight into lab scale lithium ion batteries during operation *J. Electroanal. Chem.* **772** 52–7

[43] Weber W, Kraft V, Grützke M, Wagner R, Winter M and Nowak S 2015 Identification of alkylated phosphates by gas chromatography–mass spectrometric investigations with different ionization principles of a thermally aged commercial lithium ion battery electrolyte *J. Chromatogr.* A **1394** 128–36

[44] Bondue C J, Hegemann M, Molls C, Thome E and Baltruschat H 2016 A Comprehensive study on oxygen reduction and evolution from lithium containing DMSO based electrolytes at gold electrodes *J. Electrochem. Soc.* **163** A1765–75

[45] Alwast D, Schnaidt J, Law Y T and Behm R J 2016 A novel approach for differential electrochemical mass spectrometry studies on the decomposition of ionic liquids *Electrochim. Acta* **197** 290–9

[46] Alwast D, Schnaidt J, Hancock K, Yetis G and Behm R J 2019 Effect of Li^+ and Mg^{2+} on the electrochemical decomposition of the ionic liquid 1-Butyl-1- methylpyrrolidinium bis(trifluoromethanesulfonyl)imide and related electrolytes *ChemElectroChem* **6** 3009–19

[47] Eifert L, Jusys Z, Banerjee R, Behm R J and Zeis R 2018 Differential electrochemical mass spectrometry of carbon felt electrodes for vanadium redox flow batteries *ACS Appl. Energy Mater.* **1** 6714–8

[48] Peled E 1979 The electrochemical behavior of alkali and alkaline earth metals in nonaqueous battery systems—the solid electrolyte interphase model *J. Electrochem. Soc.* **126** 2047

[49] Winter M, Besenhard J O, Spahr M E and Novák P 1998 Insertion electrode materials for rechargeable lithium batteries *Adv. Mater.* **10** 725–63

[50] Lanz M and Novák P 2001 DEMS study of gas evolution at thick graphite electrodes for lithium-ion batteries: the effect of γ-butyrolactone *J. Power Sources* **102** 277–82

[51] Michalak B, Berkes B B, Sommer H, Brezesinski T and Janek J 2017 Electrochemical cross-talk leading to gas evolution and capacity fade in $LiNi_{0.5}Mn_{1.5}O_4$/graphite full-cells *J. Phys. Chem.* C **121** 211–6

[52] Xu G *et al* 2018 Tracing the impact of hybrid functional additives on a high-voltage (5 V-class) SiO_x-C/$LiNi_{0.5}Mn_{1.5}O_4$ Li-ion battery system *Chem. Mater.* **30** 8291–302

[53] Lu D *et al* 2019 Deciphering the interface of a high-voltage (5 V-Class) Li-ion battery containing additive-assisted sulfolane-based electrolyte *Small Methods* **3** 1900546

[54] Xia H, Meng Y S, Lu L and Ceder G 2007 Electrochemical properties of nonstoichiometric $LiNi_{0.5}Mn_{1.5}O_{4-\delta}$ thin-film electrodes prepared by pulsed laser deposition *J. Electrochem. Soc.* **154** A737

[55] Santhanam R and Rambabu B 2010 Research progress in high voltage spinel $LiNi_{0.5}Mn_{1.5}O_4$ material *J. Power Sources* **195** 5442–51

[56] Kumai K, Miyashiro H, Kobayashi Y, Takei K and Ishikawa R 1999 Gas generation mechanism due to electrolyte decomposition in commercial lithium-ion cell *J. Power Sources* **81–82** 715–9

[57] Wang H, Rus E, Sakuraba T, Kikuchi J, Kiya Y and Abruña H D 2014 CO_2 and O_2 evolution at high voltage cathode materials of Li-ion batteries: a differential electrochemical mass spectrometry study *Anal. Chem.* **86** 6197–201

[58] Berkes B B, Schiele A, Sommer H, Brezesinski T and Janek J 2016 On the gassing behavior of lithium-ion batteries with NCM523 cathodes *J. Solid State Electrochem.* **20** 2961–7

[59] Michalak B, Berkes B B, Sommer H, Bergfeldt T, Brezesinski T and Janek J 2016 Gas evolution in $LiNi_{0.5}Mn_{1.5}O_4$/graphite cells studied in operando by a combination of differential electrochemical mass spectrometry, neutron imaging, and pressure measurements *Anal. Chem.* **88** 2877–83

[60] Terada Y, Yasaka K, Nishikawa F, Konishi T, Yoshio M and Nakai I 2001 *In Situ* XAFS analysis of Li(Mn, M)2O4 (M=Cr, Co, Ni) 5V cathode materials for lithium-ion secondary batteries *J. Solid State Chem.* **156** 286–91

[61] Luo M *et al* 2020 Identifying the anionic redox activity in cation-disordered $Li_{1.25}Nb_{0.25}Fe_{0.50}O_2$/C oxide cathodes for Li-ion batteries *J. Mater. Chem.* A **8** 5115–27
[62] Hong J, Lim H-D, Lee M, Kim S-W, Kim H, Oh S-T, Chung G-C and Kang K 2012 Critical role of oxygen evolved from layered li–excess metal oxides in lithium rechargeable batteries *Chem. Mater.* **24** 2692–7

IOP Publishing

Batteries
Materials principles and characterization methods
Chen Liao

Chapter 6

Electron microscopies for batteries

Weiqun Li, Nigel D Browning and B Layla Mehdi

6.1 Electron microscopy SEM and TEM

6.1.1 Basics of electron microscopy

Electron microscopies (EMs), including scanning electron microscopy (SEM) [1] and (scanning) transmission electron microscopy ((S)TEM) [2], are important to characterize the micro- to atomic-scale structural and compositional properties for battery materials. Both SEM and TEM consist of electron optical system, vacuum system, signal processing system, power system, etc. There are two different electron sources in the electron optical system: thermionic sources and field-emission sources. The thermionic sources, such as tungsten filaments or lanthanum hexaboride (LaB_6) crystals, are heated up to stimulate electron beam. The field-emission sources, such as fine tungsten needles, render higher resolution and brightness by applying high voltage, lowering the work function barrier, and tunnelling out the electrons.

The electron beam is firstly converged through the condenser lenses. In SEM, the movement of electron beam is controlled by the scanning coil, and the electron beam is focused on the sample surface through the objective lenses. The secondary electrons, emitted from the sample surface by inelastic scattering of electron beam, are used for imaging the morphology of the sample. In TEM, the parallel electron beam is concentrated by the condenser lenses, transmits through the sample, and is magnified by the intermediate, diffraction, and projector lenses. In both SEM and TEM, the imaging and signal processing system is used to display and analyze the images.

Due to the shorter wavelength of electron than visible light, EMs have much higher resolution than optical microscopy. The resolution is generally increased with the decrease of electron wavelength and the increase of accelerating voltage. SEM has up to nanoscale resolution, while TEM has the resolution of sub-angstrom. Such high resolution is still affected by astigmatism, chromatic aberration, and spherical aberration. Astigmatism occurs when the electrons sense the nonuniform magnetic field, which can be easily corrected using stigmators. The chromatic aberration

doi:10.1088/978-0-7503-2682-7ch6

occurs due to the variation of the electron energy, which can be solved using very good high-tension suppliers. The spherical aberration, which is caused by the lens field acting inhomogeneously on the off-axis rays, is the main limitation for higher resolution nowadays. Therefore, aberration-corrected STEM has been developed to provide higher resolution images.

In both SEM and TEM, the energy-dispersive x-ray spectroscopy (EDS), which is analyzed based on the characteristic x-rays produced by the high-energy electron beam, is used to determine the chemical composition of the battery materials. In TEM, by analyzing the selected area diffraction pattern (SAED) and measuring the inter-distance of lattice plane, the chemical components of the materials can be approximately confirmed. The electron energy loss spectroscopy (EELS), which is produced by the inelastic scattering of electron beam, is used to determine the composition, valence state, and thickness of the samples.

6.1.2 Electron beam effect

Electron beam dose has a significant influence on imaging quality of EMs, especially for (S)TEM. Generally, the effect from electron beam can be minimized by using lower accelerating voltages and shorter electron beam exposure time. The electron beam effect becomes even more severe for *in situ* and *operando* EMs, as beam-sensitive electrolyte is used and long exposure time is needed. In order to control the electron beam dose, avoid unnecessary side reactions and increase the image frame rate, subsampling image acquisition and image inpainting have been studied and used. Subsampling acquisition is the sampling of the fully sampled image. In the subsampling imaging, the image is analyzed by acquiring only a small random portion of the pixels in the image. By using image inpainting, the missing information of the images can be restored through reconstructing the sparse distribution of pixels in the image based on the subsampling image acquisition. This method allows independent control of both spatial and temporal dose envelope for *ex situ*, *in situ*, and *operando* experiments. The quantitative information can be acquired from images with fewer pixels under subsampling [3–8]. Compressive sensing (CS) [9, 10] uses an image inpainting algorithm to restore the missing information and reconstruct the incomplete image for the beam sensitive materials.

Figures 6.1(a)–(c) show the full range of various electron beam configurations, which control the number of electrons on picosecond-to-microsecond timescales [11]. The radiolysis occurs mainly in the beam-exposed area when the electron beam hits the thin liquid sample. The chemical species become inhomogeneous, and diffusion takes place as a result of the localized radiolysis process, as shown in figure 6.1(a). The STEM configuration in figure 6.1(b) considers each of these cells to be an individual pixel, and all the pixels are exposed to the electron beam. While subsampling with CS is adopted, random pixels are exposed to the electron beam for a certain time period, as shown in figure 6.1(c).

Figures 6.1(d)–(f) and (g)–(i) show two applications of subsampling image acquisition with image inpainting. As shown in figures 6.1(d)–(f), a set of sub-sampled images are taken, including 100%, 25% and 12.5% sampling levels, for the

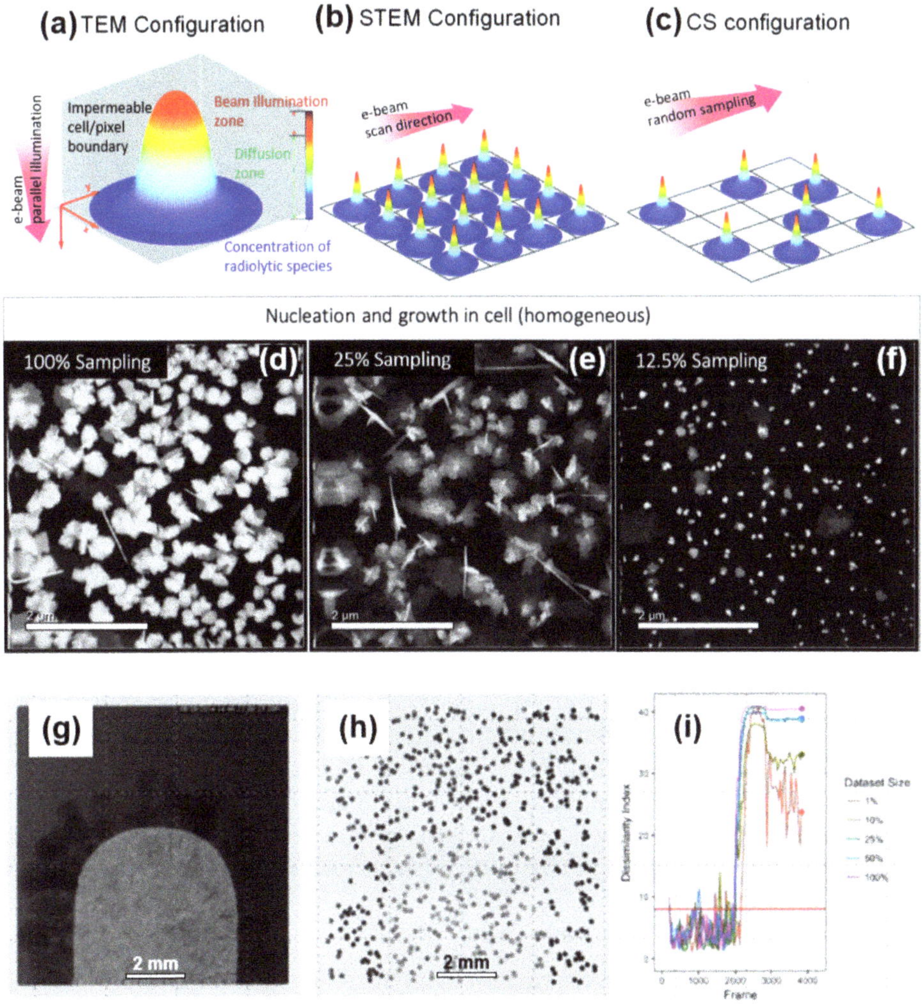

Figure 6.1. Schematics of the electron/liquid interaction zones in the EMs for (a) a single illumination spot, (b) multiple illumination spots, and (c) random-sampling of illumination spots. Reprinted with permission from [11]. Copyright 2021 Royal Chemistry Society. Subsampling STEM images of the *in situ* liquid cell for the nucleation and growth of Ag nanoparticles at (d) 100%, (e) 25% and (f) 12.5% sampling rates. Adapted with permission from [12]. Copyright 2019 AIP Publishing LLC. (g) 100% and (h) 1% subsampled images for Li deposition in a STEM liquid cell. (i) Comparison of different subsampling levels for Li deposition amount. Reprinted with permission from [13]. Copyright 2019 Cambridge University Press.

in situ nucleation and growth of Ag nanoparticles [12]. Such sampling rates have significant effect on the morphology of Ag nanoparticles, indicating unavoidable electron beam effect on the nucleation and growth processes. At high sampling levels, the Ag nanoparticles are mostly rafts and pillars; while at low sampling levels, the particles tend to be regular spherical shapes.

The dendrite deposition in an *operando* liquid cell is shown in figures 6.1(g)–(i), in which the electron beam dose is $\leqslant 0.3$ e Å^{-2} s^{-1} (less than 0.3 electrons per square

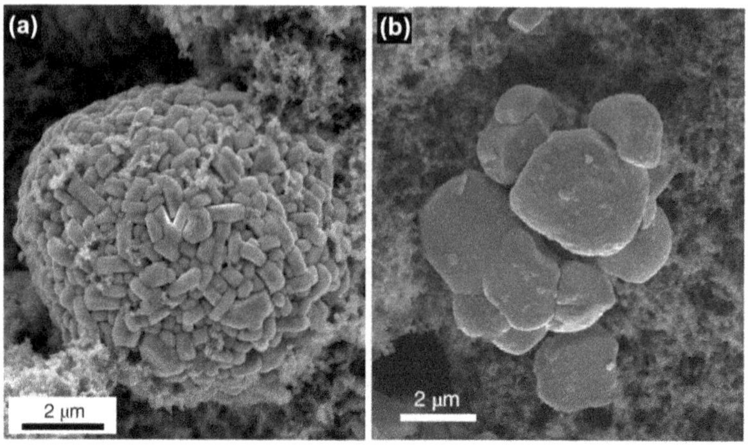

Figure 6.2. SEM images of (a) pristine polycrystalline NMC811 electrode and (b) pristine single-crystal NMC electrode. Reprinted with permission from [14]. Copyright 2020 Springer Nature Limited.

angstrom per second). In order to achieve high speed of imaging and avoid severe electron beam effect, subsampling has been used through an external scan generator that randomly hops between lines of the scan at a rate designed to minimize the effects of scan hysteresis [5]. Subsequently, reconstruction is used to fill in the gaps in the adaptive sub-sampling data so that high speed and high resolution images can be achieved. Figures 6.1(g)–(i) show a practical application of this method to study the Li deposition process in the liquid cell battery [13]. Figures 6.1(g) and (h) are the 100% and 1% subsampled images, respectively. Figure 6.1(i) shows the comparison among different subsampling levels, indicating that only 1% of the image can provide precise measurement of the Li dendrite amount during the cycling.

6.1.3 Applications of electron microscopy

EMs are the most effective characterization facilities to image the nano- and structure and morphology of battery materials, including cathode, anode, separator, current collector, etc. For example, SEM images in figures 6.2(a) and (b) show the structure and morphology of pristine polycrystalline $LiNi_{0.8}Mn_{0.1}Co_{0.1}O_2$ (NMC811) and single-crystal nickel-manganese-cobalt (NMC) cathode materials [14]. The size, uniformity, morphology and surface from micro- to nano-scale can be clearly seen for both materials. The secondary polycrystalline NMC811 has spherical morphology of several micrometers, and is composed of several hundreds of nanometers of primary particles, as shown in figure 6.2(a).

A nano-to-angstrom-scale dark-field STEM (DF-STEM) image of the aged NMC811 cathode is shown in figure 6.3(a), in which the lattice planes of layered $R\bar{3}m$ and rock salt $Fm\bar{3}m$ can be distinguished from the green and blue dashed squared regions, respectively. The corresponding fast Fourier transform (FFT) images in figures 6.3(b) and (c) show the lattice planes for the two regions, respectively. Through comparing the structural and morphological changes before and after cycling, the fading and degradation mechanisms of the battery materials

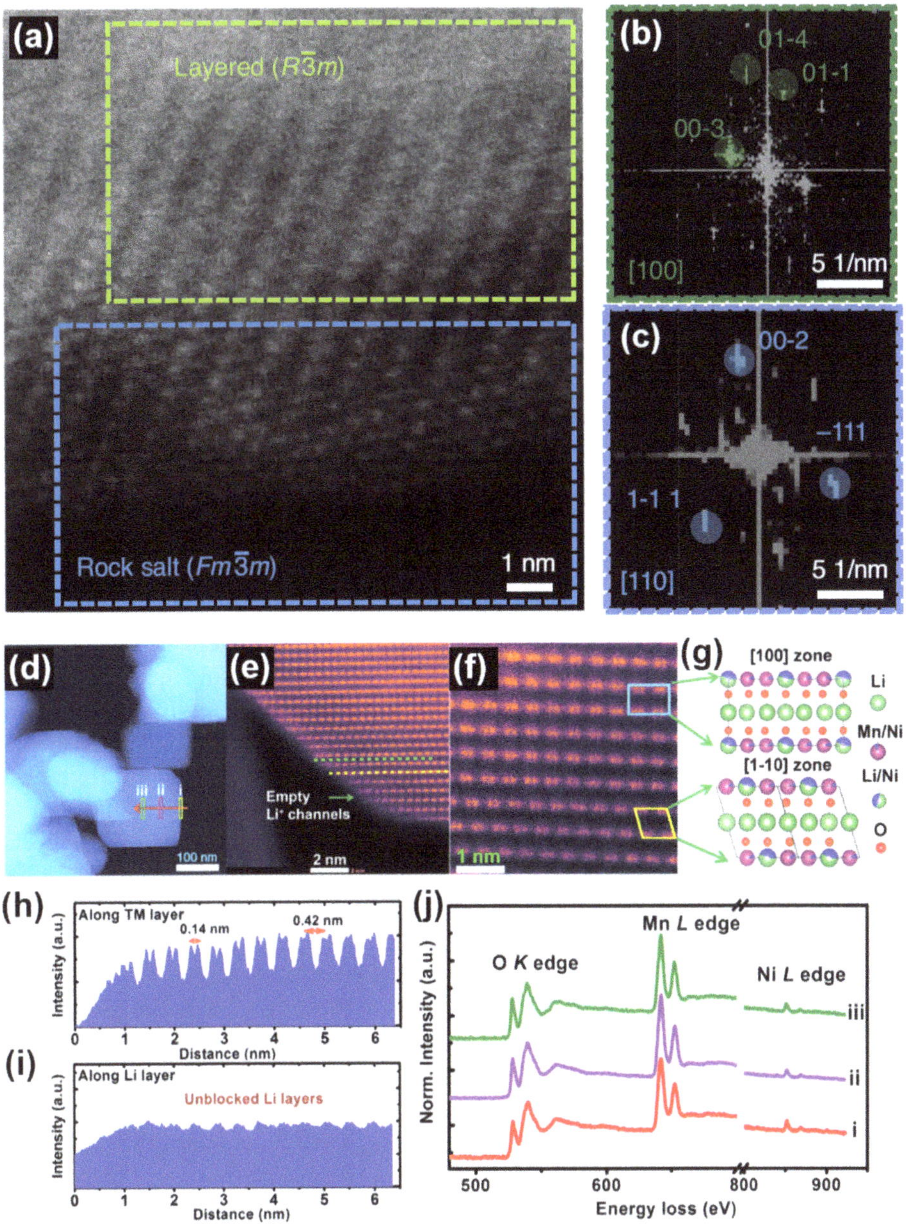

Figure 6.3. (a) DF-STEM images of aged NMC811. FFTs of the (b) upper and (c) lower regions in (a). Reprinted with permission from [14]. Copyright 2020 Springer Nature Limited. (d) Aberration-corrected HAADF-STEM image, (e, f) high resolution Z-contrast images, and (g) atomic model showing the crystal structure of LMR cathode material. Intensity profiles along (h) the yellow dashed line (TM layer) and (i) the green dashed line (Li layer) shown in (e). (j) EELS spectra showing the O-*K*, Mn-*L*, and Ni-*L* edges at locations i, ii, and iii in (d). Reprinted with permission from [15]. Copyright 2015 American Chemical Society.

can be figured out. The aberration-corrected STEM images in figures 6.3(d)–(f) show the atomic-scale structure of $Li[Li_{0.2}Ni_{0.2}Mn_{0.6}]O_2$ (LMR) cathode materials [15]. The high-angle annular dark-field STEM (HAADF-STEM) image in figure 6.3(d) shows the basic morphology of the LMR nanoparticles. The particle surface region has the *C2/m* phase, as shown in the high resolution Z-contrast image in figure 6.3(e). Higher magnification Z-contrast image in figure 6.3(f) shows that there are mixed [100] (cyan) and [1 $\bar{1}$ 0] (yellow) zones in the inner bulk region, compared with the atomic model in figure 6.3(g). The distances between two transition metal (TM) atoms can be precisely measured by the plotted intensity profile (figure 6.3(h)) along the yellow dashed line in figure 6.3(e). The Li ion diffusion pathway, which is labelled by the green dashed line in figure 6.3(e), is not blocked as shown by the plotted intensity profile in figure 6.3(i). The EELS spectra (figure 6.3(j)), performed along the red arrow in figure 6.3(d), show the identical peak shapes of O-*K* edge, Mn-*L* edge and Ni-*L* edge, indicating unchanged valence states of the TM ions from particle surface to bulk.

In order to minimize and eliminate the external influence during material process and transfer, *in situ* and *operando* SEM and (S)TEM have been used to provide high spatial resolution and precise temporal resolution for battery materials with non-invasive and diagnostic characterization capabilities. It was only in recent years that *in situ* and *operando* experiments have become more practical, owing to the development of specialized holders and instruments, including open cell [16–18], liquid cell [19–23], environmental cell [24–26], etc. Here the *in situ* EMs are defined as that the live videos are uninterrupted throughout the whole electrochemical process, while *operando* EMs are those that only parts of the live videos are taken or only snapshots are taken at different states of charge and discharge cycles.

In situ and *operando* open cell provides researchers with fundamental insights into the structural and compositional changes during electrochemical reactions [16–18]. The open cell has been used to study the structural and compositional change of SnO_2 nanowire upon Na ion insertion/extraction. In the standard open cell configuration (figure 6.4(a)), the electrode (SnO_2 in this case) contacts with either solid (Na_2O and NaOH in this case) or liquid electrolyte through a point electrical contact. Such point contact alters the diffusion or transportation patterns and local reaction conditions occurring in the real systems, which is not necessarily representative of the real situations in batteries. Figures 6.4(b)–(d) show the pristine, sodiated and desodiated states of SnO_2 nanowire, respectively (with schematic of each step in figure 6.4(g)). Upon sodiation, the SnO_2 nanowire electrode is expanded and transformed to armorphous Na_xSn and crystalline Na_2O. With further sodiation, the Na_xSn is crystallized into $Na_{15}Sn_4$. Figures 6.4(e) and (f) show the higher resolution of TEM image of the sodiated nanowire with its SAED pattern. Upon desodiation, the Na_xSn is transformed to Sn nanoparticles which are dispersed in the Na_2O hollow matrix. The formation of hollow matrix leads to the increase of electrical impedance and poor cycle life of SnO_2 electrode for sodium ion batteries.

Figure 6.4(h) shows the open cell experimental setup of *in situ* and *operando* SEM characterization for $Li(Ni_{0.8}Co_{0.15}Al_{0.05})O_2$ (NCA) cathode materials [18]. The secondary NCA electrode particle is attached on the tip of the manipulator, which

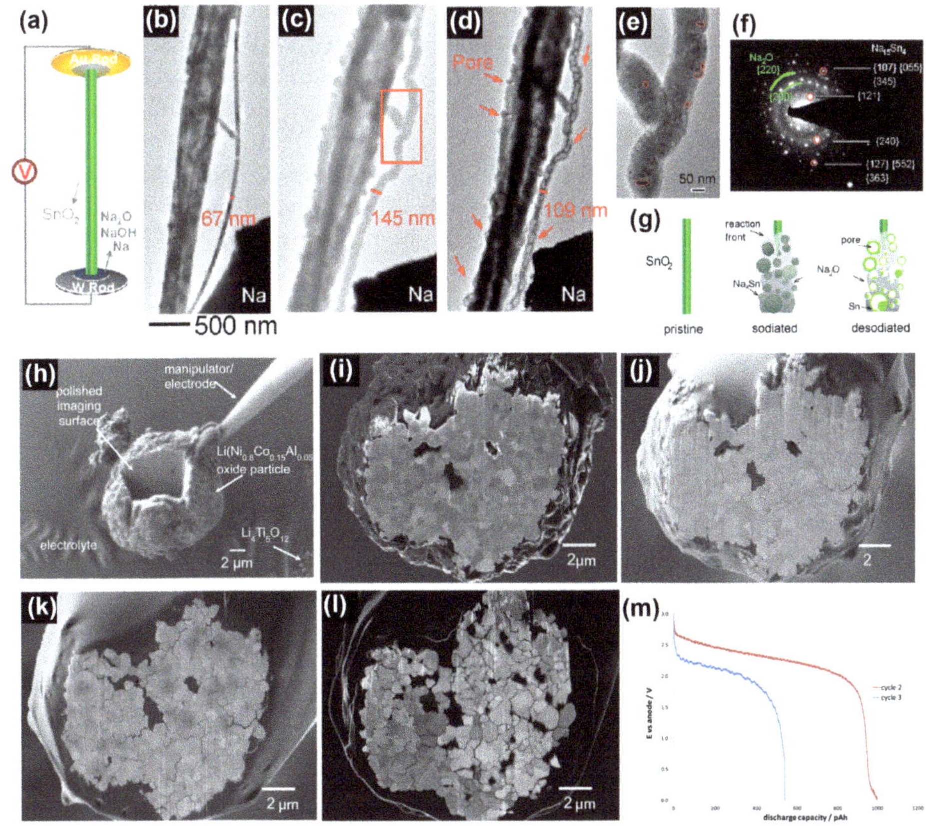

Figure 6.4. (a) Schematic of the open cell system. TEM images showing the (b) pristine, (c) sodiated, and (d) desodiated states of the SnO_2 nanowire electrode. (e) Higher resolution TEM image of the sodiated nanowire, indicated by red square in (c), with its (f) SAED pattern. (g) Schematic showing the pristine, sodiated and desodiated states of the SnO_2 nanowire electrode. Reprinted with permission from [17]. Copyright 2013 American Chemical Society. (h) Experimental setup of the *in situ* and *operando* SEM micro-battery. SEM images showing (i) the pristine NCA particle, and the same particle after (j) first, (k) second, and (l) third cycle. (m) Discharge capacity and voltage profiles for the second and third cycles. Reprinted with permission from [18]. Copyright 2013 John Wiley & Sons, Inc.

can be controlled and moved freely. The electrode particle is subsequently immersed or partially immersed in the low-vapor pressure electrolyte droplet (P_{13} TFSI + 0.5 mol kg^{-1} LiTFSI). The $Li_4Ti_5O_{12}$ anode is covered by the electrolyte droplet. Figures 6.4(i)–(l) show the fracture of an NCA particle in the first three cycles by using the *operando* SEM open cell. The intergranular cracks form upon electrochemical cycling, leading to the loss of electrical contact among the primary NCA particles. The impedance is therefore increased, and the discharge capacity fades, as shown by the capacity loss from cycle 2 to cycle 3 in figure 6.4(m).

In order to obtain the dynamic information which involves liquid environments, the closed liquid cell has been developed [19–23]. The liquid cell provides rapid and direct characterization of the reactions under controlled conditions directly relevant to the practical situations. In the standard liquid cell, the materials are immersed in the liquid electrolyte, forming a conformal contact between the tested materials and the environment. Such conformal liquid environment mimics the real environment where the electrochemical reactions occur. Liquid cell accomplishes the direct observation of the structural and chemical evolution of battery materials and how these phenomena relate to the materials' intrinsic properties. The electron beam rate must be carefully controlled to avoid the side effect induced by the electron beam itself. The strong electron beam can result in the formation of bubbles and/or precipitates, the breakdown of the solution, or change in local chemistry or pH value of the solution [11, 27]. *Operando* STEM liquid cell has been used to study the

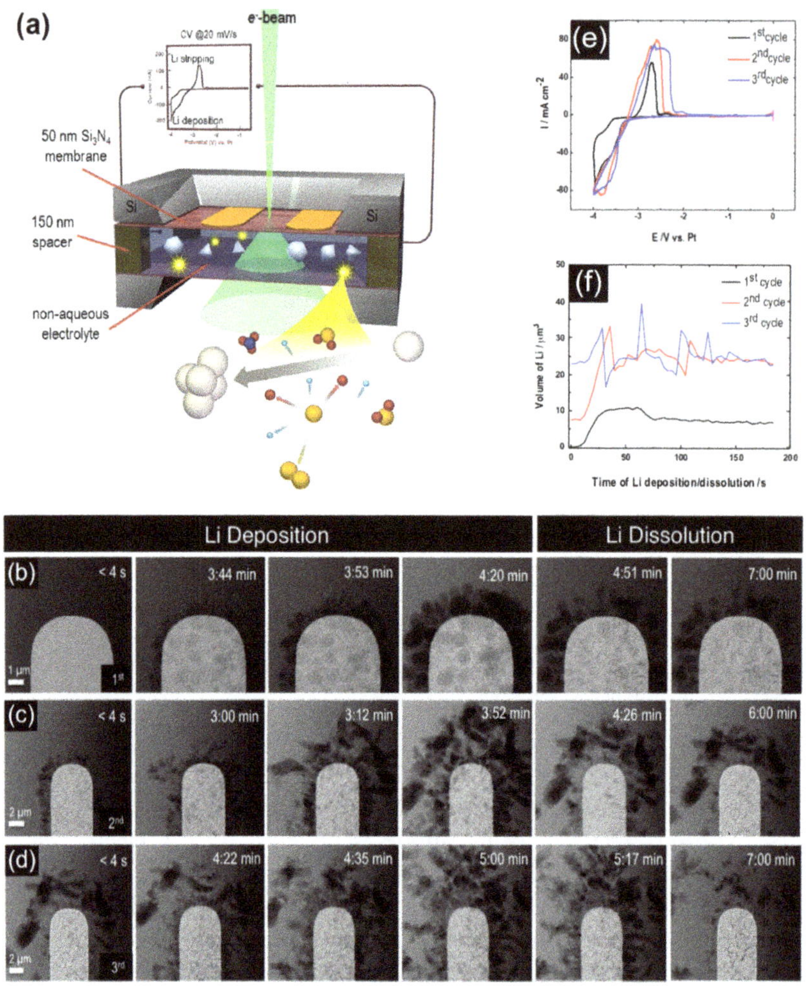

Figure 6.5. (a) Schematic of the *operando* liquid cell. Reprinted with permission from [20]. Copyright 2016 Springer Nature Limited. HAADF STEM images of Li deposition and dissolution during (b) first, (c) second, and (d) third charge/discharge cycles. (e) Cyclic voltammetry and (f) growth rate of Li in the first three cycles. Reprinted with permission from [19]. Copyright 2015 American Chemical Society.

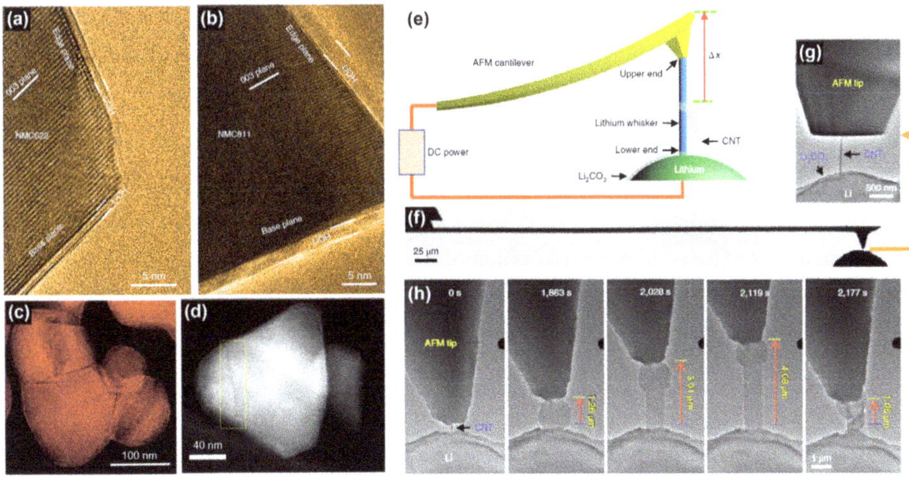

Figure 6.6. TEM images showing the passivation layers on the surface of (a) NMC811 and (b) NMC622 particles after exposing samples to H_2O vapor ($P_{H_2O} = 5 \times 10^{-2}$ Torr) for 30 min. Reprinted with permission from [24]. Copyright 2020 Springer Nature Limited. STEM images showing (c) intergranular and (d) intragranular cracks of the NMC nanoparticle upon thermal activation. Reprinted with permission from [25]. Copyright 2018 American Chemical Society. (e) Schematic of the AFM-ETEM setup. (f) TEM image of the AFM cantilever and the Li metal electrode. (g) TEM image of a carbon nanotube attaching to a AFM tip. (h) TEM images showing the growth of a Li whisker. Reprinted with permission from [26]. Copyright 2020 Springer Nature Limited.

electrochemical processes of Li dendrite [19, 20]. Figure 6.5(a) shows the schematic of an *operando* liquid cell. The liquid cell is composed of two silicon microchips containing electron beam transparent silicon nitride (Si_3N_4) membranes. Three platinum (Pt) microelectrodes are aligned parallel with each other, serving as reference, working, and counter electrodes. The electron beam passes through the membranes and electrolyte. The electrochemical reaction is monitored by an electrochemical work station. Figures 6.5(b)–(d) show the first three cycles of Li deposition and dissolution at the surface of Pt electrode. Due to solid Li being less dense than the surrounding liquid electrolyte and the electrode itself, the deposition and dissolution can be clearly seen, and 'dead Li' forms at the end of each cycle. Quantitative studies of the corresponding cyclic voltammetry and Li growth rate can be simultaneously recorded, as shown in figures 6.5(e) and (f). The nucleation, growth and dissolution mechanisms of battery materials can be thoroughly analyzed by using the *in situ* and *operando* liquid cell.

Environmental EMs have been used to study the atmosphere effect, thermal stability, mechanical property, etc, for battery materials [24–26]. For example, environmental gas cell has been used to study the cathode-air interfacial reactions for NMC cathode materials [24]. The cathode-H_2O reactions produce a conformal layer of LiOH at the surface of NMC811 and $LiNi_{0.6}Mn_{0.2}Co_{0.2}O_2$ (NMC622) cathodes, as shown in figures 6.6(a) and (b), respectively. There is a critical thickness of LiOH beyond which the other interfacial reactions can be arrested. The thermal stability of $LiNi_{0.4}Mn_{0.4}Co_{0.2}O_2$ (NMC442) cathode has been studied through a

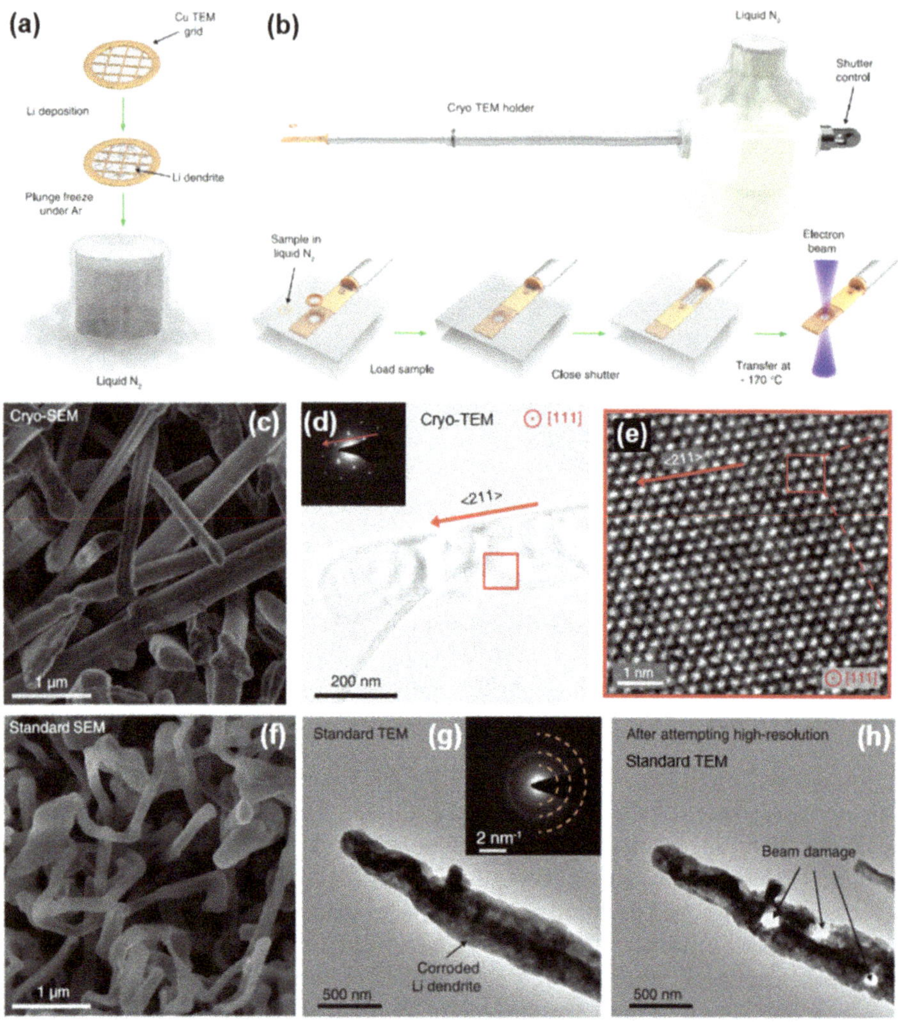

Figure 6.7. (a) Schematic showing the assembly process of Li dendrite for cryo-EMs. (b) Schematic showing the cryo-TEM holder. (c) Cryo-SEM, (d) cryo-TEM and SAED pattern, and (e) high resolution cryo-TEM images of Li dendrite. (f) Standard SEM of Li dendrite after brief air exposure and (g) standard TEM of Li dendrite after ~1 s air exposure. (h) Standard TEM image of Li dendrite after ~1s electron beam exposure (dose rate of ~500 e Å^{-2} s^{-1}). Reprinted with permission from [30]. Copyright 2017 American Association for the advanced of Science.

heating TEM holder [25]. The delithiated NMC particles were loaded on a silicon nitride passivated silicon chip, and subsequently heated up. Both intergranular and intragranular cracks can be observed by the *in situ* TEM images, as shown in figures 6.6(c) and (d), respectively. The intragranular cracks are induced by local stress build-up from the layered to rock salt phase transformation. The intergranular cracks are caused by the accumulation of oxygen vacancies and continuous phase changes.

An atomic force microscope (AFM) cantilever was combined with an environmental TEM (ETEM) to study the growth and mechanical properties of Li dendrite, as shown in figures 6.6(e)–(g) [26]. After applying a voltage, the Li dendrite grows against the AFM tip, as shown in figure 6.6(h). The growth stress (up to 130 MPa) is substantially higher than that of bulk or micrometer Li reported elsewhere [28, 29]. The yield strength of Li dendrite was reported as high as 244 MPa. This technique provides a precise measurement of the intrinsic mechanical properties of the battery materials in the native environment.

Due to the volatility and electron beam-sensitivity of liquid electrolyte, cryogenic (cryo) EMs have been used to vitrify the liquid electrolyte, and visualize the structural components and chemical mapping of the solid–liquid interfaces in their native state. The cryo-SEM and cryo-(S)TEM have been used to provide high resolution structural and chemical information about dendrites and solid–electrolyte interphase layers [30, 31]. Figures 6.7(a) and (b) show the operation process of a cryo-EM with a normal Cu grid. By comparing the cryo-SEM/TEM images (figures 6.7(c)–(e)) and the standard SEM/TEM images (figures 6.7(f)–(h)), the facet morphology is well preserved by using the cryo-transfer method. The atomic-resolution TEM image (figure 6.7(e)) shows that the Li dendrite is stable in the cryo state under electron beam rate ~1000 e Å^{-2} s^{-1} for ~30 s. In contrast, the Li dendrite is corroded after exposing under electron beam in the standard condition (figures 6.7(g) and (h)).

SEM and (S)TEM are important and powerful characterization approaches for imaging the micro- to atomic-scale morphological and structural properties of battery materials. Compositional information can be analyzed through the EDS, EELS, SAED, FFT, etc. *In situ* and *operando* EMs provide not only as high as atomic level spatial resolution, but also precise temporal resolution for electrochemical analysis. Cryo-EMs accomplish the direct observation of beam sensitive materials in their native environments. The beam effect, which is unavoidable, can be minimized by using subsampling and inpainting techniques.

References

[1] Reimer L 1998 *Scanning Electron Microscopy: Physics of Image Formation and Microanalysis* (Berlin: Springer)

[2] Williams D B and Carter C B 2009 *Transmission Electron Microscope—A Textbook for Materials Science* (Boston, MA: Springer)

[3] Leary R, Saghi Z, Midgley P A and Holland D J 2013 Compressed sensing electron tomography *Ultramicroscopy* **131** 70–91

[4] Stevens A, Yang H, Carin L, Arslan I and Browning N D 2014 The potential for Bayesian compressive sensing to significantly reduce electron dose in high-resolution STEM images *Microscopy* **63** 41–51

[5] Kovarik L, Stevens A, Liyu A and Browning N D 2016 Implementing an accurate and rapid sparse sampling approach for low-dose atomic resolution STEM imaging *Appl. Phys. Lett.* **109** 164102

[6] Béché A, Goris B, Freitag B and Verbeeck J 2016 Development of a fast electromagnetic beam blanker for compressed sensing in scanning transmission electron microscopy *Appl. Phys. Lett.* **108** 93103

[7] Ferroni M, Signoroni A, Sanzogni A, Masini L, Migliori A, Ortolani L, Pezza A and Morandi V 2016 Biological application of compressed sensing tomography in the scanning electron microscope *Sci. Rep.* **6** 33354
[8] Nicholls D, Lee J, Amari H, Stevens A J, Mehdi B L and Browning N D 2020 Minimising damage in high resolution scanning transmission electron microscope images of nanoscale structures and processes *Nanoscale* **12** 21248–54
[9] Candes E J, Romberg J and Tao T 2006 Robust uncertainty principles: exact signal reconstruction from highly incomplete frequency information *IEEE Trans. Inf. Theory* **52** 489–509
[10] Donoho D L 2006 Compressed sensing *IEEE Trans. Inf. Theory* **52** 1289–306
[11] Lee J, Nicholls D, Browning N D and Mehdi B L 2021 Controlling radiolysis chemistry on the nanoscale in liquid cell scanning transmission electron microscopy *Phys. Chem. Chem. Phys.* **23** 17766–73
[12] Mehdi B L *et al* 2019 Controlling the spatio-temporal dose distribution during STEM imaging by subsampled acquisition: in-situ observations of kinetic processes in liquids *Appl. Phys. Lett.* **115** 63102
[13] Mehdi B L, Browning N D, Lee J, Amari H, Johnson N, Nicholls D and Stevens A J 2019 Low-dose and high-speed observations of battery processes by operando STEM *Microsc. Microanal.* **25** 1454–5
[14] Xu C *et al* 2020 Bulk fatigue induced by surface reconstruction in layered Ni-rich cathodes for Li-ion batteries *Nat. Mater.* **20** 84–92
[15] Zheng J, Xu P, Gu M, Xiao J, Browning N D, Yan P, Wang C and Zhang J-G 2015 Structural and chemical evolution of Li- and Mn-rich layered cathode material *Chem. Mater.* **27** 1381–90
[16] Huang J Y *et al* 2010 *In Situ* observation of the electrochemical lithiation of a single SnO_2 nanowire electrode *Science (80-.).* **330** 1515–20
[17] Gu M, Kushima A, Shao Y, Zhang J-G, Liu J, Browning N D, Li J and Wang C 2013 Probing the failure mechanism of SnO_2 nanowires for sodium-ion batteries *Nano Lett.* **13** 5203–11
[18] Miller D J, Proff C, Wen J G, Abraham D P and Bareño J 2013 Observation of microstructural evolution in Li battery cathode oxide particles by *in situ* electron microscopy *Adv. Energy Mater.* **3** 1098–103
[19] Mehdi B L *et al* 2015 Observation and quantification of nanoscale processes in lithium batteries by operando electrochemical (S)TEM *Nano Lett.* **15** 2168–73
[20] Mehdi B L, Stevens A, Qian J, Park C, Xu W, Henderson W A, Zhang J-G, Mueller K T and Browning N D 2016 The impact of Li grain size on coulombic efficiency in Li batteries *Sci. Rep.* **6** 34267
[21] Gu M *et al* 2013 Demonstration of an electrochemical liquid cell for operando transmission electron microscopy observation of the lithiation/delithiation behavior of Si nanowire battery anodes *Nano Lett.* **13** 6106–12
[22] Abellan P *et al* 2014 Probing the degradation mechanisms in electrolyte solutions for Li-ion batteries by *in situ* transmission electron microscopy *Nano Lett.* **14** 1293–9
[23] Hou J, Freiberg A, Shen T-H, Girod R, Gonthier J, Kim S-J, Maglia F, Gasteiger H A and Tileli V 2020 Charge/discharge cycling of $Li_{1+x}(Ni_{0.6}Co_{0.2}Mn_{0.2})_{1-x}O_2$ primary particles performed in a liquid microcell for transmission electron microscopy studies *J. Phys.: Energy* **2** 34007

[24] Zou L *et al* 2020 Unlocking the passivation nature of the cathode–air interfacial reactions in lithium ion batteries *Nat. Commun.* **11** 3204
[25] Mu L *et al* 2018 Oxygen release induced chemomechanical breakdown of layered cathode materials *Nano Lett.* **18** 3241–9
[26] Zhang L *et al* 2020 Lithium whisker growth and stress generation in an *in situ* atomic force microscope–environmental transmission electron microscope set-up *Nat. Nanotechnol.* **15** 94–8
[27] Schneider N M, Norton M M, Mendel B J, Grogan J M, Ross F M and Bau H H 2014 Electron–water interactions and implications for liquid cell electron microscopy *J. Phys. Chem.* C **118** 22373–82
[28] Schultz R P 2002 *Lithium: Measurement of Young's Modulus and Yield Strength* (Batavia, IL: Fermi National Accelerator Laboratory)
[29] Xu C, Ahmad Z, Aryanfar A, Viswanathan V and Greer J R 2017 Enhanced strength and temperature dependence of mechanical properties of li at small scales and its implications for Li metal anodes *Proc. Natl. Acad. Sci.* **114** 57–61
[30] Li Y *et al* 2017 Atomic structure of sensitive battery materials and interfaces revealed by Cryo–electron microscopy *Science (80–.).* **358** 506–10
[31] Boyle D T, Huang W, Wang H, Li Y, Chen H, Yu Z, Zhang W, Bao Z and Cui Y 2021 Corrosion of lithium metal anodes during calendar ageing and its microscopic origins *Nat. Energy* **6** 487–94

Chapter 7

Nuclear magnetic resonance as an analytical tool in battery materials science

Mounesha N Garaga, David Clarkson and Steve Greenbaum

Nuclear magnetic resonance (NMR) spectroscopy has been employed productively for many years to characterize a wide variety of phenomena that are central to the functioning of batteries. The predominance of lithium-ion batteries as power sources in electronic devices and more recently in electric vehicles has also led to an abundance of NMR-based investigations of Li insertion and deinsertion mechanisms in electrode materials, ion transport in electrolytes, and even interfacial phenomena that are crucial to battery performance, lifetime, and safety.

This chapter begins with a review of background and implementation of standard NMR methods including 1-D and 2-D spectroscopy, relaxometry, and diffusometry, and some specialized techniques including imaging and *operando* methods. Most of the investigations reviewed here, some from our own laboratory, concern structural and dynamical aspects of Li-ion battery materials, in particular carbonaceous anodes, transition metal oxide cathodes, and both liquid and solid electrolytes.

7.1 Introduction

The lithium-ion battery (LIB) is one of the most ubiquitous components in modern day portable electronics and in increasingly pervasive application in larger scale technologies such as electric vehicles. Recognition of the enormous impact of LIBs on daily lives was undoubtedly a major factor in the awarding of the 2019 Nobel Prize in Chemistry to Goodenough, Whittingham, and Yoshino. Despite its enormous commercial success, the basic design of the LIB has changed little since its market introduction in the 1990s. The essential ingredients are two intercalation electrodes—a carbonaceous anode and a transition metal oxide or phosphate cathode—and a liquid electrolyte consisting of a Li salt dissolved in a liquid carbonate solvent. There has been much materials development on new cathode materials, essentially replacing Co with other transition metals, which eliminates or

doi:10.1088/978-0-7503-2682-7ch7

reduces the use of this strategic metal, but also generally raises the battery voltage. The resulting increase in energy density is offset by electrolyte stability issues at the higher voltage. Recent efforts are also geared toward enabling alternatives to graphite such as Si or Sn [1, 2]. However, remarkably few changes have been made in the electrolyte beyond the introduction of small quantity additives, including fluorinated compounds, or expanding from binary to tertiary or quaternary solvent mixtures. Nonetheless, there is still a very active research area in these electrolytes with an eye toward improving battery stability, lifetime, and safety [3, 4].

In parallel fashion, the last three decades have seen growing interest in solid electrolytes, which include solvent-free polymer electrolytes, ceramic ion conductors, and polymer/ceramic composites. The obvious advantage of a solid electrolyte is improved safety, especially for large format vehicle propulsion and grid storage batteries, but progress has been limited by low ionic conductivity and mechanical interface integrity upon cycling, among other issues [5, 6]. Although there is also considerable interest in hybrid or gel electrolytes which are essentially liquid electrolyte immobilized in a polymer matrix [7]. Nuclear magnetic resonance (NMR) spectroscopy is one of a few very powerful methods for probing the immediate environments of ions such as Li^+ in both the liquid and solid states, especially in disordered materials. The NMR spectrum is sensitive to a number of physical parameters that are directly related to the efficacy of a working battery. These include Li-ion mobility, electronic conductivity, and the changes of valence or electronic structure of the cations and transition metal ions that are involved in the redox processes. Furthermore, NMR is element-specific (i.e., nuclear-specific) and may be used to directly examine the species that are involved in the working battery. As a probe of the immediate environment of the nucleus under investigation, NMR is an ideal method for investigating the small changes in local structure that occur on doping, or for elucidating lithium insertion and extraction mechanisms. An important feature of NMR is that the integrated spectroscopic signal is proportional to the number of resonant NMR nuclei in the sample. Thus, NMR analysis can be used to ascertain the relative phase content of a material in cases where individual spectral contributions can be resolved.

Though there are several excellent reviews of NMR in battery materials science, especially in solid electrode materials [8], electrode–electrolyte interface [9] and both liquid and solid electrolyte [10–12], this chapter describes many of the important NMR techniques (figure 7.1) in detail, with several literature references serving as examples of the use of these techniques in battery science. We also include a brief discussion of the solid–electrolyte interphase (SEI), which forms as a result of electrolyte decomposition processes at the electrodes. Unfortunately, space limitations do not allow an exhaustive survey of other important battery chemistries under development, such as those based on Mg or Al.

7.2 Methods

7.2.1 Introduction to NMR methodology

NMR comes with a suite of techniques which are ideal for examining the structure of both liquid and solid electroactive materials, as well as measurements which can

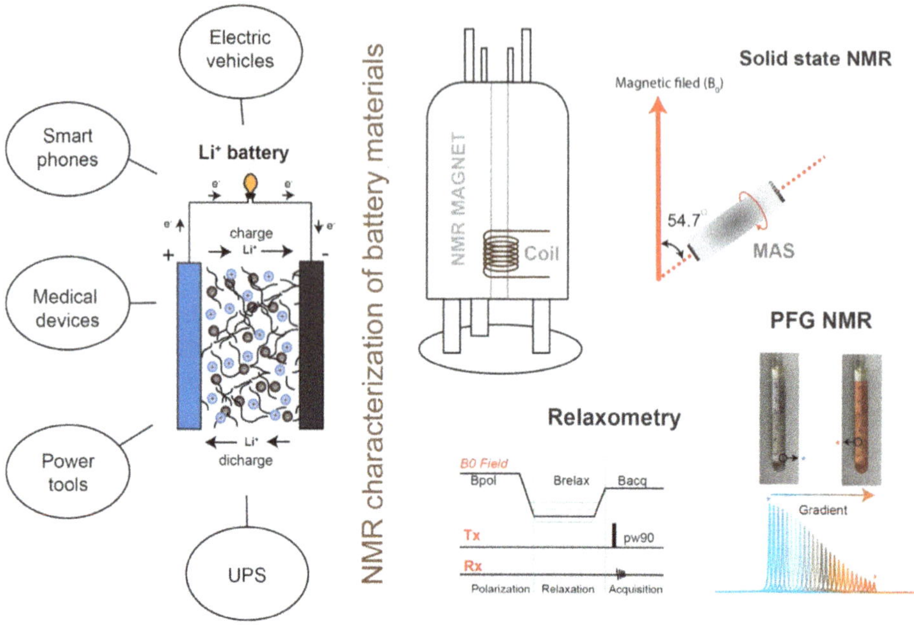

Figure 7.1. Schematics of Li^+ battery and NMR characterization tool.

elucidate both short and long-range dynamics on timescales from hundreds of millisecond range to (in some cases) the picosecond range.

Isotopes which have a spin quantum number $I > 0$ have a nonzero magnetic dipole moment and are NMR active (table 7.1). There are a number of high abundance NMR active isotopes which are common in battery materials—notably, given the wide usage of lithium ion battery chemistries, it is convenient that there are

Table 7.1. Properties of selected nuclei relevant to NMR of electroactive materials, values from Stone (2005) [126] and Pyykko (2008) [127].

Isotope	Spin	Gyromagnetic ratio γ (Mrad $T^{-1}s^{-1}$)	Quadrupole moment Q (barn)	Natural abundance (%)
^{1}H	1/2	267.5100		99.9855
^{6}Li	1	39.3699	−0.000 806	7.59
^{7}Li	3/2	103.9720	−0.04	92.41
^{13}C	1/2	42.2700		1.07
^{19}F	1/2	251.8050		100
^{23}Na	3/2	67.2800	0.104	100
^{27}Al	5/2	70.8021	0.1466	100
^{29}Si	1/2	29.4250		4.6832
^{31}P	1/2	108.3910		100

two NMR active lithium isotopes, ^{6}Li and ^{7}Li. Protons and deuterons also make possible a number of experiments looking at the behaviors of polymers and organic solvents, and abundant NMR isotopes such as ^{23}Na, ^{31}P, ^{27}Al, make measurements of materials for battery chemistries beyond Li-ion, including Li-polymer, Na ion, Al, etc. Somewhat less abundant nuclei such as ^{13}C and ^{29}Si, are also readily accessed by NMR.

When samples containing NMR active nuclei are placed in a strong static magnetic field (B_0), the resulting population difference in spins aligned and anti-aligned with the magnetic field follow a Boltzmann distribution, resulting in a net polarization of nuclei. The net magnetization is therefore described by [13, 14],

$$\frac{N_m}{N} = e^{-E_m/k_BT} \sum_{n=-I}^{I} e^{\frac{-E_n}{k_BT}} \approx \frac{1}{2I+1}\left(1 + \frac{m\hbar\gamma B_0}{k_BT}\right) \text{at high } T$$

The resulting net magnetization scales with the gyromagnetic ratio, and is aligned with B_0, which is by convention taken to point in the z direction. However, when perturbed, the polarization of the nuclei precesses about the magnetic field at the Larmor frequency [13],

$$\omega_L = \gamma B_0$$

By illuminating the sample with pulsed radio-frequency electromagnetic fields at the Larmor frequency, the nuclear magnetization can be manipulated from its equilibrium position along the z-axis into the observation xy-plane, and the resulting free induction decay (FID) can be measured using an inductive coil, in practice often the same coil that delivers the rf pulse. A basic 1D NMR experiment consists of simply applying a single resonant radio pulse and observing this FID, which takes the form

$$FID = S_0 e^{-(i\omega + R_2)t}$$

FIDs are then Fourier transformed to produce spectra. The utility of NMR as a probe of the environment of the nucleus is due to nuclear coupling to its surroundings. The most obvious form of this coupling is through the electronic shielding of the magnetic field, causing nuclei at different molecular or lattice sites, or different domains of the material, to experience static magnetic fields of slightly different strength, resulting in a chemical shift in the frequency of nuclei at the site. Chemical shifts are typically measured according to a reference frequency ω_{ref} of a highly symmetric compound in ppm:

$$\delta = \frac{\omega - \omega_{\text{ref}}}{\omega_{\text{ref}}}$$

The benefit of this reporting of chemical shifts is that to first order, the chemical shift of a given site is the same at any field strength, allowing comparison between measurements taken at different field strengths. Chemical shifts in crystalline systems are rank two tensors whose value depends strongly on the crystal orientation, resulting in powder patterns in powdered crystalline systems—we will return to this in our discussion of MAS and solid structural determination techniques.

Nuclear spins also may be coupled to one another directly, either via through space interactions or through coupling to shared electrons (J-coupling), and in paramagnetic samples may be coupled to unpaired electron spins, or to conduction band electrons in metals (Knight shift). Isotopes with $I>1/2$ have a nonzero electric quadrupole moment which may be coupled with the local electric field gradient, which results in spectral broadening. In isotropic liquids, rapid molecular tumbling results in chemical shift anisotropy (CSA), direct dipolar coupling, and first-order quadrupolar coupling averaging to 0 and having no resulting impact on spectra—however, they continue to play a role in relaxation [13].

Because of these couplings, spins relax to equilibrium according to two separate time constants: T_1, the spin-lattice or longitudinal relaxation time which is a measure of how long it takes for the polarization to relax to its equilibrium along the z-axis, and T_2, a measure of the decay of polarization in the xy-plane. Beyond providing inherent experimental limitations, both constants can provide insight into the dynamics of the species being investigated. They may be related to the autocorrelation function of the stochastic process driving the relaxation $G(t)$ and its Fourier transform the spectral density function $J(\omega)$. Examining T_1 at a range of temperatures and frequencies can provide both information about species translational diffusion, rotation, and other motional processes across timescales on the order of picoseconds to microseconds.

7.2.2 Structural investigations and MAS

Spectral information can be difficult to gather from solids due to the strength and asymmetry of couplings in crystalline systems. In general, chemical shift, dipolar, and quadrupolar couplings all depend strongly on orientation in crystalline systems. It is possible to gather useful information from a single crystal system by taking measurements at different orientations of the crystal principal axis system with respect to B_0 [15], but this is frequently unfeasible or inconvenient. Powdered crystals exhibit powder NMR patterns, whose enormous breadth can make it difficult to distinguish between sites. Magic angle spinning NMR (MAS NMR) is a technique which can significantly improve resolution, making sites resolvable, by rotating the sample in such a way that the motion averages out anisotropic effects and reduces the effect of the dipolar and quadrupolar couplings.

The chemical shift interaction is due to the electronic shielding of the nucleus from the external magnetic field—generally, it includes both paramagnetic and diamagnetic contributions. Paramagnetic effects in cathode materials containing transition metal ions (e.g. Fe^{2+}, $Mn^{2+,4+}$, Ni^{3+}) can be rather large, giving very broad spectra that can be challenging to capture in their entirety [16]. Paramagnetic contributions due to conduction electrons in metals, called the Knight shift, are conventionally treated separately from the rest of the chemical shift tensor, due to the magnitude and sign of their effect and general importance (for example, measuring the shift from conductor to semiconductor as Li is intercalated into Li_xCoO_2) [17]. The chemical shift tensor is typically represented as:

$$\begin{pmatrix} \delta_{xx} & \delta_{xy} & \delta_{xz} \\ \delta_{yx} & \delta_{yy} & \delta_{yz} \\ \delta_{zx} & \delta_{zy} & \delta_{zz} \end{pmatrix}$$

The chemical shift tensor is typically assumed to be symmetric—in fact, chemical shift tensors are not generally symmetric, but spectra are not sensitive to their skew symmetric components, although they may play a role in relaxation. Dealing only with the symmetric component of the matrix, it can be diagonalized into its principal axis system (PAS) with three orthogonal principal axes:

$$\begin{pmatrix} \delta_{XX}^{PAS} & 0 & 0 \\ 0 & \delta_{YY}^{PAS} & 0 \\ 0 & 0 & \delta_{ZZ}^{PAS} \end{pmatrix}$$

Chemical shifts are frequently reported in terms of the isotropic shift $\delta_{iso} = \frac{\delta_{XX}^{PAS} + \delta_{YY}^{PAS} + \delta_{ZZ}^{PAS}}{3}$, the chemical shift anisotropy (CSA), and the biaxiality η. By convention, the PAS is selected such that $| \delta_{ZZ}^{PAS} - \delta_{iso} | \geqslant | \delta_{XX}^{PAS} - \delta_{iso} | \geqslant | \delta_{YY}^{PAS} - \delta_{iso} |$. The CSA is defined as the maximal deviation from the isotropic chemical shift $\delta_{aniso} = \delta_{ZZ}^{PAS} - \delta_{iso}$. The biaxiality, $\eta = \frac{\delta_{YY}^{PAS} - \delta_{XX}^{PAS}}{\delta_{aniso}}$, is a measure of the asymmetry between the two axes not captured by the CSA.

Nearby nuclear dipoles interact with each other in two relevant ways: direct dipole–dipole coupling through space, and coupling through interactions with shared electron bonds (J-coupling). Direct dipole–dipole coupling takes the following form [18]:

$$b_{jk}\left(3\cos^2\left(\theta_{jk}\right) - 1\right)\hat{I}_{jz}\hat{I}_{kz}$$

Where b_{jk} is the dipole–dipole coupling constant between spins j and k, θ_{jk} is the polar angle of the displacement between the two spins, and $\hat{I}_{jz}$ $\hat{I}_{kz}$ are the spin angular momentum operators of spins j and k.

J-coupling generally takes the form of a coupling through a rank two tensor J:

$$2\pi \boldsymbol{\hat{I}}_j \boldsymbol{J}_{jk} \boldsymbol{\hat{I}}_k$$

Quadrupolar coupling is in general more complicated. The nuclear quadrupolar moment Q is coupled to the electric field gradient (EFG), which is represent by a second rank, symmetric, traceless tensor

$$\begin{pmatrix} V_{xx} & V_{xy} & V_{xz} \\ V_{yx} & V_{yy} & V_{yz} \\ V_{zx} & V_{zy} & V_{zz} \end{pmatrix}$$

Like the chemical shift tensor, an orthogonal PAS can also be found for the EFG, and principal axes are chosen such that $| V_{ZZ}^{PAS} | \geqslant | V_{YY}^{PAS} | \geqslant | V_{XX}^{PAS} |$, a slightly different ordering from that of the chemical shifts. Since the EFG is traceless, $V_{ZZ}^{PAS} + V_{YY}^{PAS} + V_{XX}^{PAS} = 0$. The largest principal component is called $eq = V_{ZZ}^{PAS}$, and the biaxiality parameter is defined $\eta_Q = \frac{V_{XX}^{PAS} - V_{YY}^{PAS}}{V_{ZZ}^{PAS}}$ and is always a value between 0 and 1. The first order correction can be written in terms of a uniaxial and biaxial term

$$\frac{e^2qQ}{4I(I+1)\hbar}\left((3\cos^2(\theta) - 1) + \eta_Q \sin^2(\theta)\cos(2\varphi)\right)$$

Similarly to direct dipole–dipole coupling, the uniaxial term depends on $(3\cos^2(\theta) - 1)$. Uniaxial field gradients are common under highly symmetric coordinations (for example, octahedra with Jahn– Teller distortions). When analyzing quadrupolar coupling, first order perturbative corrections to the Zeeman eigenstates are frequently insufficient for nuclei with large quadrupole moments, notably ^{23}Na and heavier nuclei such as ^{59}Co and ^{55}Mn. Other nuclei found in battery materials including the halogens $^{35,37}Cl$ and ^{127}I exhibit very large quadrupolar couplings, making lines too broad to be observed except under highly symmetric conditions.

The appearance of $(3\cos^2(\theta) - 1)$ in both the direct dipolar and quadrupolar coupling terms suggests a way to reduce the broadening of spectra due to these terms. The magic angle, $\theta_M \approx 54.7°$, is the angle such that $3\cos^2(\theta_M) - 1 = 0$ [19]. Magic angle spinning NMR (MAS-NMR) consists of rotating the sample about an axis at the magic angle at speeds on the order of the strength of the coupling. The motional averaging removes these terms, provided the rotation is on the order of the coupling strengths. Depending on the rotors size, where realistic maximum spinning speed is inversely proportional to the rotor dimension, current state of the art commercially available NMR probes achieve faster than 100 kHz MAS, resulting in unprecedented sensitivity and resolution similar to that of liquid-state NMR [20]. This advanced ultra-fast MAS technology is often used to understand the complex molecular system in biomacromolecules or inorganic materials with large heterogeneities, which demands high-resolution [21].

7.2.3 NMR relaxometry

Relaxometry studies the relaxation of nuclear spins to equilibrium after being perturbed by radio pulses. Phenomenologically, spins relax according to two primary time constants: T_1, the spin-lattice or longitudinal relaxation constant, which measures the rate at which the magnetization relaxes towards thermal equilibrium along the z-axis, and T_2, the spin–spin or transverse relaxation constant, which measures the decay of polarization in the xy observation plane [22]. They are also measured by their respective relaxation rates, $R_1 = \frac{1}{T_1}$ and $R_2 = \frac{1}{T_2}$. A number of pulse sequences exist to measure T_1 and T_2, most commonly T_1 inversion recovery (T_1IR) and spin-echo pulse sequences.

Relaxation plays a role in all measurements by placing inherent limitations on experiments. Faster relaxation may result in broader spectra (through lifetime broadening), lower resolution and difficulty distinguishing peaks. They also place limitations on the duration of pulse sequences, which must be kept short relative to relaxation. Samples with large relaxation rates (frequently due to strong quadrupolar coupling) can relax faster than the receiver delay in a single pulse experiment and the digitization rate of the spectrometer, making spectral measurements impossible.

But relaxation measurements also contain valuable information about the dynamics of the species being observed. Spontaneous photon emission from typical proton NMR energy levels occur on the order of 10^{21} seconds, meaning nuclear relaxation occurs primarily through coupling to their surroundings [14]. Relaxometry measurements can provide information about different transport processes and molecular relaxation in the nucleus' surroundings. Spin relaxation theory is generally complex, depending both on the spin quantum number and the type of coupling dominant in relaxation, and requires some *a priori* knowledge of dynamic processes underlying the system. In the framework of Bloch/Wangsness/Redfield theory [23], the effect on transitions due to stochastic motions which couple the spin to its environment are related to the autocorrelations of spherical harmonic functions, and their related spectral densities, which can be used to determine transition rates between nuclear spin eigenstates, thereby relating relaxation rates with relevant molecular motion. Isotropic liquid motion is well represented by Bloembergen–Pound–Purcell theory, which is still very much in use since first formulated in 1948 [24].

Collecting variable temperature data can provide invaluable information about the underlying molecular relaxation processes in both liquid and solid electrolytes. Correlation times τ_c are typically expected to follow Arrhenius or Vogel–Fulcher–Tammann (VFT) behavior in liquids and polymers, and far from the T_1 minimum, T_1 scales with the correlation time allowing for measurements of activation energies of molecular processes. At the T_1 minimum itself, $\omega\tau_c = 1$, allowing for measurement of the underlying relaxation process.

Other important relaxation terms are $T_{1\rho}$ and T_2*, which basically are spin-lattice relaxation time during the spin–lock and spin–spin relaxation affected by magnetic inhomogeneity, respectively. Any spatially varying sources of magnetic field resulting in field-inhomogeneity leads to a similar dephasing phenomenon. It should be noted that $T_2^* < T_2$. The signal intensity measured as a function of spin–lock duration can be used to measure the $T_{1\rho}$,

$$M = M_\infty + (M_0 - M_\infty)e^{-\tau/T1\rho}$$

Where, M_0 is the initial magnetization and M_∞ is the magnetization when the spin system and the lattice reach a quasi-equilibrium position during the spin–lock. $T_{1\rho}$ is known as the relaxation time in the rotating frame, because the spin–lock sequence provides an effective field much lower than the laboratory static field B_0. In this way, $T_{1\rho}$ provides information on processes with a significantly lower frequency than the NMR Larmor frequency.

Fast field cycling relaxometry (FFCR) also provides information about the underlying relaxation processes, but across an even broader length of time scales by measuring relaxation at lower magnetic fields. A typical field cycling experiment has three parts—after allowing the spins to equilibrate and applying a 180° pulse at a higher field, the *B*-field is rapidly decreased to a lower relaxation field, and the spins are allowed to relax. After relaxing, the *B*-field is increased to an observation field and a 90° observation pulse is applied, and the resulting signal intensity is measured. Very generally, this gives access to dynamics measurements across a broader range of timescales between the millisecond dynamics accessible via diffusometry and nanosecond range accessible via conventional relaxometry, by measuring the spectral density function $J(\omega)$ by varying ω. Field cycling is limited by poor resolution, and most commonly applied to nuclei with large gyromagnetic ratios like protons—however, recent studies have found success measuring ^{7}Li relaxation [25]. For a more detailed discussion of underlying theories and techniques, see Kimmich and Anoardo [26].

7.2.4 NMR diffusometry

Perhaps the most useful information NMR can provide about electrolytes is the measurement of distinct diffusion coefficients for cation and anion species. While ionic conductivities can be measured using impedance spectroscopy, these measurements lack specific information about the current contributions of individual cation and anion species which can be provided by NMR diffusometry. The benefit of this approach is that one does not only learn about the conductivity of the electrolyte, but can specifically determine the contribution of electrochemically active cation to this current, and can measure a cation transference number t^+. Further, comparisons of diffusion coefficient measurements with impedance spectroscopy can provide information about the correlated motion of cation and anion diffusion, providing information about the solvation of the salt [27].

Pulsed field gradient NMR (PFG-NMR) works by applying strong pulsed *B*-field gradients G_z for a time δ to the sample after a 90° pulse, causing nuclear spins to accumulate phase according to their position along the z-axis, causing spins to accumulate phase according to their position. After being allowed to diffuse for some diffusion time Δ, a second mirrored gradient pulse is applied to undo this phase accumulation. The diffusion of the nuclei results in a difference in the phases accumulated in the first and second pulse, and the resulting echo is attenuated according to the Stejskal–Tanner equation [28–30]:

$$S = S_0 e^{-\gamma^2 G_z^2 D \delta^2 (\Delta - \delta/3)}$$

where D is the self-diffusion coefficient of the nucleus under observation. By varying the gradient strength, the logarithm of the signal intensity varies linearly with $G_z^{\ 2}$, allowing the diffusion coefficient of the nucleus to be fit. Diffusion can be asymmetric, and can have multiple diffusion components for a single line.

Access to the self-diffusion coefficients of individual cation and anion species provides additional invaluable data on electrolytes, because the information provided about individual cationic and anionic motion across millisecond-to-second timescales, and ~ one to several μm length scales, gives better understanding about their individual contributions to the current in the electrolyte. The cationic transference number in the dilute limit may be calculated from

$$t^{+} = \frac{D_{+}}{D_{+} + D_{-}}$$

The cationic transference numbers are critical to the development of new electrolytes—anion contributions to ionic currents result in anion concentration gradients, limiting the rate of charge and discharge and creating a concentration overpotential which limits the operating voltage of the cell. Ionic conductivity Λ can be calculated from the NMR self-diffusion coefficients using the Nernst–Einstein equation [27]

$$\Lambda_{\mathrm{NMR}} = \frac{z^{2}F^{2}}{RT}(D_{+} + D_{-})$$

An objection may be raised about the above methodology that while self-diffusion coefficients measure ion diffusion in a homogeneous chemical environment, the transference number and Nernst–Einstein equations are calculated with mutual diffusion coefficients [31] which describe diffusion against a concentration gradient, and that the two are equivalent only in the limit of infinite dilution. However, the self-diffusion coefficient should accurately characterize transport across tens of milliseconds, and micron length scales, for most electrolytes. These length and time scales ought to accurately characterize the transport of the ion in the bulk solvent—deviations from this transport at longer length and time scales (for example, due to depletion of unoccupied lattice sites across a concentration gradient in solids, or due to cation–anion interactions and correlated motion) are expected to be small.

The Nernst–Einstein relation further assumes total ion dissociation. Thereby, a measure of ion dissociation can be calculated from the NMR conductivity calculated from the Nernst–Einstein equation and comparing it with direct DC conductivity measurements using impedance spectroscopy or similar techniques

$$\%\ \mathrm{Dissociation} = \frac{\Lambda_{\mathrm{Impedance}}}{\Lambda_{\mathrm{NMR}}}$$

All PFG-NMR rely on the measurement of an echo, in which an FID dephases due to magnetic field inhomogeneities, and then rephases after the application of a 180° pulse flips the spins exactly cancelling the effects. These sequences fall broadly into two categories: Spin echo and stimulated echo. Spin echo experiments are the simplest experiment, in which after applying a field gradient pulse, a time τ is waited for before applying a π pulse and waiting an additional τ and applying the same pulsed field gradient, which this time acts in the opposite direction, for a total diffusion time $\Delta = 2\tau$. However, frequently in solid materials, T_2 relaxation times are too short, and a stimulated echo is used instead, in which after pulsing the field gradient, a $\pi/2$ pulse is applied to store the magnetization along the z-axis so that it relaxes according to T_1, not T_2, relaxation. The maximum diffusion time is then limited by the longer T_1 relaxation time, and after diffusing for a time delta an additional $\pi/2$ pulse is applied bringing the stored magnetization into the xy plane, at which time a second identical pulsed field gradient is applied just as in the spin echo experiment. However, in addition to storing the magnetization, these additional $\pi/2$ pulses create additional coherences, which can be destroyed by the application of small spoiler gradient pulses.

Two experimental considerations which must be taken into account in order to measure accurate self-diffusion coefficients are the presence of eddy currents after gradient application, and the effects of coherent flow, particularly in low viscosity samples which contain convective currents. Eddy currents in conducting elements in the probe result from the rise in currents in the magnetic field gradient coils during PFG-NMR. There are a number of techniques to mitigate these effects, both using specialized hardware and special pulse techniques. The use of shielded gradient coils is now standard, and gradient pre-emphasis techniques make it possible to correct eddy current effects on gradient shape by changing the shape of the input pulse. More commonly, bipolar pulse pairs are used, in which gradients of opposite polarities are applied in pulse pairs, causing their eddy current effects to cancel. In the event that such methods fail, an eddy current delay may be utilized, applying a 90° pulse at the end of the sequence to store the magnetization along the z-axis for a longitudinal eddy current delay time. These techniques are all capable of compensating for eddy current effects, but careful attention must be paid while running and interpreting PFG-NMR experiments—eddy currents can produce phase effects and artificial signal attenuation in PFG-NMR spectra, resulting in measured diffusion coefficients that are artificially fast.

Pulse sequences also exist to control for coherent flow in samples. The issue of coherent flow often arises when measuring diffusion in low viscosity liquid electrolytes. Temperature in NMR experiments is typically controlled by passing heated or cooled gas across the NMR tube, and in practice this method gives rise to thermal gradients and convection currents forming in the sample. Convection currents, and coherent flow, contribute to the measured attenuation of the signal and result in an increased apparent diffusion coefficient.

While this problem can be controlled for by using smaller sample sizes and more even temperature controls, there are also pulse sequences available which measure only non-coherent transport of nuclei. The contribution of coherent flow to the apparent diffusion coefficient can be removed through the use of a double stimulated echo DSTE sequence [32].

Diffusion measurements via NMR can provide detailed information about transport across different length scales, and in particular can provide information about the sample, the structure and porosity of porous polymer electrolytes. In the case of restricted diffusion, measured apparent diffusion coefficients will depend on diffusion time Δ. The free diffusion coefficient in the limit as $\Delta \rightarrow 0$ is generally much higher than the restricted diffusion coefficient as $\Delta \rightarrow \infty$. The ratio of the two is related to the tortuosity of the pores, providing a simple method to measure tortuosity. Other models exist for other restricted flow geometries, and generally measuring apparent diffusion coefficients at different diffusion times delta can provide information about transport within the 0.01 micron–100 micron range.

7.2.5 Other NMR methods useful in material characterization

Solid state NMR is a powerful tool to characterize materials with different range of molecular order, from amorphous to crystalline, for instance. It should be noted that some of the NMR active nuclei such as ^{1}H, ^{31}P etc are highly abundant and the corresponding spectra are much easier to obtain. However, low natural abundance and low gyromagnetic ratio, for example in the case of ^{13}C or ^{29}Si, makes it difficult for adequate signal detection. Large anisotropic interactions further add to the poor sensitivity. Nevertheless, many advanced NMR technical and methodological improvements make it possible to obtain the high-resolution NMR spectra of low sensitive nuclei [33]. As explained in detail earlier, one of the most often used techniques is MAS, which averages the dipolar and first-order quadrupole couplings, and CSA.

Another technique is cross polarization MAS (CPMAS) [34] where the magnetization is transferred from highly abundant nuclei to less sensitive nuclei through dipolar couplings, for instance from ^{1}H to ^{13}C. This experiment not only enhances the signal of low gamma nuclei by a factor on the order of gyromagnetic ratio ($\eta \propto \frac{\gamma_I}{\gamma_S}$) but also reduces the experimental time where relaxation of magnetization depends on the high gamma nuclei. The key to achieve signal enhancement in CPMAS experiment is to fulfill the so called Hartman–Hahn condition [35].

$$\gamma_I B_{1I} = \gamma_S B_{1S}$$

Where, B_{1I} and B_{1S} are the applied RF field strength on each channel. Basically, the precession frequencies of I and S nuclear spins are matched by adjusting the corresponding applied RF field strength. The resonant transfer of energy from I to S spin is succeeded through strong heteronuclear dipolar couplings when they are in close proximity. There are several varieties of CP-techniques in existence in order to fulfill the Hartman–Hahn condition such as ramped-CP [36] where the amplitude of the contact pulse of either I or S spin is steadily varied and adiabatic-CP [37] that uses adiabatic passage modulations of the RF profiles.

The sensitivity and resolution of the NMR spectra is further achieved by heteronuclear spin decoupling, in most of the cases ^{1}H, during data acquisition of X nuclei (X = ^{13}C, ^{29}Si, ^{7}Li etc). The broad spectral features in solid state NMR spectrum are often dominated by heteronuclear dipolar couplings. Even high-speed MAS is unable to improve the resolution if the nuclear spins X coupled through strong dipolar couplings (^{1}H–X), often seen in protonated solid materials. Such interactions can be greatly reduced by applying a train of rf pulses on the ^{1}H channel. This averages out the heteronuclear dipolar coupling associated with spins coupled to X nuclei through space. Hence the resultant ^{1}H magnetic moment averages to zero, which improves the resolution of X NMR spectral features. Most widely used ^{1}H decoupling schemes include continuous wave (CW), TPPM, SPINAL64, etc [38, 39].

In the complex molecular system, similarity in the local environments and the corresponding shifts may overlap heavily in the 1D spectra make it difficult to interpret the individual contributions. This can be addressed by acquiring NMR signals in the second dimension in two-dimensional (2D) NMR experiments [40]. 2D NMR provides detailed structural information on coupled homonuclear or heteronuclear spins through dipolar or *J*-couplings. Basically, 2D NMR experiments consist of four steps: preparation, evolution (t_1), mixing and detection (t_2) periods. Optimized RF pulses are applied in the preparation period to generate the magnetization coherences, which are modulated under the relevant Hamiltonian during evolution and detected in the indirect or second dimension. In the mixing time, the nuclear spins can be manipulated by applying carefully designed pulses during which the magnetization either evolves under a different Hamiltonian or is transferred to different nuclear spins depending on their type of spin–spin interactions. The signal obtained with the manipulation of nuclear spins through complex pulse sequence is collected in the detection period. Overall, the signal output in the 2D NMR spectrum is the result of Fourier transformation of FIDs in both direct (F_2) and indirect (F_1) dimension.

$$S(t_1,\ t_2) \rightarrow S(\omega_1,\ \omega_2)$$

The manipulation of pulses that leads to the coherence transfer can be done through-bond or through-space homo- or heteronuclear *J*-couplings or dipolar couplings. A specific type of 2D NMR experiments can be designed to discover structural and dynamical information of the molecular system. Some of the useful solid state NMR experiments to study battery materials are dipolar heteronuclear correlation (HETCOR), exchange correlation spectroscopy (EXSY), homonuclear dipolar correlation, and multiple-quantum MAS (MQMAS) NMR experiments.

While varieties of multidimensional NMR experiments provide important structural information, achieving higher sensitivity has still been an issue in some important battery science situations, even with ultrafast MAS or other advanced NMR technology. An important example is that detailed structural and chemical information about the SEI in lithium ion batteries (or CEI on the cathode side) is limited by the low number of nuclei in this thin region. In this context, dynamic nuclear polarization (DNP) NMR spectroscopy has been a breakthrough technique boosting the signal intensity of low sensitive nuclei by several orders of magnitude where spin polarization transferred from electrons to the surrounding nuclear spins by means of microwave irradiation at electron paramagnetic resonance (EPR) transition. This technique was proposed initially by Albert E Overhauser [41] and experimentally confirmed by Carver and Slichter [42]. Theoretical enhancement factor (ε) is the ratio between gyromagnetic ratios of electron spin (S) and nuclear spin (I),

$$\varepsilon_{max} = \frac{P_S}{P_I}\left(\approx\frac{\gamma_S}{\gamma_I}\text{ for }\gamma_S\hbar B_0 \ll k_B T\right)$$

This results in much larger non-thermal population differences than those determined by Boltzmann factor and hence a dramatic NMR signal enhancement by a factor of 660 in the case of protons [41]. This eliminates the need for extensive signal averaging, achieving sufficiently high-signal-to-noise ratio with novel NMR possibilities. Over the past few decades, a large amount of research has been done in materials science and structural biology by utilizing DNP NMR technique at high magnetic fields under MAS conditions [43, 44]. DNP NMR requires specialized instrumentation such as, source of microwave irradiation, DNP MAS NMR probe and cryogenic gas supply and polarizing agent with paramagnetic centers to be incorporated into the sample of interest. Several review articles about MAS-DNP including dissolution DNP NMR technique can be found in the literature [43–45]. MAS-DNP NMR has been a very efficient approach in solid state battery research to boost the NMR sensitivity of low gamma nuclei, low natural abundance and without isotopic enrichment [46–48].

7.3 NMR of battery materials: examples

We now turn to some examples from the literature of specific materials investigations using the NMR tools described above. Again, this is not an exhaustive review, and we have chosen a limited set of studies that we believe best illustrate the power of the methods.

7.3.1 Electrode materials

7.3.1.1 Anode materials

NMR studies have proved useful both in further characterizing well known Li-ion anode materials and in probing potential alternatives, providing detailed structural information at different stages of material processing, and characterizing materials at different charge states both before and after cycling. While the graphite anode has been the standard anode since the commercialization of the Li-ion battery, its specific energy capacity is limited by a maximum reversible lithium content being LiC_6, motivating a search for alternative carbonaceous materials with higher Li:C intercalation ratios. Hayes, Wullen, and Eckert *et al* [49] conducted CPMAS studies to investigate the synthesis and lithium deposition on poly(acrylonitrile/divinylbenzene) (PAN/DVB) and poly(methacrylonitrile/divinylbenzene) (PAN/DVB) copolymers to assess their stability at different pyrolyzation times and temperatures.

CPMAS was utilized to examine both ^{13}C and ^{15}N peaks both with cross polarization to enhance sites near protons and with selective dead time before signal observation without proton decoupling so that ^{13}C and ^{15}N sites would be selectively reduced due to dipolar interaction with protons. Besides the well-known signature of the Knight-shifted lithium resonance in LiC_6 [50], lithium motion in LiC_6 can be characterized via time-domain ^7Li NMR [51].

A wide range of anode materials have been studied to improve the capacity of the Li-ion battery [52]. Because of its high specific capacity, about a factor of 10 as compared to graphite, silicon-based anodes are being developed for Li-battery applications. Silicon monoxide (a-SiO) is a promising anode material which has been considered superior to the silicon (Si) metal anode with its exceptional 1500 mAh g^{-1} specific capacity. Freytag and coworkers [53] reported *in situ* ^7Li and *ex situ* ^{29}Si NMR a comparison of the lithiation behavior of a-SiO and Si metal anode materials.

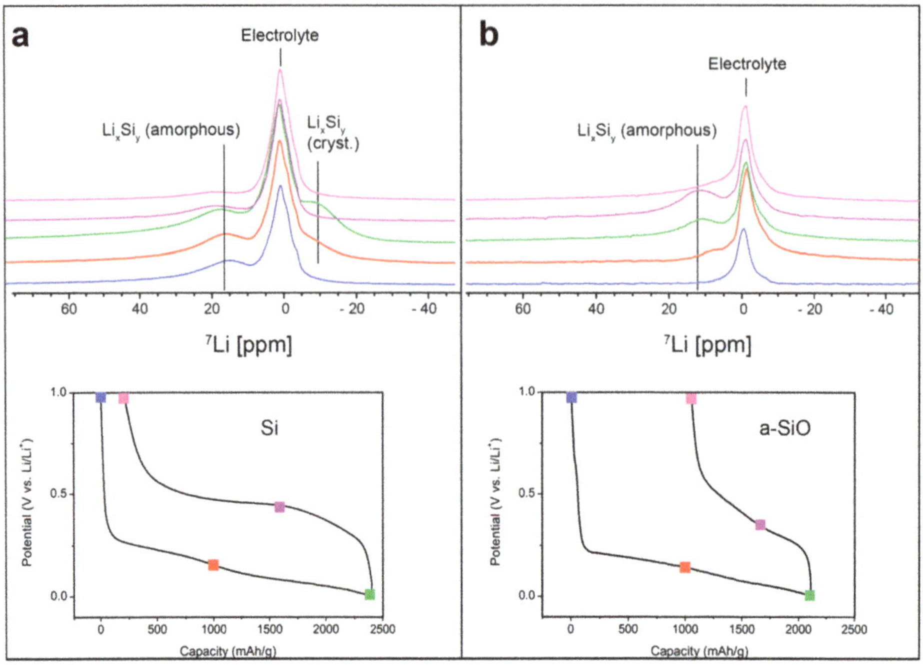

Figure 7.2. A change in ^{7}Li chemical shift during charging/discharging process was compared between (a) cellulose-based silicon metal anodes and (b) a-SiO. Reprinted with permission from [53]. Copyright (2019) American Chemical Society.

In situ ^{7}Li NMR spectra of both a-SiO and Si metal anode collected during the first cycle of charging/discharging process are shown in figure 7.2. A slight increase in the ^{7}Li chemical shift upon lithiation substantiate the similarity in the initial cycling behavior. Contrarily, the deviation of ^{7}Li peaks show opposite trend at the end of the lithiation. For Si metal anode, ^{7}Li peak centered at 16 ppm was assigned to amorphous lithium silicide, which deviates ±1 ppm up until the formation of crystalline phase. However, in the case of a-SiO, ^{7}Li peak deviates in the range of 4–11 ppm upon lithiation. The chemical shift difference between two anode materials, Si metal and a-SiO, was attributed to the apparent difference in the distribution of Li atoms and their electronic environments in the amorphous phase. An additional peak at ≈−10 ppm observed in the case of Si metal was assigned to the crystalline $Li_{15}Si_4$ phase, which is absent in the a-SiO ^{7}Li spectrum. Because the amorphous to crystalline transition is correlated with capacity fade, the absence of a crystalline ^{7}Li signal in the case of a-SiO highlights its potential advantage as an anode material.

To support the insights obtained from *in situ* ^{7}Li NMR, the authors have also reported the *ex situ* ^{29}Si MAS NMR collected for a-SiO at different points during charging/discharging process [53]. Two ^{29}Si features observed for pristine a-SiO at −107 and −63 ppm were assigned to the SiO_2 and metallic silicon, respectively. It is found that the SiO_2 peak disappeared, and Si peak intensity reduced significantly upon lithiation (300 mAh g^{-1}). At 800 mAh g^{-1}, a new narrow peak centered at −70 ppm and a broad peak centered at −117 ppm were observed. The intensity of −117 peak increased and shifted towards to −30 ppm as a function of lithiation. The opposite trend with decrease of intensity and peak shift towards lower ppm was noticed upon delithiation. These results revealed the existence of at least two ^{29}Si domains, where the narrow peak at −70 ppm was referred to the Q_0 sites of Li_4SiO_4 irreversible lithiation product. Considering the width and chemical shift trend with reference to the previous studies, authors designated the broad peak as a collection of amorphous lithium silicides, Li_xSi_y.

NMR relaxation measurements, on the other hand, can provide information on translational diffusion in some solid materials with complex structures, in cases where diffusion coefficients cannot be measured directly by PFG NMR. For example, Kuhn and coworkers [54] discovered the fastest motions of Li ions in crystalline $Li_{12}Si_7$ and at low temperature with the help of ^7Li NMR spin-lattice relaxation (SLR) measurements. The Li jump rate was found to be ~ 10^5 s^{-1} at 150 K reaching 10^9 s^{-1} at 425 K. In addition, three distinct 1D diffusion processes were established based on the variable-temperature (VT) and -frequency NMR measurements. Thus, information on the structure of $Li_{12}Si_7$, in particular, its dimensionality and that of the Li^+ jump process could be obtained by variable temperature and frequency SLR measurements.

The ^7Li spectra of $Li_{12}Si_7$ show a broad feature with complex line shape at 125 K becomes narrower as the temperature increases, figure 7.3(a). The broad line shape was attributed to the large dipole–dipole interactions, chemical shift anisotropies and interaction of ^7Li quadrupole moment with electric field gradient at different Li sites. Motional narrowing of the lineshape begins even at very low temperature, 130 K (figure 7.3(b)). The inflection point $\Delta\omega(T)$ obtained from the $\Delta\omega$ versus T plot estimates the Li jump rate (τ^{-1}), which is on the order of 10^5 s^{-1} at 150 K. A clear distinction of quadrupolar powder pattern, features consisting of central transition and quadrupolar satellite intensities, can be seen in the ^7Li spectra recorded at 188, 258, and 723 K. The absence of gradual change of δ_Q with increasing temperature was ascribed to stepwise averaging of electric quadrupolar interactions because of Li jump process between different crystallographic sites in $Li_{12}Si_7$. This explains the existence of at least three different jump processes where each one leads to a specific δ_Q.

The ^7Li SLR measurements (SLR rates in the Laboratory—T_1^{-1} and rotating-$T_{1\rho}^{-1}$ frame of references), as shown in figure 7.4, provide further valuable insights into the Li ion dynamics. The SLR rates were collected at different resonance and locking frequencies. The SLR rates measured with respect to temperature at different resonant frequencies give an absolute jump rate $\tau^{-1}(T_{max})$. The global maxima of the two $T_1^{-1}(1/T)$ curves and the low-temperature $T_{1\rho}^{-1}(1/T)$ curve reflect the same ultrafast diffusion phenomenon. The maximum conditions such as $\omega_0 \approx \tau^{-1}$ and $2\omega_1 \approx \tau^{-1}$ determine the Li jump rates (τ^{-1}), where τ_0^{-1} (pre-exponential factor) and found to be in the order of 10^{11} s^{-1}. Further detailed analysis of SLR rates show the existence of quasi-1D or at least strongly anisotropic diffusion process. On the other hand, the well-separated ^7Li SLR rate peaks at higher temperatures confirms the two additional jump processes in $Li_{12}Si_7$. Overall,

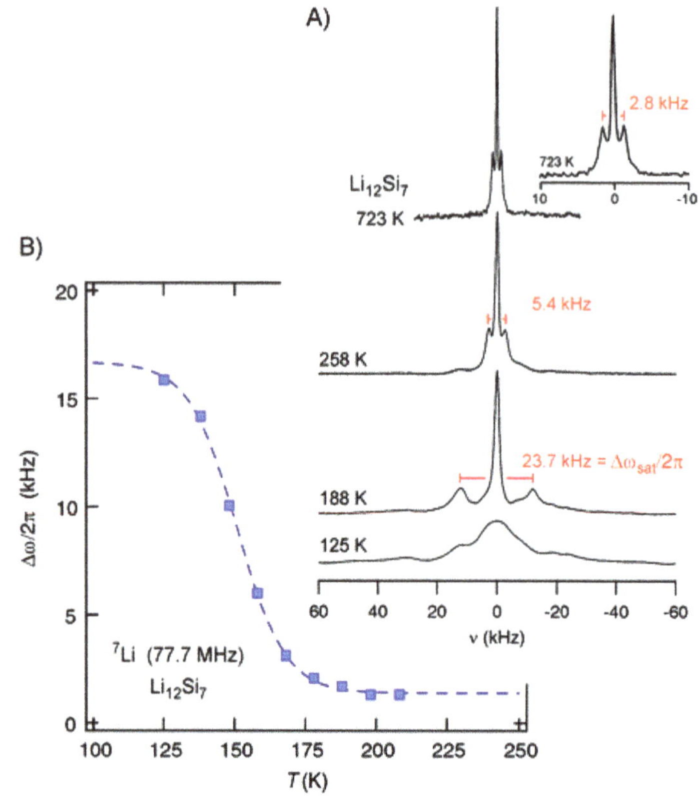

Figure 7.3. (A) ^{7}Li NMR spectra of $Li_{12}Si_7$ collected at different temperatures. (B) Full width at half maximum (FWHM) of central ^{7}Li transition plotted against the temperature. Reprinted with permission from [54]. Copyright (2011) American Chemical Society.

the reported diffusion-induced NMR SLR measurements are consistent with the three distinct Li jump processes.

Some of the low sensitive NMR nuclei, for instance ^{17}O, can provide valuable structural information in the study of anode materials, but measurements in the solid state cannot be performed at natural abundance, which is only 0.037%. A recent study by Wolf *et al* used MAS-DNP NMR technique to enhance the ^{17}O signal without ^{17}O isotopic enrichment to study $Li_4Ti_5O_{12}$ (LTO) and $Li_2ZnTi_3O_8$ anodes and other materials doped with Mn(II) ions [47].

7.3.1.2 Cathode materials

Transitional metal-based cathode materials containing paramagnetic iron, nickel, or manganese, present significant challenges to acquisition of undistorted broad ^{7}Li NMR spectra. Several strategies to reduce the paramagnetic broadening are: (i) observe the lower γ ^{6}Li nucleus, which sometimes requires isotopic enrichment; (ii) run the measurement at lower magnetic field, which proportionally reduces the extent of the paramagnetic broadening. A few examples discussed below highlight

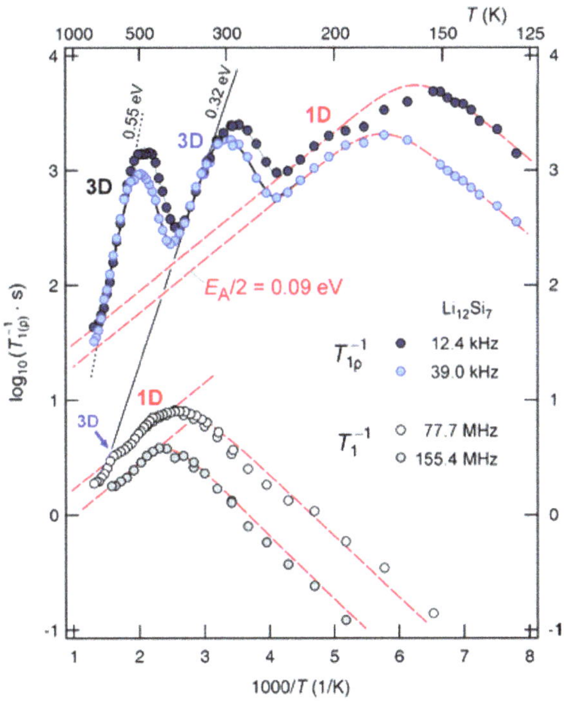

Figure 7.4. Arrhenius plots ^{7}Li T_1 relaxation rates measured in the Laboratory- (T_1^{-1}) and rotating frame ($T_{1\rho}^{-1}$). Reprinted with permission from [54]. Copyright (2011) American Chemical Society.

the importance of NMR technique elucidating the structural changes caused during cycling and the corresponding dynamics at NMR time scale. Detailed review articles focusing on the use of NMR technique in cathode materials can be found in the literature [17, 55].

$LiCoO_2$ has been used as a cathode material in commercially available batteries for decades, first introduced by SONY Energytec in 1990. Goodenough and coworkers showed ability of $LiCoO_2$ to reversibly deintercalate lithium [56]. The NMR study of deintercalated Li_xCoO_2 was initially reported by Greenbaum and coworkers [57]. The ^{7}Li feature obtained for Li_xCoO_2 ($0.4 \leqslant x < 1$) was interpreted as the localized electron spins. In another study, Delmas and coworkers [58] performed a detailed study of Li_xCoO_2 to further understand the process involved in the lithium deintercalation. A drastic change in the electronic interactions experienced by lithium was clearly noticed in the NMR experiment. In the quest to find alternative low-cost cathode materials, lithium manganese oxides $Li_xMn_2O_4$ was found to be the most promising material [59]. Furthermore, layered $LiMnO_2$ [60]; structurally analogous to $LiCoO_2$, has been extensively investigated.

Grey and coworkers utilized solid state NMR as a main characterization tool to unravel the structural changes of lithium manganese cathode materials prepared under different conditions [17, 61]. ^{6}Li NMR was chosen as the probe nucleus because of the limitations associated with broad ^{7}Li ($I = 3/2$) spectral features;

attributable primarily to interactions with unpaired electron spins in this paramagnetic material. Homonuclear dipolar coupling and quadrupolar interactions also contribute to the broadening in ^{7}Li spectra. Contrarily, ^{6}Li (I = 1) with its smaller quadrupolar interactions and weaker magnetic couplings gives better resolution. The ^{6}Li MAS NMR spectra of studied $LiMn_2O_4$ prepared at different temperatures are shown in the figure 7.5. The existence of spinel and the absence of impurities were confirmed by XRD. An intense ^{6}Li peak observed for LiMn850 sample was assigned to the Li in normal position (8a) of the spinel structure. Additional ^{6}Li peaks observed for samples prepared at lower temperature were attributed to the existence of multiple local environments, even though the spinel structure contains one crystallographic site for lithium. It was also found that the line width of the main ^{6}Li peak increases and its intensity decreases as the synthesis temperature decreases. In addition, the MAS (40 kHz) ^{7}Li spectrum showed a weak signal at 2300 ppm for Mn550 and Mn600 [55]. The spinning side band pattern of

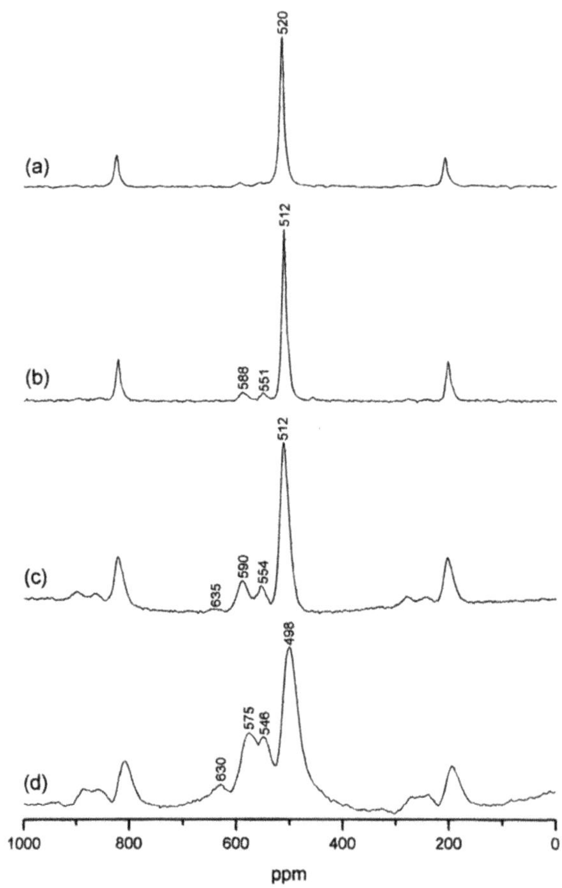

Figure 7.5. ^{6}Li MAS NMR spectra of $LiMn_2O_4$ as a function of synthesized sample temperatures (LiMn850, LiMn650, LiMn600, LiMn550). Reprinted with permission from [55]. Copyright (2003) American Chemical Society.

this peak was correlated to that of the spinel $Li_{0.5}Zn_{0.5}[Mn_{0.5}Li_{0.5}]O_4$, confirming the Li substitution in the octahedral site (16d). This indicates that the samples prepared at lower temperature contain excess lithium. The additional spectral features of nominally stoichiometric material $LiMn_2O_4$ were assigned to the tetrahedral Li sites that are close to the Mn^{4+}. Further detailed and more recent [15] NMR analysis of cathode materials can be found in these references [17, 55, 61].

7.3.2 Electrolytes

The electrolyte is the medium that separates anode and cathode and transports the Li^+ ions between the electrodes in Li-ion batteries. In addition to the liquid electrolytes in commercial use, gel electrolytes and solid electrolytes are also under development. Solid electrolytes include inorganic oxides and sulfides, which could be crystalline or amorphous, and solvent-free polymer electrolytes. Polymer electrolytes containing a liquid solvent component are commonly referred to as gels. Here, we discuss several such examples where NMR technique provides important structural and dynamical information.

7.3.2.1 Liquid electrolytes, ionic liquids electrolytes

The most used liquid electrolytes include $LiPF_6$ dissolved in organic solvents consisting of a mixture of ethylene carbonate (EC) and dimethyl carbonate (DMC) or other linear carbonates. There are many limitations associated with the carbonate electrolyte because of thermal degradation at higher temperatures or electrochemical decomposition at higher potentials. NMR plays a key role in elucidating the structure of electrolyte additives and their decomposition products as well as their interactions [62–66]. Recent research has also been focused on improving the safety of liquid electrolytes by adding ionic liquids or by simply mixing Li salt in ionic liquids [67, 68]. Multinuclear NMR spectroscopy has been used in a number of ionic liquids-based electrolytes to probe the structure and dynamics at local level [69–73].

The low natural abundance of some isotopes impedes NMR characterization e. g. ^{17}O, which has a natural abundance of 0.038%. Furthermore, it is a quadrupolar nucleus, resulting in low resolution due to its broad signal. However, accumulating ^{17}O signal at natural abundance is still possible for liquid samples. For instance, Bogle and coworkers were able to identify specific solvation lithium structures in binary carbonate mixtures containing $LiPF_6$ [74]. In particular, ^{17}O NMR spectra of EC, DMC and EC/DMC mixture with Li-salt were reported, and the resultant spectral changes were analyzed to probe the local environments of structural units that are directly responsible for Li^+-solvation. As shown in figure 7.6, the ^{17}O NMR spectra of studied samples, ^{17}O peaks corresponding to the carbonyl oxygens experience upfield shift because of the shielding effect of the lithium ion on the lone pair electrons, while ethereal oxygens experience downfield. The apparent change in chemical shifts as compared to their neat solvent counterpart was attributed to the ion–dipole interactions between Li^+ and solvent molecules. Apparently, Li^+ interacts more strongly with carbonyl oxygens than ethereal

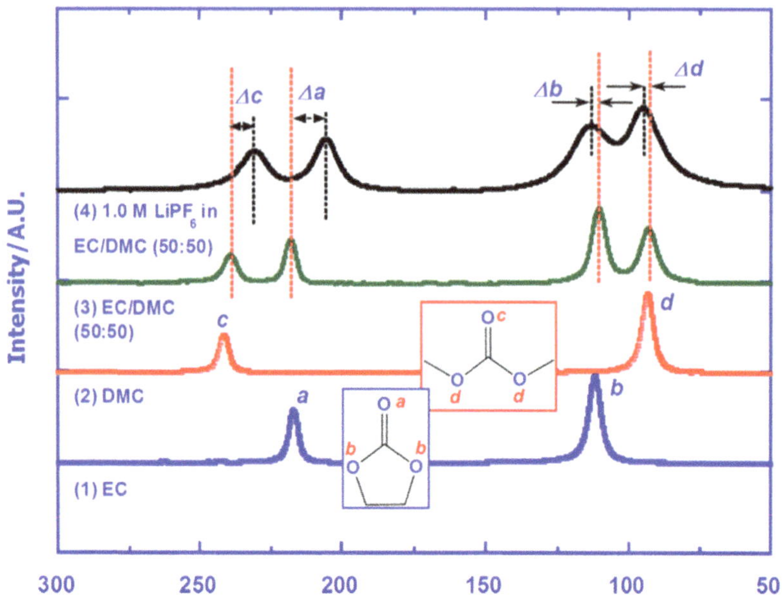

Figure 7.6. ^{17}O NMR spectra of EC, DMC, EC/DMC with and without $LiPF_6$. Reprinted with permission from [74]. Copyright (2013) American Chemical Society.

oxygens as one can observe from the corresponding larger ^{17}O shift differences. In addition, with its higher donicity and dielectric permittivity, EC undergoes a larger shift than DMC with addition of Li-salt, highlighting its status as the stronger Li^+ solvator. Further understanding of primary and secondary structures of Li^+-solvation has been achieved through analyzing the ^{17}O NMR as a function of EC/DMC solvent ratio.

In another study, Wan *et al* [10] investigated the solvation structures of LiFSI in DME electrolytes by ^{17}O NMR in addition to ^6Li, which is also a low abundance nucleus. The observed changes in ^{17}O and ^6Li chemical shifts were supplemented with computational results to understand how the coordination of Li^+ and FSI^- increases while association between Li^+ and DME molecule decreases with increasing LiFSI concentration. Figure 7.7 shows the ^{17}O and ^6Li NMR spectra of LiFSI-DME electrolyte at various concentrations of LiFSI collected at ultrahigh magnetic field of 20 T.

As shown in the figure 7.7 two ^{17}O peaks of the studied samples, at 160 ppm and −24 ppm, were assigned to the sulfonyl oxygen atoms (Os) of FSI- and ethereal oxygen (Oe) atoms of DME. The Oe ^{17}O peak shifts to lower ppm values with increasing LiFSI concentration suggesting increasing coordination of DME molecules with increased Li^+ concentration. Contrarily, the Os ^{17}O peak shifts to higher ppm from 0.05 to 2 M concentration and then shifts to lower ppm above 3 M to saturated concentration. The later trend where Os peak shifts to lower ppm was attributed to the ion–dipole interaction between Li^+ and Os of FSI^- at higher LiFSI concentration. In the case of ^6Li, as shown in figure 7.7, multiple ^6Li peaks were

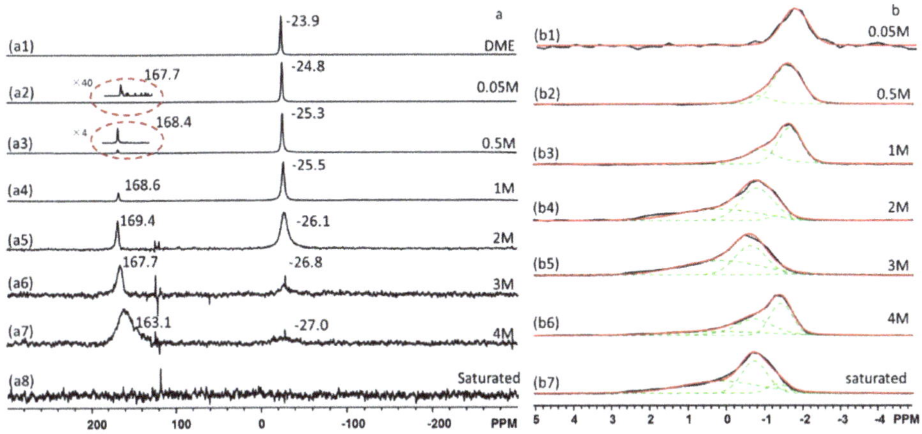

Figure 7.7. Natural abundance ^{17}O and ^{6}Li spectral analysis of LiFSI-DME liquid electrolytes at various concentrations of LiFSI performed to determine the solvation structure of Li ions. Reprinted with permission from [75]. Copyright (2016) American Chemical Society.

observed while changing the concentration of LiFSI. The variation in the ^{6}Li chemical shift, peak line width and the relative ratios of peak intensity was attributed to the corresponding difference in number of Li^+ cluster species or in Li^+ solvation structures. Guided by these ^{17}O and ^{6}Li NMR parameters, authors performed the DFT calculations for a series of possible model solvation structures. It was shown that calculated NMR parameters match the NMR experimental results, thereby predicting the evolution of Li^+ solvation as a function LiFSI concentration.

7.3.2.2 Solid electrolytes

To alleviate limitations associated with liquid electrolytes, for instance leakage and flammability, major research efforts have been focused on solid electrolytes [6]. An interesting class of solid electrolytes include ceramic-based sulfide materials, to which a large amount research has been devoted in order to understand the structure and dynamics of these materials. Gobet *et al* [76] report that the disordered beta phase of Li_3PS_4 lithium-ion conductor shows improved ionic conductivity by three orders of magnitude as compared to the crystalline gamma phase. The thermal treatment of the Li_3PS_4–3THF precursor that produces the nanoporous beta- Li_3PS_4 phase was monitored by multi-nuclear NMR spectroscopy (^{31}P, $^{6,7}Li$, ^{1}H). The dynamics of Li^+ ions were further studied through PFG NMR by measuring the lithium diffusion coefficients. As shown in figure 7.8, a single sharp ^{31}P peak in gamma-Li_3PS_4 sample was attributed to the unique P site of isolated PS_4^{3-} tetrahedra. However, three narrow ^{31}P peaks observed for the Li_3PS_4–3THF sample corresponds to the three distinct chemical environments associated with PS_4^{3-} tetrahedra induced by co-crystallization of THF. In addition to the three ^{31}P peaks, a broad ^{31}P peak corresponding to the metastable amorphous or glassy phase belongs to 45% of total P content emerges after the 70 °C treatment. Furthermore, ^{31}P NMR spectrum of Li_3PS_4–100 °C sample shows a broad Gaussian signal

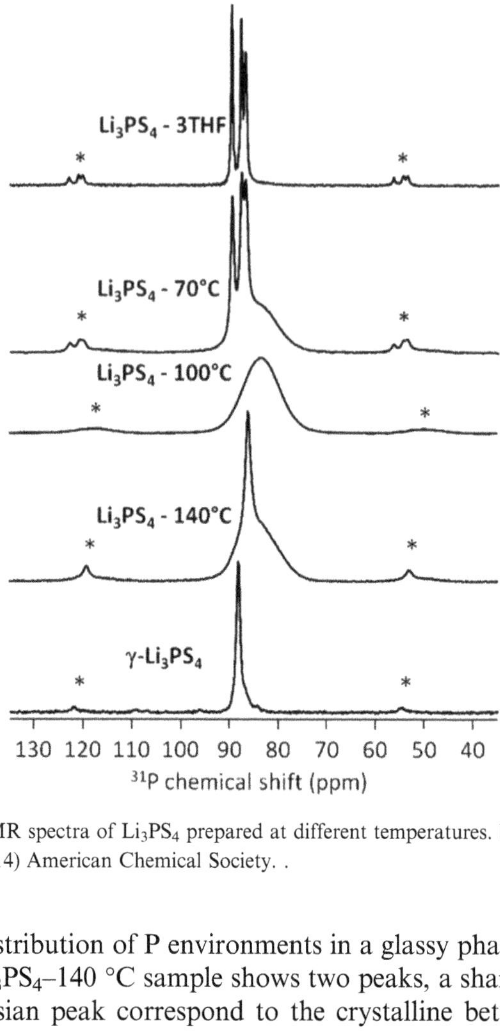

Figure 7.8. ^{31}P MAS NMR spectra of Li_3PS_4 prepared at different temperatures. Reprinted with permission from [76]. Copyright (2014) American Chemical Society. .

attributed to the distribution of P environments in a glassy phase. Interestingly, ^{31}P NMR spectrum Li_3PS_4–140 °C sample shows two peaks, a sharp peak at 86.3 ppm and a broad Gaussian peak correspond to the crystalline beta-Li_3PS_4 and glassy phase, respectively. Further morphological changes to these intermediate samples during the process of NMR experiments have been reported in detail [76]. It has been shown that a controlled heat treatment of Li_3PS_4–3THF produces crystalline beta-phase, where the amorphous phase possesses transitory unorganized structure established during the desolvation of the precursor. The higher conductivity of beta phase is supported by PFG NMR where the average D_{Li+} was found to be 5.4×10^{-13} m^2 s^{-1}.

Amorphous, glassy solid electrolytes can achieve ionic conductivities ranging from 10^{-7} to 10^{-2} S cm^{-1} [77, 78]. The disordered structure of glassy electrolytes also eliminates the grain boundaries associated with polycrystalline materials which are susceptible to dendrite formation. Recently, Kaup and coworkers explored the possibility of achieving superionic conductivity in lithium oxythioborate halide or LIBOSS (aLi_2S-$B_2S_{3-x}SiO_{2-z}LiI$) glass system [79]. It was found that the incorporation of SiO_2 increases the dissolution of LiI, in the glass matrix upon re-restructuring the thioborate network. While XRD lacks provision of structural information due to the amorphous nature of sample, solid state NMR plays a significant role in probing the local structure and thus insight into the fast ion conduction mechanism.

To understand the maximum in conductivity (2.1 mS cm^{-1}) at the 0.25 SiO_2 composition, high-field ^{11}B, ^{29}Si and ^{7}Li MAS NMR spectra were collected for LIBOSS with 0, 0.25 and 0.5 SiO_2 content, figure 7.9. ^{11}B NMR spectrum shows multiple peaks corresponding to BS_4 tetrahedra and trigonal B units. At $x = 0$, the

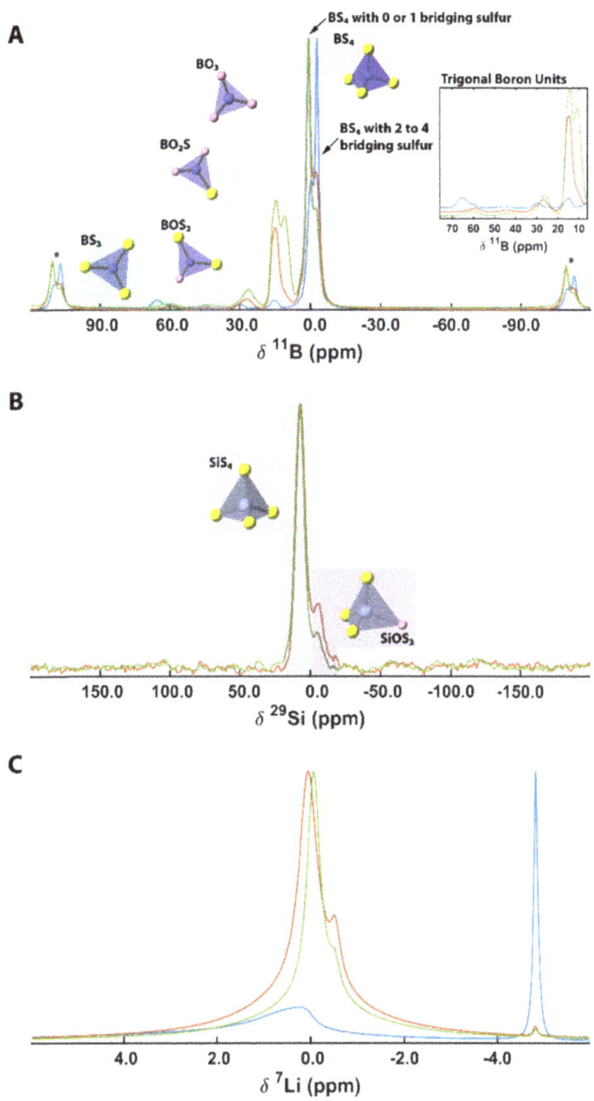

Figure 7.9. ^{11}B, ^{29}Si and ^{7}Li NMR spectra of $LiB_{0.5}Si_xO_{2x}SI_{0.5}$ where $x = 0$ (blue), 0.25 (orange), and 0.5 (green) [79], copyright (2020) John Wiley & Sons.

^{11}B peaks at −2.9 ppm and 0 ppm were attributed to BS_4 with two to four bridging sulfur and isolated BS_4 tetrahedra or BS_4 with one bridging sulfur, respectively. The opposite trend of intensity of these ^{11}B peaks, with the addition of SiO_2, reveals that BS_4 units favors the isolated or network terminating tetrahedra. Consequently, the non-bridging sulfur species gives smaller inorganic units, which further increases the solubility of LiI as well as Li ion density in glass matrix. Addition of SiO_2 generates BOS_2 or BO_2S, which subsequently forms isolated BO_3 units and BO_3 with multiple bridging oxygen anion groups at high SiO_2 content.

Another interesting class of solid electrolytes is the garnet-type oxide electrolytes, $Li_{3+y}La_3M_2O_{12}$ (M = Nb, Ta, Sb, Zr, etc) [80], which exhibit cubic structure, achieve high ionic conductivities (10^{-4} S cm^{-1}) for example as in LLZO, and maintain reasonably high working voltage against Li [81–83]. In this regard, Larraz and coworkers [84] reported a detailed NMR study $^6Li_{7x}H_xLa_3Zr_2O_{12}$ ($0 \leqslant x \leqslant 5$) garnets probing the distribution of Li-ions. LLZO garnet contains both tetrahedral and octahedral Li sites with relative populations that depend on Li content. The dynamic exchange between octahedral and tetrahedral sites and relative distribution of protons and lithium ions were investigated through variable temperature (VT) ^1H-^6Li CPMAS NMR experiments. The lack of resolution obtained in ^7Li spectrum is addressed successfully by using the ^6Li isotope where narrow peaks clearly distinguish different Li sites of the garnet structure. Hence, ^6Li spectra were recorded for $^6Li_7La_3Zr_2O_{12}$ at different ageing and annealing or washing process, figure 7.10, which helps in identifying different Li sites. A small peak at 0.1 ppm was assigned to the Li_2CO_3 or $LiOH.H_2O$ species. On the other hand, with support of Raman data, all ^6Li peaks observed for the washed and aged sample were assigned to the specific Li sites of the garnet cubic structure. Further spectral analysis of

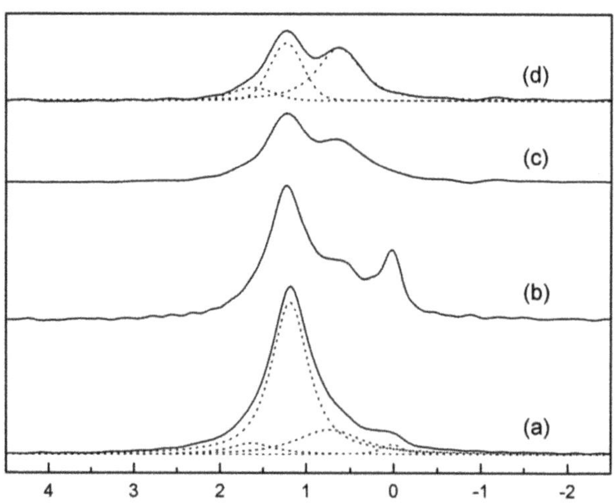

Figure 7.10. ^6Li NMR spectra of $^6Li_7La_3Zr_2O_{12}$ prepared at different experimental condition, (a) as-prepared, (b) aged at room temperature for $^6Li_{3.8}H_{3.2}La_3Zr_2O_{12}$ before and after washing step without (c) and with ^1H decoupling (d). Reproduced from [84] with permission of The Royal Society of Chemistry.

$^6Li_{3.8}H_{3.2}La_3Zr_2O_{12}$ identifies three Li environments which can be compared to the parent tetragonal $^6Li_7La_3Zr_2O_{12}$ compound. These three Li environments were classified as less distorted octahedral, distorted octahedra and tetrahedral sites. In addition, T_1 relaxation measurements and VT NMR experiments provide insight into the dynamics of octahedral and tetrahedral Li sites and their apparent exchange process.

7.3.2.3 Solvent-free polymer electrolytes

Polymer electrolytes have attracted considerable interest in both industry and academia with their prominent advantages such as suppression of dendrite growth, enhanced endurance to varying volume during cycling, improved safety and better shape flexibility and manufacturing integrity [85]. Again, NMR spectroscopy has played a significant role in elucidating the structure and dynamics of polymer elctrolytes [86]. Polyethylene oxide (PEO)-based polymer electrolytes are the most widely studied materials for batteries [87–89]. The ion transport mechanism is a cooperative one, involving polymer segmental motion in the amorphous phase of the usually heterogeneous amorphous/crystalline morphology of the polymer–salt complex. Consequently, significant effort has been devoted to reducing the crystalline phase for better performance at ambient temperature. Although there has been evidence presented for a conductivity mechanism in the crystalline phase under chain alignment conditions [90], the vast majority of research on PEO-based electrolytes involves suppression of the crystalline phase.

In a recent example, increased conductivity was observed when the PEO polymer is combined with polycarbonates. Meabe and coworkers [91] studied the properties of poly(ethylene oxide carbonates) by solid state NMR ^1H, ^{13}C, ^7Li and ^{19}F. NMR experiments were performed for a series of samples with different ratios of PEO, PC and Li–salt. The local dynamics of ions was investigated through ^7Li and ^{19}F NMR analysis, in particular, for PEO_{34}-PC as a function of LiTFSI concentration (20 wt% to 80 wt%), figure 7.11. Multiple ^7Li and ^{19}F peaks for 80 wt% sample indicated that there are two different Li sites. The presence of additional Li sites was confirmed by

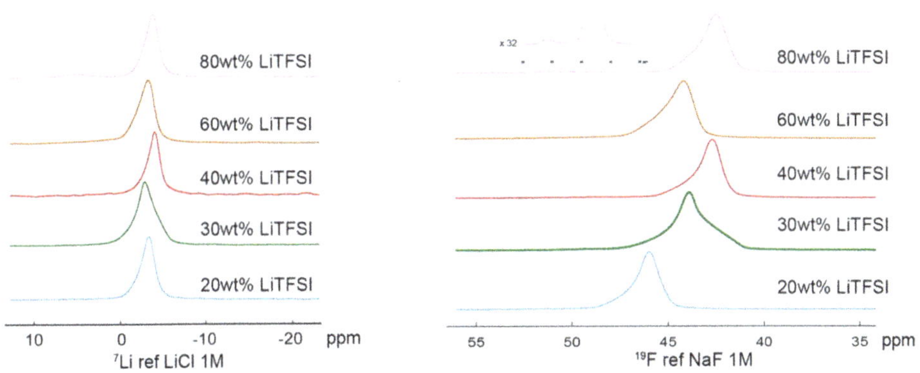

Figure 7.11. ^7Li (left) and ^{19}F (right) NMR spectra of PEO_{34}-PC with different wt% of LiTFSI collected at 343 K. Reprinted from [91], Copyright (2018), with permission from Elsevier.

^{7}Li relaxation measurements, which was attributed to the formation of LiTFSI aggregates.

Longer correlation times obtained from the relaxation measurements for ion aggregates are consistent with the slower dynamics *and less mobility*. ^{7}Li and ^{19}F diffusion coefficients were also measured for PEO_{34}-PC with different concentration of LiTFSI and analyzed to unravel the ion dynamics at longer time/length scales. In another study, Becher *et al* [92] reported the selective dynamics of polymer, cation and anion in polymer electrolyte by measuring ^{1}H, ^{7}Li and ^{19}F diffusion coefficients at various time and length scales. The disordered polymer electrolytes were achieved by mixing the PPG polymer with $LiClO_4$ or LiTFSI salt. The local dynamics and mobilities of all ions and polymer chain decrease with increase of salt concentration with more pronounced effect on the ClO_4 than on the TFSI anions. It was found that the cations diffuse at slower rate than anions, and follow the VFT temperature dependence. Field cycling spin-lattice relaxation studies were employed to investigate the short-range dynamics of ions and polymer chain. A bimodal behavior of correlation times observed for intermediate salt concentrations was attributed to the presence of salt-rich and salt-depleted regions. The obtained NMR results suggest the presence of the majority of lithium ions in salt-rich regions. By relating the diffusion coefficients with correlation time, the authors showed that the segmental motion of the polymer strongly affects the lithium ion transport even in polymer electrolytes with micro-heterogeneous salt distributions.

Polymer electrolytes are often studied through high-resolution solid-state NMR by spinning the sample packed in a rotor. In an interesting but usually overlooked aspect of standard MAS studies, one must consider that the physical state of the electrolyte can change under rapid rotation. A recent study by Messinger *et al* [93] reviews the physical state of PEO polymer containing LiTFSI salt at static and MAS condition. The distribution of stresses as well as the local orientation of the polymer chain within the rotor changes with MAS condition, resulting in corresponding line shape changes. MAS averages the inhomogeneities associated with dipolar couplings, which improves the spectral resolution and helps in studying the dynamics of slow diffusing components. At the same time, the dynamics of ions are sensitive to the MAS-induced centrifugal pressures, because of the chain alignment.

An example shown in figure 7.12 illustrates the effect MAS and static condition at different temperatures on the ^{1}H line shape of PEO(LiTFSI) sample. A single broad peak of static ^{1}H NMR spectrum collected at 25 °C is assigned to the CH_2 of amorphous PEO domains. After heating the sample for 1 h at 97 °C above the T_M, a narrow ^{1}H peak is observed. Further cooling down to 25 °C results in the same line shape of ^{1}H peak depicting the original state of the polymer. On the other hand, spinning the sample at 10 kHz at 97 °C enhances the ^{1}H resolution. Further spinning the sample under 10 kHz and halting MAS subsequently while cooling down to 25 °C shows a characteristic doublet line shape with a splitting of 2.2 kHz at static condition. The obtained broad ^{1}H doublet feature is attributed to the local ordering and partial chain alignment induced by rapid rotation of the sample. The diffusion coefficients of Li and TFSI ions measured at static and MAS condition were found to be in the same range, however, a slight difference observed in the obtained NMR

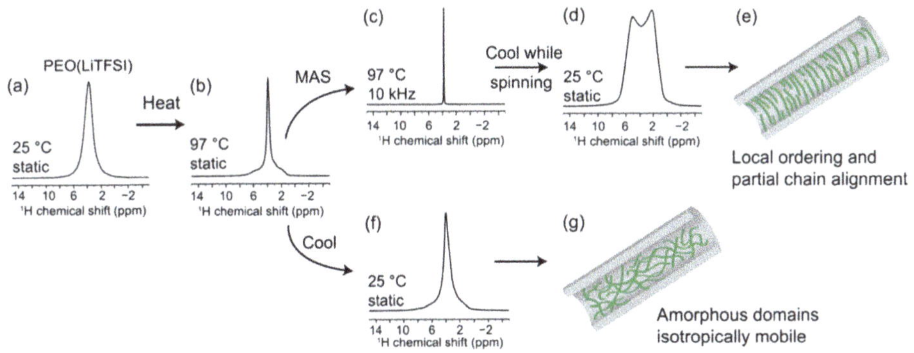

Figure 7.12. The physical state of PEO-LiTFSI electrolyte is analyzed through solid state NMR at MAS and static condition [93]. John Wiley & Sons, copyright (2020).

activation energy. This was attributed, again, to the local ordering of the polymer chain due to the rapid rotation of the sample at MAS condition that alters the local magnetic susceptibility tensor. Hence, it has been recommended that the samples should be spun at possible slowest MAS rate that is necessary to achieve the maximum resolution and compare with the static NMR spectra to establish the stress-induced local ordering.

7.3.2.4 Ionic liquid–polymer electrolytes (gel electrolytes)

Confining liquid electrolytes in a solid polymer is a common strategy to improve the mechanical stability (compared to liquid imbibed in a porous separator, as is the current practice in commercial LIBs). In this regard, confined ionic liquid (IL)-based polymer electrolytes are of great interest in energy-related applications, especially considering the enhanced safety aspects of ILs [94]. Again, NMR spectroscopy plays a vital role in understanding the structure and dynamics of polymer gel electrolytes [7]. A recent work from Garaga *et al* reports the molecular level interactions and dynamics of ions with polymer in phosphonium- and ammonium-based TFSI ILs with PMMA by PFG and solid-state NMR as well as FFC relaxometry.

At higher polymer concentration, the diffusivity of ions is largely affected by their interaction with the polymer chain. Such molecular interactions of ions with polymer network can be probed through CPMAS solid state NMR. For instance, in the case of PMMA-ILs polymer electrolyte the interaction of TFSI ions with PMMA polymer are established through $^{13}C(^{19}F)$ CPMAS NMR [7]. As mentioned in the methods section, the basic principle is transferring the magnetization from ^{19}F of TFSI to ^{13}C of TFSI or PMMA by fulfilling the Hartmann–Hahn condition. As shown in figure 7.13 of PMMA polymer gel electrolyte, ^{13}C peaks at 120 ppm and small intense peaks at 10–60 ppm were assigned to the ^{13}C of TFSI and PMMA, respectively. This confirms that the TFSI ions interact with the polymer chain. These types of NMR experiments probe the molecular-level interactions through spatial proximities between ions and the polymer chain up to a certain distance, which could help in designing novel ILs-polymer electrolytes.

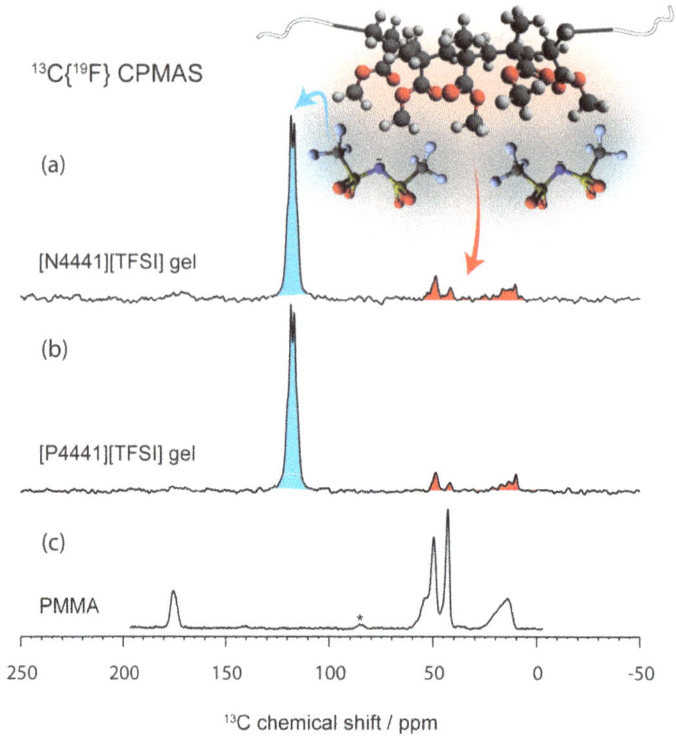

Figure 7.13. $^{13}C(^{19}F)$ CPMAS NMR spectra of N4441- and P4441-gel electrolytes. Reprinted from [7], copyright (2021), with permission from Elsevier.

It was found that the ions diffuse faster in [P4441]- as compared to the [N4441]-ILs and gel electrolytes. Fast field cycling relaxometry was further employed to understand the dynamics of $P4441^+$ or $N4441^+$ and $TFSI^-$ by measuring ^1H and ^{19}F spin-lattice relaxation rate (R_1) profiles, respectively [7]. figure 7.14 shows an example of ^1H spin-lattice relaxation rate profiles of pure ILs, ILs electrolyte and polymer gel electrolyte. These R_1 profiles were analyzed by using an appropriate model to unravel the rotational and translational dynamics of ions. Because of higher viscosity, ions in [N4441][TFSI] diffuse more slowly than those in [P4441][TFSI], which was confirmed unambiguously with the large relaxation rate observed for [N4441][TFSI] in both the ILs and polymer gel electrolyte. Recall that above the temperature of the T_1 minimum (or R_1 maximum), slower dynamics leads to higher relaxation rate. The slowing down of ions with the addition of LiTFSI or PMMA was also observed with the corresponding larger relaxation rates. In addition, ^{19}F spin lattice relaxation has also been reported. Analysis of FFC data supports the PFG NMR results regarding the dynamics of ions, which also confirms that the $TFSI^-$ has a stronger interaction with polymer chain than the $P4441^+$ or $N4441^+$. Another important parameter associated with dynamics of ions and polymer could be calculated from FFC data is rotational correlation time (τ_R), where it was shown

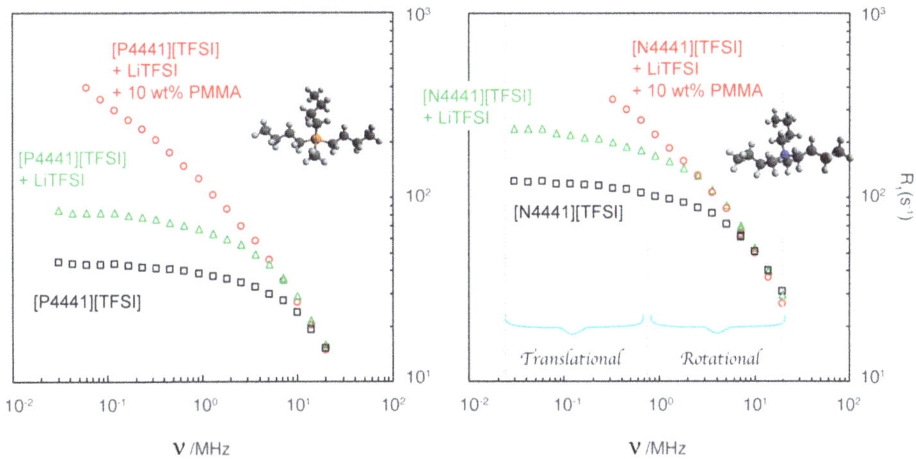

Figure 7.14. [1]H spin-lattice relaxation rate profiles of pure ILs, ILs electrolyte and polymer gel electrolytes. Reprinted from [7], copyright (2021), with permission from Elsevier.

that the τ_R increases with increasing Li–salt concentration and decreases with increasing temperature.

Another example of quasi solid electrolyte based on ionic liquids described here is the well-studied IL BMIM TFSI, mixed with LiTFSI and confined in a silica matrix, in which both standard PFG diffusion and FFC relaxometry were made [72]. Because broadband relaxometry profiles contain contributions from both translational diffusion and molecular rotation, in cases where it is possible to measure diffusion directly by PFG NMR, these measured values also serve the useful purpose of constraining the FFC relaxometry fits. The analysis of the experimental data highlight the existence of two fractions of the liquid: a bulk fraction (at least several ionic radii from the silica particles) and a surface fraction (close to the silica particles).

7.3.2.5 Composite ceramic polymer electrolytes

All-solid-state Li batteries containing solid electrolytes, for instance, garnet and argyrodite electrolytes, show improved energy density and safety. However, complications associated with physical/electrical contact at the electrode–electrolyte interface and a large interfacial resistance limit their practical use in batteries. Polymer electrolytes show better mechanical stability and good electrode–electrolyte contact, however, lack the desired conductivity. In this regard, the addition of ceramic materials, Li^+ insulating oxides and oxide/sulfide Li^+ conductors, to the polymer electrolytes improves the overall Li^+ conductivity. PEO/perovskite $Li_{3/8}Sr_{7/16}Ta_{3/4}Zr_{1/4}O_3$ composite electrolytes reported by Xu and coworkers, found to be more suitable for all-solid-state Li batteries showing 5.4×10^{-5} S cm^{-1} conductivity at room temperature [95]. The improved mobility of Li-ions in the studied PEO-perovskite electrolyte was ascribed to the stronger interaction between F^- of TFSI and Ta^{5+} of the perovskite. In addition, the anode SEI prevents the formation lithium dendrites. In another study, with the aim of investigating Li^+ local environments and Li^+ conduction mechanism in polymer/Li^+-conducting filler composite electrolytes, Wu and coworkers [96] used ^7Li relaxation time and $^6Li \rightarrow {}^7Li$ trace-exchange NMR measurements. NASICON $LiZr_2(PO_4)_3$ was used as a filler in the PEO-based composite polymer electrolyte. It was found that the salt-concentration and ratio of ceramic filler significantly influence the population of Li^+ in two local environments.

7.3.3 Electrode and electrolyte interface

Despite the importance of the SEI in battery performance and safety, the paucity of nuclei in the thin SEI layer presents challenges to their investigation by NMR. Early studies involved identification of irreversible Li (which may also include electrolyte decomposition products atop the SEI) aided by the substantial chemical shift difference between intercalated Li and surface species [97, 98]. Additional information on the chemical compound makeup of the SEI could also be accomplished by selective carbon-13 enrichment of the electrolyte solvent [99].

Conventional 1D and 2D NMR experiments are insightful in solving major problems in battery science but still insufficient because of the poor sensitivity and selectivity. In this context, DNP NMR has been used as a main characterization tool to boost the signal of low sensitive nuclei in number of occasions probing the local structure and dynamics of battery materials and interfaces [48, 100]. As explained in the methods section, a typical DNP NMR experiment uses organic biradicals that requires cryogenic sample cooling, which leads to severe limitations as discussed in Hope *et al* [46]. Instead of using organic biradical, authors showed the advantage of using conduction electrons of lithium metal to enhance the ^7Li, ^1H, ^{19}F NMR signals of SEI species, which helps in understanding their chemical structure and spatial distribution.

The basic theory involves using the Pauli paramagnetism of the metallic electrons as the source of polarization to enhance the nuclear magnetization via the Overhauser effect [41], instead of the localized radicals. In this work, a significant hyperpolarization of ^7Li signal of cycled lithium metal anode achieved at room temperature by using Overhauser DNP. Furthermore, this technique has been used to probe the selective proximity of different species to the Li metal at the interface by acquiring enhanced diamagnetic ^7Li, ^1H, ^{19}F NMR spectra. Signatures of SEI with polymer organic species or LiF are then revealed through double-resonance $^7Li \rightarrow {}^1H$ and $^7Li \rightarrow {}^{19}F$ CPMAS, and $^7Li(^{19}F)$ rotational echo double resonance (REDOR) NMR experiments.

A large chemical shift (~250 ppm) of lithium metal, as shown in figure 7.15, arises from the well–known metallic Knight shift. A factor of 7.9 signal enhancement was observed with the microwave irradiation. In addition, distribution of NMR frequencies was observed in the lower frequency region because of the partial saturation of conduction ESR (CESR) transition that decreases the moment as well as the Knight shift. As demonstrated in the article, the DNP NMR mechanism is further confirmed by performing the experiment via Field sweep [46]. A detailed analysis of DNP NMR experiments performed at low temperature and lower field

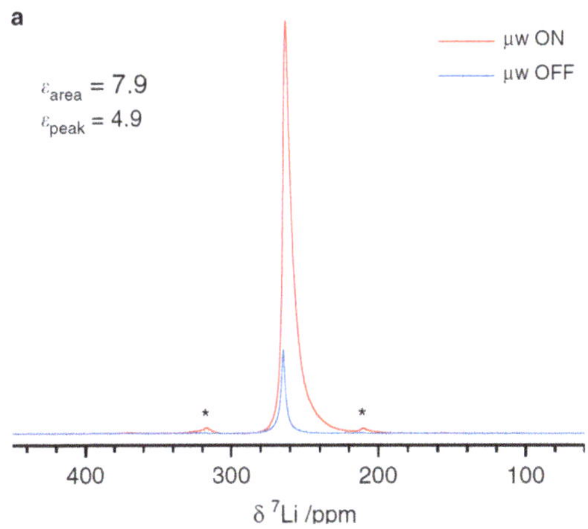

Figure 7.15. ^{7}Li MAS NMR spectra of microstructural lithium metal collected with and without 15.6 W microwave irradiation (at 395.29 GHz) at a magnetic field of 14.1 T and 12.5 kHz MAS spinning, and at 300 K using Hahnecho pulse sequence. Asterisks indicate the spinning sidebands. Reproduced from [46], copyright (2020) with permission from Springer.

strength provide valuable insights in understanding the details of ^{7}Li signal enhancement upon microwave irradiation. This hyperpolarization technique has further been applied to investigate lithium metal and solid–electrolyte interface. As shown in figure 7.16, the DNP NMR experiments were performed for several samples with a standard electrolyte (LP30), 1:10 volume ratio of additive FEC in standard electrolyte (LP30+FEC). The achieved ^{7}Li signal enhancement indicated that the observed hyperpolarization is sufficient to probe the SEI.

In the case of LP30, the ^{7}Li enhancement was found to be on the same order as obtained in the case of metallic lithium, and was assigned to lithium compounds in the SEI, such as Li_2CO_3 and LiOH. The absence of signal enhancement for the sharp Li peak at −0.8 ppm is consistent with its assignment to Li in residual electrolyte. However, in the case of LP30+FEC, the broad ^{7}Li signal at −0.9 ppm was found to be enhanced upon microwave irradiation indicating the close proximity of residual electrolyte to the metal surface. A large enhancement of small peak at −0.8 ppm was assigned to the LiF obtained during the reduction of FEC or decomposition of $LiPF_6$.

In the ^{1}H NMR spectra of LP30, a large signal enhancement observed for 3.8 ppm ^{1}H peak was assigned to the poly-ethylene oxide (PEO)-like species. Another narrow peak at 4.7 ppm that showed enhancement of ~2 was assigned to the mobile EC molecules and/or small organic lithium carbonates close to the metal surface. The unenhanced broad signal at ~4.6 ppm and signals at 0.1–0.2 and 7.2–7.3 were attributed to the solid residual electrolyte and impurities, respectively. In contrast, the ^{1}H NMR spectra of LP30+FEC showed the signature of poly-vinylene carbonate as was assigned to the broad ~4.7 signal with enhancement of 1.8. Further detailed investigation, including static DNP-enhancement for microstructural

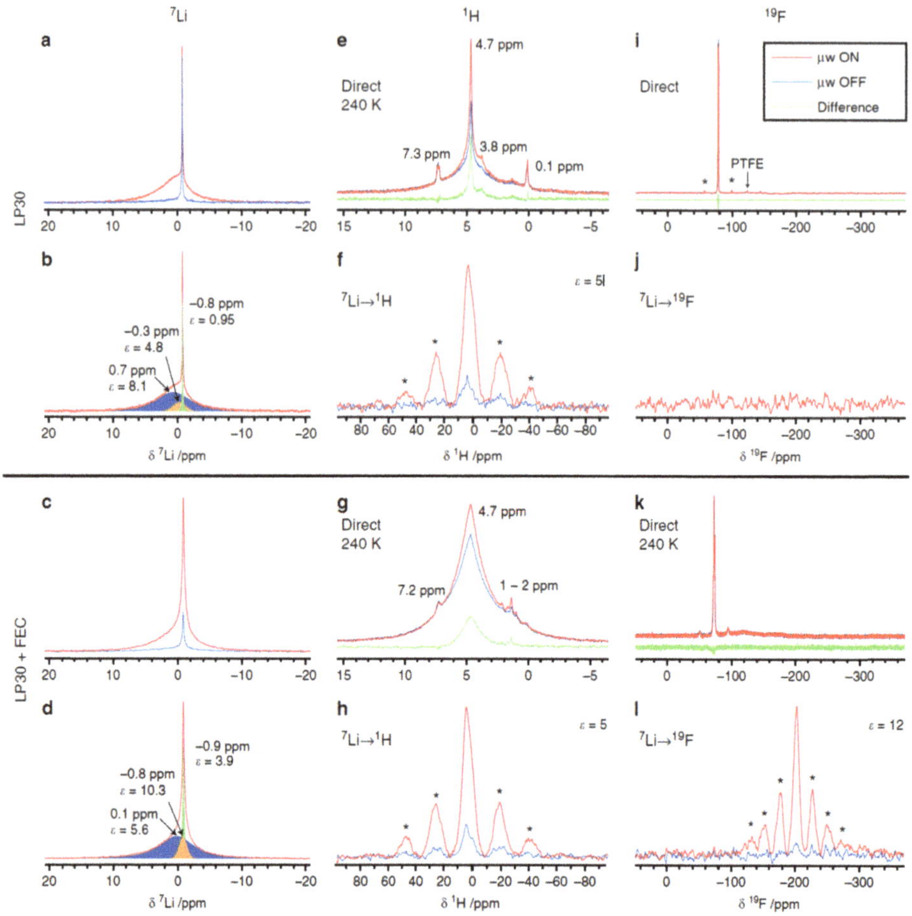

Figure 7.16. ^{7}Li, ^{1}H and ^{19}F DNP NMR spectra of SEI samples. Diamagnetic (a-d) ^{7}Li, (e, g) ^{1}H, (i, k) ^{19}F NMR spectra of LP30 and LP30+FEC electrolytes. Reproduced from [46], copyright (2020) with permission from Springer

lithium, enable the understanding of structural and physical insights by probing buried/internal metal–diamagnetic solid–solid interfaces.

As a final example of advanced methods to probe interfacial phenomena, direct lithium-ion exchange at the electrode–electrolyte interfaces in all-solid-state batteries [101–103] has been reported. Solid state NMR offers a number of experiments to study the lithium-ion exchange process between different Li-ion containing electrode/electrolyte combinations. Recently, Yu and coworkers evaluated the selective charge transfer between the cathode–electrolyte phases in argyrodite Li_6PS_5Cl-Li_2S [104] and Li_6PS_5Br-Li_2S [105] all-solid-state Li-sulfur battery by using multi-dimensional solid-state NMR. In particular, 2D exchange spectroscopy (EXSY) was used to probe the lithium-ion exchange between electrode and electrolyte.

1D and 2D EXSY ^{7}Li NMR spectra were recorded for Li_6PS_5Br-Li_2S cathode mixtures with different Li_2S particle size (micron to nanosized), as shown in figure 7.17. It was found that the smaller sized Li_2S reduces the kinetic restriction by minimizing the charge transport distances. While 1D ^7Li MAS NMR distinguish the local environments of Li ions in Li_6PS_5Cl and Li_2S phases, 2D EXSY NMR probes the spontaneous lithium-ion exchange between these phases. The upfield shift of Li_6PS_5Br peak was attributed to the local effect of Br on the Li ions. The well-resolved ^7Li peaks in 1D NMR facilitates the acquisition of 2D EXSY NMR to gather more information. The off-diagonal peaks in 2D spectra indicate the exchange of Li ions, during the time between the ^7Li signal at $t = 0$ and $t = t_{mix}$. As demonstrated in the paper [105] and shown in figure 7.17, off-diagonal peaks are observed for the thoroughly mixed Li_6PS_5Br-Li_2S sample. This means that Li ions present in Li_6PS_5Br phase at $t = 0$ diffused to Li_2S phase at $t = t_{mix}$. The intensity of the off-diagonal peaks further provides quantitative information, such as, how many Li ions transferred to the Li_2S phase and the corresponding exchange current density. In the case of thoroughly mixed sample, 20% Li ions moved from Li_6PS_5Br to Li_2S phase at 100 ms mixing time. Based on these results, authors emphasized the necessity of high-speed ball milling to achieve the intimate mixing and hence the desired significant Li ions transfer.

The effect of cycling on the Li-ions transfer was further investigated by collecting 2D EXSY NMR of cycled cathode mixtures at different mixing times, as shown in

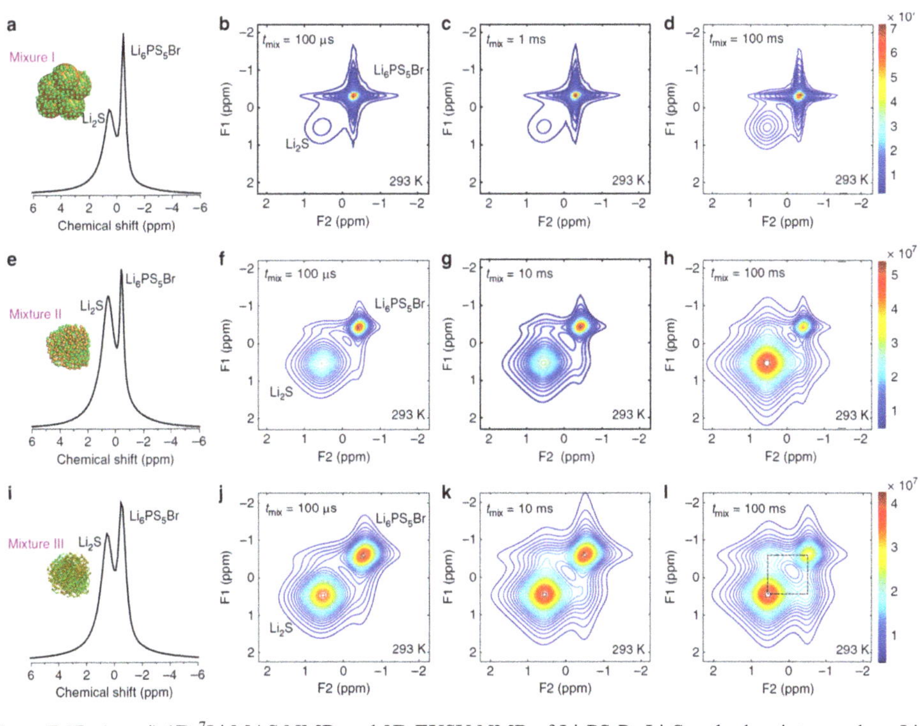

Figure 7.17. (a, e, i) 1D ^7Li MAS NMR and 2D EXSY NMR of Li_6PS_5Br-Li_2S cathode mixtures where Li_2S is nanosized (b, c, d) and micron sized (f, g, h), and nanosized Li_2S is well-mixed with Li_6PS_5Br (j, k, l) collected at different mixing times (100us, 10ms and 100ms). Reproduced from [105], copyright (2017) with permission from Springer

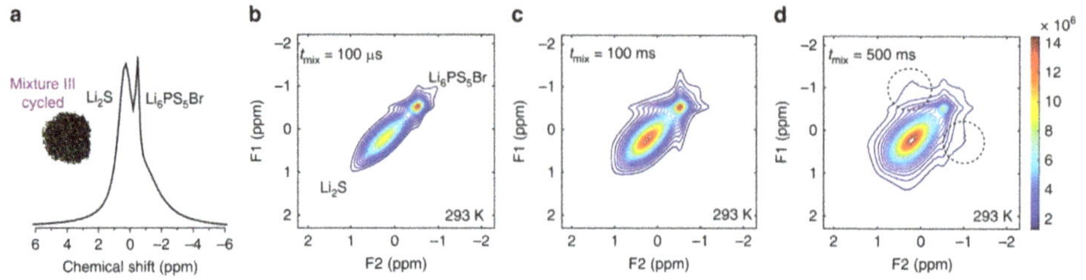

Figure 7.18. (a) 1D ^{7}Li MAS and (b, c, d) 2D ^{7}Li EXSY NMR spectra of thoroughly mixed Li_6PS_5Br–Li_2S material. Reproduced from [105], copyright (2017) with permission from Springer

figure 7.18. Only a weak intensity of off-diagonal peaks was observed and only at 500 ms, unlike those observed for uncycled mixture at 100 ms. This confirms that cycling can affect the facile transfer of Li ions over the solid–solid Li_6PS_5Br–Li_2S interface. In addition to 2D EXSY, authors performed the 1D ^{7}Li–^{7}Li exchange NMR experiment as a function of temperature and t_{mix} to assess the quantification of Li exchange between the electrode and electrolyte. These results highlight the versatility of exchange NMR experiments in quantifying the Li-ion transport at the interface in all-solid-state batteries.

7.3.4 *Operando* NMR techniques

To understand how the battery works at molecular level, one should characterize a series of materials prepared in full range of condition in an operating lithium-ion battery. Instead, much more powerful *operando*/*in situ* techniques provide detailed information about electrode structure evolution, electrochemical reaction mechanisms and side reactions in real time during the battery operation [106]. This enables developing high-performance materials for future advanced battery technology. *In situ* NMR offers different methods to probe the structural and dynamical changes at local level during, for example, cycling process, as discussed in section 7.3.1.1 (anode materials). In this section, we discuss some examples related to electrophoretic NMR (eNMR) and *in situ* magnetic resonance imaging (MRI).

7.3.4.1 Electrophoretic NMR

Electrophoretic NMR can be used to directly measure the velocity (v_i) of ionic drift in the presence of an applied electric field (E), which differentiates the charged species from neutral molecules or ion pairs [107–109]. This technique uses double stimulated echo pulse sequence with an applied electric field to measure the ionic mobility and hence the transference number. The mobilities of depend on the electric field. The transference number (t_i) is given by,

$$t_i = \frac{\sigma_i}{\Sigma_j \sigma_j}$$

Conductivity of one ionic species as a fraction of the overall ionic conductivity (summation of all partial conductivities (σ_i)) can also be expressed in terms of the electrophoretic mobility μ_i, the charge number z_i and number density N_i.

$$t_i = \frac{z_i N_i \mu_i}{\Sigma z_i N_i \mu_i}$$

Rosenwinkel and Schonhoff [110] determined the lithium transference numbers in PEO-LiTFSI electrolyte by measuring electrophoretic mobilities of ions and compared them to the PFG NMR results. As compared to the lithium, the negative sign corresponding to the TFSI ion indicates their apparent migration to the opposite electrode. In the case of studied PEO/LiTFSI electrolytes, t_{Li} can be written as,

$$t_{\mathrm{Li}} = \frac{\mu_{\mathrm{Li}}}{\mu_{\mathrm{Li}} - \mu_{\mathrm{TFSI}}}$$

The lithium transference numbers calculated by eNMR found to be in same range of 0.2, consistent with the t_{Li} obtained by PFG NMR. Because electrophoretic mobility does not track ion pairs or neutral aggregates, the corresponding transference number measured in diluted media was found to be more accurate than PFG NMR. Furthermore, eNMR has been applied to other ILs-based electrolytes to determine the Li ion transference number and their motion in Li-anion clusters, for example [111, 112]. A negative transference number, for example, provides evidence of $[\mathrm{Li(TFSI)_2}]^-$ mobile species.

7.3.4.2 In situ *MRI*

While, *in situ* NMR techniques provide subsequent structural and dynamical changes of battery materials occurring during operation, complementary to other *operando* techniques, *in situ* MRI has been a powerful alternative tool for studying spatial ion distribution in lithium-ion batteries [113–116]. Basically, the signal of spatially distributed ions that evolve during the course of time can be mapped for visualization and parametrization. This is directly related to the mass transport phenomena occurring within a battery cell. During the charging and discharging, the ions migrate towards their respective electrodes through an electrolyte while conduction electrons flow through an external circuit. An ionic concentration gradient forms during this process in the electrolyte solution. *In situ* MRI with inverse mathematical modelling can establish such spatially distributed ions by estimating the apparent diffusion coefficient and Li ion transference number [117]. A crucial finding is that the ionic diffusivity on or near the two different electrodes can differ by up to 60%.

In another study, Klett and coworkers [113] reported the transport properties under or near operating conditions by using *in situ* ^{7}Li NMR imaging. The experiments were conducted on a home-built electrochemical cell, to which 30 μA current was constantly applied, consisting of $LiPF_6$ electrolyte held in between two Li metal electrodes. The authors used the concentrated binary electrolyte mass transport model [118] to obtain the corresponding salt diffusivity and lithium transport number in spatially distributed $LiPF_6$ electrolyte.

Aiming to understand the electrode microstructures build-up, in other words lithium dendrites deposited on the anode during charging process, Grey, Jerschow and coworkers [116, 119, 120] demonstrated the efficacy of *in situ* MRI technique. This allows one to establish a direct correlation between lithium dendrites and electrolyte concentration gradient. The authors estimated the time when dendrite formation begins via ^{7}Li chemical shift images of Li metal collected during charge process, as shown in figure 7.19 [120]. A uniform distribution of Li ions, before applying current, was noticed with rectangular shape in concentration profile obtained from *z* spin-echo images. The ^{7}Li chemical shift image for the corresponding pristine cell shows peaks of lithium metal electrodes at the top and bottom of the cell resonating at ca. 250 ppm. Upon applying the current, the concentration profile increased on the positive electrode and decreased on the negative electrode (figure 7.19(b), top), whereas the peak intensity increased only on the negative electrode (figure 7.19(b)), the peak at 260 ppm, a new peak at 270 ppm emerged, the former assigned to mossy-type and later assigned to dendritic microstructures. Hence, these results establish the real-time changes observed in both Li electrolyte concentration profiles and growth of different dendritic microstructure on the electrode.

Another recent work by Bazak *et al* [121] by using *in situ* MRI technique, investigated the temperature dependence of the concentration gradient under applied current. The quantitative *in situ* MRI analysis of concentration gradient was found to be in strong agreement with *ex situ* NMR and other electrochemical techniques. One of the main limitations in NMR of battery components are metal-containing materials, which are often detected as imaging artifacts. However, recent development in MRI has enabled real-time visualization of the electrochemical process in metal-containing batteries or other materials [122, 123]. This technique has further been extended to study other battery materials [124]. For example, Xiang

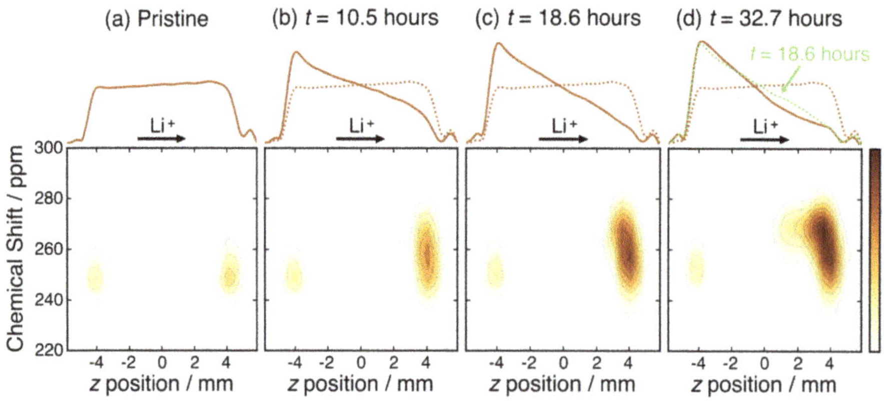

Figure 7.19. ^{7}Li MRI electrolyte concentration profiles (top) and the ^{7}Li chemical shift image of lithium metal (bottom) for the electrochemical cell charged at 0.76 mA cm^{-2}. Reprinted with permission from [120]. Copyright (2015) American Chemical Society.

et al implemented the *in situ* MRI technique detecting ^{23}Na signal to understand the growth of sodium dendrites and the SEI layer [125].

Acknowledgements

The authors have benefited from research support provided by the U.S. Office of Naval Research and the U.S. Department of Energy, Division of Basic Energy Sciences prior to and during the preparation of this chapter.

References

[1] Wu H and Cui Y 2012 Designing nanostructured Si anodes for high energy lithium ion batteries *Nano Today* **7** 414–29

[2] Ying H and Han W-Q 2017 Metallic Sn-based anode materials: application in high-performance lithium-ion and sodium-ion batteries *Adv. Sci.* **4** 1700298

[3] Goodenough J B and Kim Y 2010 Challenges for rechargeable Li batteries *Chem. Mater.* **22** 587–603

[4] Tarascon J-M and Armand M 2010 Issues and challenges facing rechargeable lithium batteries *Materials for Sustainable Energy* (Singapore: World Scientific) pp 171–79

[5] Li J, Ma C, Chi M, Liang C and Dudney N J 2015 Solid electrolyte: the key for high-voltage lithium batteries *Adv. Energy Mater.* **5** 1401408

[6] Zhao Q, Stalin S, Zhao C-Z and Archer L A 2020 Designing solid-state electrolytes for safe, energy-dense batteries *Nat. Rev. Mater.* **5** 229–52

[7] Garaga M N, Jayakody N, Fraenza C C, Itin B and Greenbaum S 2021 Molecular-level insights into structure and dynamics in ionic liquids and polymer gel electrolytes *J. Mol. Liq.* **329** 115454

[8] Pecher O, Carretero-González J, Griffith K J and Grey C P 2017 Materials' methods: NMR in battery research *Chem. Mater.* **29** 213–42

[9] Haber S and Leskes M 2018 What can we learn from solid state NMR on the electrode–electrolyte interface? *Adv. Mater.* **30** 1706496

[10] Wiemers-Meyer S, Winter M and Nowak S 2019 NMR as a powerful tool to study lithium ion battery electrolytes *Annual Reports on NMR Spectroscopy* ed G A Webb vol 97 (New York: Academic) ch 3 pp 121–62

[11] Morales D J and Greenbaum S 2020 NMR Investigations of crystalline and glassy solid electrolytes for lithium batteries: a brief review *Int. J. Mol. Sci.* **21** 3402

[12] O'Donnell L F and Greenbaum S G 2021 Review of multivalent metal ion transport in inorganic and solid polymer electrolytes *Batteries* **7** 3

[13] Abragam A 1961 *The Principles of Nuclear Magnetism* (Oxford: Oxford University Press)

[14] Cavanagh J, Fairbrother W J, Palmer A G III and Skelton N J 1995 *Protein NMR Spectroscopy: Principles and Practice.* (Amsterdam: Elsevier)

[15] Stallworth P E, Samueli R, Sideris P, Vaknin D and Greenbaum S G 2012 ^{7}Li and ^{31}P nuclear magnetic resonance studies of single crystal $LiMPO_4$ (M = Co, Fe) *Advances in Inorganic Phosphate Materials* (Columbus, OH: American Ceramic Society) pp 117–26

[16] Pell A J, Pintacuda G and Grey C P 2019 Paramagnetic NMR in solution and the solid state *Prog. Nucl. Magn. Reson. Spectrosc.* **111** 1–271

[17] Grey C P and Dupré N 2004 NMR studies of cathode materials for lithium-ion rechargeable batteries *Chem. Rev.* **104** 4493–512

[18] Levitt M H 2013 *Spin Dynamics: Basics of Nuclear Magnetic Resonance.* (New York: Wiley)

[19] Lowe I J 1959 Free induction decays of rotating solids *Phys. Rev. Lett.* **2** 285–7

[20] Nishiyama Y 2018 Solid-state NMR under ultrafast MAS rate of 40–120 kHz *Experimental Approaches of NMR Spectroscopy: Methodology and Application to Life Science and Materials Science* (Singapore: Springer) pp 171–95

[21] Polenova T, Gupta R and Goldbourt A 2015 Magic angle spinning NMR spectroscopy: a versatile technique for structural and dynamic analysis of solid-phase systems *Anal. Chem.* **87** 5458–69 0p

[22] Bloch F 1946 Nuclear induction *Phys. Rev.* **70** 460–74

[23] Redfield A G 1957 On the theory of relaxation processes *IBM J. Res. Dev.* **1** 19–31

[24] Bloembergen N, Purcell E M and Pound R V 1948 Relaxation effects in nuclear magnetic resonance absorption *Phys. Rev.* **73** 679

[25] Graf M, Kresse B, Privalov A F and Vogel M 2013 Combining ^{7}Li NMR field-cycling relaxometry and stimulated-echo experiments: A powerful approach to lithium ion dynamics in solid-state electrolytes *Solid State Nucl. Magn. Reson.* **51** 25–30

[26] Kimmich R 2018 *Field-cycling NMR Relaxometry: Instrumentation, Model Theories and Applications* (Cambridge: Royal Society of Chemistry)

[27] Price W S 2009 *NMR Studies of Translational Motion: Principles and Applications* (Cambridge: Cambridge University Press)

[28] Stejskal E O and Tanner J E 1965 Spin diffusion measurements: spin echoes in the presence of a time-dependent field gradient *J. Chem. Phys.* **42** 288–92

[29] Sinnaeve D 2012 The Stejskal–Tanner equation generalized for any gradient shape—an overview of most pulse sequences measuring free diffusion *Concepts Magn. Reson.* A **40** 39–65

[30] Kuchel P W, Pagès G, Nagashima K, Velan S, Vijayaragavan V, Nagarajan V and Chuang K H 2012 Stejskal–tanner equation derived in full *Concepts Magn. Reson.* A **40** 205–14

[31] Mehrer H 2007 *Diffusion in Solids: Fundamentals, Methods, Materials, Diffusion-Controlled Processes* vol 155 (Cham: Springer)

[32] Johnson C S 1999 Diffusion ordered nuclear magnetic resonance spectroscopy: principles and applications *Prog. Nucl. Magn. Reson. Spectrosc.* **34** 203–56

[33] Slichter C P 1978 *Principles of Magnetic Resonance* (Berlin: Springer)

[34] Pines A, Gibby M G and Waugh J S 1973 Proton-enhanced NMR of dilute spins in solids *J. Chem. Phys.* **59** 569–90

[35] Hartmann S R and Hahn E L 1962 Nuclear double resonance in the rotating frame *Phys. Rev.* **128** 2042–53

[36] Metz G, Wu X L and Smith S O 1994 Ramped-amplitude cross polarization in magic-angle-spinning NMR *J. Magn. Reson.* A **110** 219–27

[37] Hediger S, Meier B H and Ernst R R 1995 Adiabatic passage Hartmann–Hahn cross polarization in NMR under magic angle sample spinning *Chem. Phys. Lett.* **240** 449–56

[38] Bennett A E, Rienstra C M, Auger M, Lakshmi K V and Griffin R G 1995 Heteronuclear decoupling in rotating solids *J. Chem. Phys.* **103** 6951–8

[39] Fung B M, Khitrin A K and Ermolaev K 2000 An improved broadband decoupling sequence for liquid crystals and solids *J. Magn. Reson.* **142** 97–101

[40] Ernst R R, Bodenhausen G and Wokaun A 1990 *Principles of Nuclear Magnetic Resonance in One and Two Dimensions.* (New York: Oxford University Press)

[41] Overhauser A W 1953 Polarization of nuclei in metals *Phys. Rev.* **92** 411–5

[42] Carver T R and Slichter C P 1953 Polarization of nuclear spins in metals *Phys. Rev.* **92** 212–3

[43] Maly T *et al* 2008 Dynamic nuclear polarization at high magnetic fields *J. Chem. Phys.* **128** 052211

[44] Lilly Thankamony A S, Wittmann J J, Kaushik M and Corzilius B 2017 Dynamic nuclear polarization for sensitivity enhancement in modern solid-state NMR *Prog. Nucl. Magn. Reson. Spectrosc.* **102–103** 120–95

[45] Ardenkjaer-Larsen J H 2016 On the present and future of dissolution-DNP *J. Magn. Reson.* **264** 3–12

[46] Hope M A, Rinkel B L D, Gunnarsdóttir A B, Märker K, Menkin S, Paul S, Sergeyev I V and Grey C P 2020 Selective NMR observation of the SEI–metal interface by dynamic nuclear polarisation from lithium metal *Nat. Commun.* **11** 2224

[47] Wolf T, Kumar S, Singh H, Chakrabarty T, Aussenac F, Frenkel A I, Major D T and Leskes M 2019 Endogenous dynamic nuclear polarization for natural abundance ^{17}O and lithium NMR in the bulk of inorganic solids *JACS* **141** 451–62

[48] Leskes M, Kim G, Liu T, Michan A L, Aussenac F, Dorffer P, Paul S and Grey C P 2017 Surface-sensitive NMR detection of the solid electrolyte interphase layer on reduced graphene oxide *J. Phys. Chem. Lett.* **8** 1078–85

[49] Hayes S, van Wüllen L, Eckert H, Even W R, Crocker R W and Zhang Z 1997 Solid-state NMR strategies for the structural investigation of carbon-based anode materials *Chem. Mater.* **9** 901–11

[50] Ménétrier M 2000 ^{7}Li NMR in some lithium intercalation compounds *Materials for Lithium-Ion Batteries* ed C Julien and Z Stoynov (Dordrecht: Springer) pp 507–15

[51] Langer J, Epp V, Heitjans P, Mautner F A and Wilkening M 2013 Lithium motion in the anode material LiC_6 as seen via time-domain ^{7}Li NMR *Phys. Rev.* B **88** 094304

[52] Park C-M, Kim J-H, Kim H and Sohn H-J 2010 Li-alloy based anode materials for Li secondary batteries *Chem. Soc. Rev.* **39** 3115–41

[53] Freytag A I, Pauric A D, Jiang M and Goward G R 2019 ^{7}Li and ^{29}Si NMR enabled by high-density cellulose-based electrodes in the lithiation process in silicon and silicon monoxide anodes *J. Phys. Chem.* C **123** 11362–8

[54] Kuhn A, Sreeraj P, Pöttgen R, Wiemhöfer H-D, Wilkening M and Heitjans P 2011 Li ion diffusion in the anode material $Li1_2Si_7$: ultrafast quasi-1D diffusion and two distinct fast 3D jump processes separately revealed by ^{7}Li NMR relaxometry *JACS* **133** 11018–21

[55] Grey C P and Lee Y J 2003 Lithium MAS NMR studies of cathode materials for lithium-ion batteries *Solid State Sci.* **5** 883–94

[56] Mizushima K, Jones P C, Wiseman P J and Goodenough J B 1980 $LixCoO_2$ ($0< x < -1$): A new cathode material for batteries of high energy density *Mater. Res. Bull.* **15** 783–9

[57] Ouyang B, Cao X, Lin H W, Slane S, Kostov S, Denboer M and Greenbaum S G 1994 Lithium-7 NMR studies of $Li_{1-x}CoO_2$ battery cathodes *MRS Proc.* 369 59

[58] Ménétrier M, Saadoune I, Levasseur S and Delmas C 1999 The insulator-metal transition upon lithium deintercalation from $LiCoO_2$: electronic properties and ^{7}Li NMR study *J. Mater. Chem.* **9** 1135–40

[59] Gummow R J, de Kock A and Thackeray M M 1994 Improved capacity retention in rechargeable 4 V lithium/lithium-manganese oxide (spinel) cells *Solid State Ionics* **69** 59–67

[60] Armstrong A R and Bruce P G 1996 Synthesis of layered $LiMnO_2$ as an electrode for rechargeable lithium batteries *Nature* **381** 499–500

[61] Lee Y J, Wang F and Grey C P 1998 ^{6}Li and ^{7}Li MAS NMR studies of lithium manganate cathode materials *JACS* **120** 12601–13

[62] Peng J, Gobet M, Devany M, Xu K, Cresce A V W, Borodin O and Greenbaum S 2018 Multinuclear magnetic resonance investigation of cation-anion and anion-solvent interactions in carbonate electrolytes *J. Power Sources* **399** 215–22

[63] Gnanaraj J S, Zinigrad E, Asraf L, Gottlieb H E, Sprecher M, Schmidt M, Geissler W and Aurbach D 2003 A detailed investigation of the thermal reactions of $LiPF_6$ solution in organic carbonates using ARC and DSC *J. Electrochem. Soc.* **150** A1533

[64] Kim J-K, Matic A, Ahn J-H and Jacobsson P 2010 An imidazolium based ionic liquid electrolyte for lithium batteries *J. Power Sources* **195** 7639–43

[65] Shah F U, Gnezdilov O I, Gusain R and Filippov A 2017 Transport and association of ions in lithium battery electrolytes based on glycol ether mixed with halogen-free orthoborate ionic liquid *Sci. Rep.* **7** 16340

[66] Eftekhari A, Liu Y and Chen P 2016 Different roles of ionic liquids in lithium batteries *J. Power Sources* **334** 221–39

[67] Sakaebe H, Matsumoto H and Tatsumi K 2007 Application of room temperature ionic liquids to Li batteries *Electrochim. Acta* **53** 1048–54

[68] Watanabe M, Thomas M L, Zhang S, Ueno K, Yasuda T and Dokko K 2017 Application of ionic liquids to energy storage and conversion materials and devices *Chem. Rev.* **117** 7190–239

[69] Herriot C, Khatun S, Fox E T, Judeinstein P, Armand M, Henderson W A and Greenbaum S 2012 Diffusion coefficients from ^{13}C PGSE NMR measurements—fluorine-free ionic liquids with the DCTA– anion *J. Phys. Chem. Lett.* **3** 441–4

[70] Castiglione F, Ragg E, Mele A, Appetecchi G B, Montanino M and Passerini S 2011 Molecular environment and enhanced diffusivity of Li^+ ions in lithium-salt-doped ionic liquid electrolytes *J. Phys. Chem. Lett.* **2** 153–57

[71] Frömling T, Kunze M, Schönhoff M, Sundermeyer J and Roling B 2008 Enhanced Lithium transference numbers in ionic liquid electrolytes *J. Phys. Chem.* B **112** 12985–90

[72] Jayakody N K, Fraenza C C, Greenbaum S G, Ashby D and Dunn B S 2020 NMR Relaxometry and diffusometry analysis of dynamics in ionic liquids and ionogels for use in lithium-ion batteries *J. Phys. Chem.* B **124** 6843–56

[73] Girard G M A, Hilder M, Zhu H, Nucciarone D, Whitbread K, Zavorine S, Moser M, Forsyth M, MacFarlane D R and Howlett P C 2015 Electrochemical and physicochemical properties of small phosphonium cation ionic liquid electrolytes with high lithium salt content *Phys. Chem. Chem. Phys.* **17** 8706–13

[74] Bogle X, Vazquez R, Greenbaum S, Cresce A V W and Xu K 2013 Understanding Li+–solvent interaction in nonaqueous carbonate electrolytes with ^{17}O NMR *J. Phys. Chem. Lett.* **4** 1664–8

[75] Wan C *et al* 2016 *J. Power Sources* **307** 231–43

[76] Gobet M, Greenbaum S, Sahu G and Liang C 2014 Structural evolution and Li dynamics in nanophase Li_3PS_4 by solid-state and pulsed-field gradient NMR *Chem. Mater.* **26** 3558–64

[77] Zheng F, Kotobuki M, Song S, Lai M O and Lu L 2018 Review on solid electrolytes for all-solid-state lithium-ion batteries *J. Power Sources* **389** 198–213

[78] Cao C, Li Z-B, Wang X-L, Zhao X-B and Han W-Q 2014 Recent advances in inorganic solid electrolytes for lithium batteries *Front. Energy Res.* **2** 25

[79] Kaup K, Bazak J D, Vajargah S H, Wu X, Kulisch J, Goward G R and Nazar L F 2020 A lithium oxythioborosilicate solid electrolyte glass with superionic conductivity *Adv. Energy Mater.* **10** 1902783

[80] Thangadurai V, Narayanan S and Pinzaru D 2014 Garnet-type solid-state fast Li ion conductors for Li batteries: critical review *Chem. Soc. Rev.* **43** 4714–27

[81] Murugan R, Thangadurai V and Weppner W 2007 Fast lithium ion conduction in garnet-type $Li_7La_3Zr_2O_{12}$ *Angew. Chem. Int. Ed.* **46** 7778–81

[82] Lian P-J, Zhao B-S, Zhang L-Q, Xu N, Wu M-T and Gao X-P 2019 Inorganic sulfide solid electrolytes for all-solid-state lithium secondary batteries *J. Mater. Chem.* A **7** 20540–57

[83] Liu Q, Geng Z, Han C, Fu Y, Li S, He Y.-b., Kang F and Li B 2018 Challenges and perspectives of garnet solid electrolytes for all solid-state lithium batteries *J. Power Sources* **389** 120–34

[84] Larraz G, Orera A, Sanz J, Sobrados I, Diez-Gómez V and Sanjuán M L 2015 NMR study of Li distribution in $Li_{7-x}H_xLa_3Zr_2O_{12}$ garnets *J. Mater. Chem.* A **3** 5683–91

[85] Song J Y, Wang Y Y and Wan C C 1999 Review of gel-type polymer electrolytes for lithium-ion batteries *J. Power Sources* **77** 183–97

[86] Munoz S and Greenbaum S 2018 Review of recent nuclear magnetic resonance studies of ion transport in polymer electrolytes *Membranes (Basel)* 8 120

[87] Fenton D E, Parker J M and Wright P V 1973 Complexes of alkali metal ions with poly (ethylene oxide) *Polymer* **14** 589

[88] Berthier C, Gorecki W, Minier M, Armand M B, Chabagno J M and Rigaud P 1983 Microscopic investigation of ionic conductivity in alkali metal salts-poly(ethylene oxide) adducts *Solid State Ionics* **11** 91–5

[89] Kovarsky R, Golodnitsky D, Peled E, Khatun S, Stallworth P E, Greenbaum S and Greenbaum A 2011 Conductivity enhancement induced by casting of polymer electrolytes under a magnetic field *Electrochim. Acta* **57** 27–35

[90] Christie A M, Lilley S J, Staunton E, Andreev Y G and Bruce P G 2005 Increasing the conductivity of crystalline polymer electrolytes *Nature* **433** 50–3

[91] Meabe L, Huynh T V, Lago N, Sardon H, Li C, O'Dell L A, Armand M, Forsyth M and Mecerreyes D 2018 Poly(ethylene oxide carbonates) solid polymer electrolytes for lithium batteries *Electrochim. Acta* **264** 367–75

[92] Becher M, Becker S, Hecht L and Vogel M 2019 From local to diffusive dynamics in polymer electrolytes: NMR studies on coupling of polymer and ion dynamics across length and time scales *Macromolecules* **52** 9128–39

[93] Messinger R J, Huynh V, Bouchet T, Sarou-Kanian R and Deschamps V M 2020 Magic-angle-spinning-induced local ordering in polymer electrolytes and its effects on solid-state diffusion and relaxation NMR measurements *Magn. Reson. Chem.* **58** 1118–29

[94] Ye Y-S, Rick J and Hwang B-J 2013 Ionic liquid polymer electrolytes *J. Mater. Chem.* A **1** 2719–43

[95] Xu H, Chien P-H, Shi J, Li Y, Wu N, Liu Y, Hu Y-Y and Goodenough J B 2019 High-performance all-solid-state batteries enabled by salt bonding to perovskite in poly(ethylene oxide) Proc. of the National Academy of Sciences 116 18815–21

[96] Wu N, Chien P-H, Li Y, Dolocan A, Xu H, Xu B, Grundish N S, Jin H, Hu Y-Y and Goodenough J B 2020 Fast Li^+ conduction mechanism and interfacial chemistry of a NASICON/polymer composite electrolyte *JACS* **142** 2497–505

[97] Wang Y, Guo X, Greenbaum S, Liu J and Amine K 2001 Solid electrolyte interphase formation on lithium-ion electrodes: a ^{7}Li nuclear magnetic resonance study *Electrochem. Solid-State Lett.* **4** A68

[98] Meyer B M, Leifer N, Sakamoto S, Greenbaum S G and Grey C P 2005 High field multinuclear NMR investigation of the SEI layer in lithium rechargeable batteries *Electrochem. Solid-State Lett.* **8** A145

[99] Leifer N, Smart M C, Prakash G K S, Gonzalez L, Sanchez L, Smith K A, Bhalla P, Grey C P and Greenbaum S G 2011 ^{13}C solid state NMR suggests unusual breakdown products in SEI formation on lithium ion electrodes *J. Electrochem. Soc.* **158** A471

[100] Jin Y *et al* 2017 Identifying the structural basis for the increased stability of the solid electrolyte interphase formed on silicon with the additive fluoroethylene carbonate *JACS* **139** 14992–5004

[101] Banerjee A, Wang X, Fang C, Wu E A and Meng Y S 2020 Interfaces and interphases in all-solid-state batteries with inorganic solid electrolytes *Chem. Rev.* **120** 6878–933

[102] Yu X and Manthiram A 2018 Electrode–electrolyte interfaces in lithium-based batteries *Energy Environ. Sci.* **11** 527–43

[103] Gauthier M *et al* 2015 Electrode–electrolyte interface in Li-ion batteries: current understanding and new insights *J. Phys. Chem. Lett.* **6** 4653–72

[104] Yu C, Ganapathy S, de Klerk N J J, Roslon I, van Eck E R H, Kentgens A P M and Wagemaker M 2016 Unravelling Li-ion transport from picoseconds to seconds: bulk versus interfaces in an argyrodite Li_6PS_5Cl–Li_2S all-solid-state li-ion battery *JACS* **138** 11192–201

[105] Yu C, Ganapathy S, Eck E R H V, Wang H, Basak S, Li Z and Wagemaker M 2017 Accessing the bottleneck in all-solid state batteries, lithium-ion transport over the solid-electrolyte-electrode interface. *Nat. Commun.* **8** 1086

[106] Li H, Guo S and Zhou H 2021 In-situ/operando characterization techniques in lithium-ion batteries and beyond *J. Energy Chem.* **59** 191–211

[107] Dai H and Zawodzinski T A 1996 Determination of lithium ion transference numbers by electrophoretic nuclear magnetic resonance *J. Electrochem. Soc.* **143** L107–9

[108] Zhang Z and Madsen L A 2014 Observation of separate cation and anion electrophoretic mobilities in pure ionic liquids *J. Chem. Phys.* **140** 084204

[109] Stilbs P and Furó I 2006 Electrophoretic NMR *Curr. Opin. Colloid Interface Sci.* **11** 3–6

[110] Rosenwinkel M P and Schönhoff M 2019 Lithium transference numbers in PEO/LiTFSA electrolytes determined by electrophoretic NMR *J. Electrochem. Soc.* **166** A1977–83

[111] Gouverneur M, Schmidt F and Schönhoff M 2018 Negative effective Li transference numbers in Li salt/ionic liquid mixtures: does Li drift in the 'Wrong' direction? *Phys. Chem. Chem. Phys.* **20** 7470–8

[112] Brinkkötter M, Giffin G A, Moretti A, Jeong S, Passerini S and Schönhoff M 2018 Relevance of ion clusters for Li transport at elevated salt concentrations in [Pyr12O1][FTFSI] ionic liquid-based electrolytes *Chem. Commun.* **54** 4278–81

[113] Klett M, Giesecke M, Nyman A, Hallberg F, Lindström R W, Lindbergh G and Furó I 2012 Quantifying mass transport during polarization in a li ion battery electrolyte by *in situ* ^{7}Li NMR imaging *JACS* **134** 14654–7

[114] Mohammadi M and Jerschow A 2019 *In situ* and operando magnetic resonance imaging of electrochemical cells: a perspective *J. Magn. Reson.* **308** 106600

[115] Krachkovskiy S A, Foster J M, Bazak J D, Balcom B J and Goward G R 2018 Operando mapping of Li concentration profiles and phase transformations in graphite electrodes by

magnetic resonance imaging and nuclear magnetic resonance spectroscopy *J. Phys. Chem.* C **122** 21784–91

[116] Ilott A J, Mohammadi M, Chang H J, Grey C P and Jerschow A 2016 Real-time 3D imaging of microstructure growth in battery cells using indirect MRI Proc. of the National Academy of Sciences 113 10779–84

[117] Sethurajan A K, Krachkovskiy S A, Halalay I C, Goward G R and Protas B 2015 Accurate characterization of ion transport properties in binary symmetric electrolytes using *In situ* NMR imaging and inverse modeling *J. Phys. Chem.* B **119** 12238–48

[118] Nyman A, Behm M and Lindbergh G 2008 Electrochemical characterisation and modelling of the mass transport phenomena in $LiPF_6$–EC–EMC electrolyte *Electrochim. Acta* **53** 6356–65

[119] Chandrashekar S, Trease N M, Chang H J, Du L-S, Grey C P and Jerschow A 2012 ^{7}Li MRI of Li batteries reveals location of microstructural lithium *Nat. Mater.* **11** 311–5

[120] Chang H J, Ilott A J, Trease N M, Mohammadi M, Jerschow A and Grey C P 2015 Correlating microstructural lithium metal growth with electrolyte salt depletion in lithium batteries using 7Li MRI *JACS* **137** 15209–16

[121] Bazak J D, Krachkovskiy S A and Goward G R 2017 Multi-temperature *in situ* magnetic resonance imaging of polarization and salt precipitation in lithium-ion battery electrolytes. *J. Phys. Chem.* C **121** 20704–13

[122] Britton M M, Bayley P M, Howlett P C, Davenport A J and Forsyth M 2013 *In situ*, real-time visualization of electrochemistry using magnetic resonance imaging *J. Phys. Chem. Lett.* **4** 3019–23

[123] Britton M M 2014 Magnetic resonance imaging of electrochemical cells containing bulk metal *ChemPhysChem* **15** 1731–6

[124] Liu X, Liang Z, Xiang Y, Lin M, Li Q, Liu Z, Zhong G, Fu R and Yang Y 2021 Solid-state NMR and MRI spectroscopy for Li/Na batteries: materials, interface, and *in situ* characterization *Adv. Mater.* 2005878

[125] Xiang Y *et al* 2020 Visualizing the growth process of sodium microstructures in sodium batteries by *in situ* 23Na MRI and NMR spectroscopy *Nat. Nanotechnol.* **15** 883–90

[126] Stone N 2005 Table of nuclear magnetic dipole and electric quadrupole moments *At. Data Nucl. Data Tables* **90** 75–176

[127] Pyykkö P 2008 Year-2008 nuclear quadrupole moments *Mol. Phys.* **106** 1965–74

Chapter 8

X-ray photoelectron spectroscopy for batteries

Justin G Connell

8.1 Principles and operation

As existing Li-ion battery systems begin to approach the theoretical limits of their performance, solving problems related to existing battery materials and, in particular, efforts dedicated to the discovery of new systems that move beyond current Li-ion technologies are becoming increasingly reliant on detailed, atomic- and molecular-level understanding of surface chemistry and its impact on electrochemical response. Due in large part to this shift in research focus, the x-ray photoelectron spectroscopy (XPS) technique is enjoying a renaissance in usage in the battery field, with a significant and increasing fraction of papers in the field utilizing XPS to understand a variety of materials phenomena including charge/oxidation state, materials composition and surface speciation. With this significant increase in usage has also come a significant increase in often conflicting or contradictory results that point to a need for improved awareness of best practices for utilizing the XPS technique to analyze battery systems.

This chapter will outline the important operating principles, applications and limitations of XPS as applied to battery materials. Rather than a comprehensive overview, this chapter is intended as an operator's guide for how to practically (and skeptically) approach the XPS technique when investigating battery systems and ensure that the analysis of XPS data is presented fairly and with appropriate qualifications and/or complementary measurements that support the proposed interpretation. A brief overview of the underlying physics of XPS is provided for baseline understanding of the technique, and a brief summary of other complementary methods is given to contextualize XPS in terms of other surface analysis and spectroscopy tools. Several excellent books can be consulted by readers interested in a more detailed discussion of different surface spectroscopy techniques [1] as well as the XPS method specifically [2, 3]. Multiple examples of the types of battery-relevant questions XPS is well-suited to answering are discussed, including descriptions of emerging *in situ* and *operando* lab- and synchrotron-based techniques

doi:10.1088/978-0-7503-2682-7ch8

that are pushing the frontier of what can be done with XPS. Finally, some practical considerations for measuring and interpreting x-ray photoelectron spectra are provided as a roadmap to aid with the appropriate interpretation, and perhaps more importantly, qualification of the conclusions drawn from XPS measurements of battery materials.

8.1.1 Photoelectric effect and x-rays

The photoelectric effect is the basis for the XPS measurement and derives from the fact that surfaces will emit electrons when irradiated by sufficiently energetic photons (figure 8.1(a)). This effect had been observed by others previously but was first explained by Einstein in 1905 and was the work for which he was awarded the Nobel Prize in 1921. The XPS measurement, also referred to as electron spectroscopy for chemical analysis (ESCA), relies upon measuring the distribution of electron energies that are photoemitted from a surface irradiated by x-ray photons.

The physical process underlying the photoelectric effect is illustrated in figure 8.1(b). Incident photons with sufficient kinetic energy (KE) result in the ejection of core level electrons whose binding energy (BE) is characteristic both of the specific element as well as the chemical environment of that species within the sample being studied. BE is related to KE of the photoemitted electron through equation (8.1):

$$KE = h\nu - BE \tag{8.1}$$

where h is Planck's constant and ν is the frequency of the incident radiation, together defining the photon KE (generally $\geqslant$1 keV). The irradiated atom relaxes from this excited state by filling the vacant core-hole with an outer shell electron. This relaxation process releases energy and results in either the emission of secondary, Auger electrons (the basis of Auger electron spectroscopy, discussed in section 8.1.5) or the emission of an x-ray photon with $h\nu$ less than that of the incident x-rays (the basis of a variety of x-ray spectroscopies). In both cases, multiple pathways are available to enable relaxation depending on the orbital structure of the atomic species, and as a result the energy of electrons and photons released via core level relaxation is defined by the specific core and valence energy levels that participate in

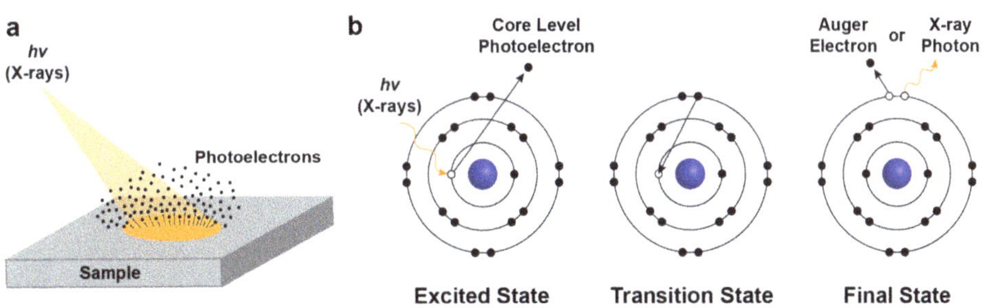

Figure 8.1. (a) The photoelectric effect. (b) Schematic of the photoemission process.

the relaxation process. Clear from equation (8.1) is the fact that if the KE of the incident photons is less than the BE of the core level electron, no excitation takes place. Furthermore, element-specific cross-sections for x-ray absorption mean that certain core levels are more susceptible to photoexcitation than others, leading to differences in peak intensities for different core level excitations [4].

8.1.2 Inelastic mean free path and information depth

Incident x-rays penetrate quite deeply into the sample (> 1 μm for ~1 keV x-rays), resulting in photoemission of electrons throughout the illuminated volume. However, these photoemitted electrons are strongly scattered (both elastically and inelastically) by the electron density of adjacent atoms in the sample and are therefore attenuated as they pass through the material. It is this attenuation that gives photoelectron spectroscopies their surface sensitivity, as only photoemitted electrons that are generated close to the sample surface are able to escape and reach the detector. The intensity of photoelectron emission from a solid material is defined by Beer's Law (equation (8.2)):

$$I_k = I_0 e^{-d/\lambda \cos\theta} \tag{8.2}$$

where I_k is the measured intensity, I_0 is the incident intensity (i.e., the intensity of photoemitted electrons of a given energy at depth d), d is the sample depth from which the photoelectron is generated, λ is the inelastic mean free path (IMFP), and θ is the angle of incidence (i.e. the angle of the spectrometer relative to the sample normal). These parameters are illustrated schematically in figure 8.2. Of these coefficients, the IMFP is most important to understand as it defines the depth from which measurable signal is emitted for a given sample.

The IMFP is an attenuation coefficient that depends sensitively on the KE of the photoelectron and is defined as the average distance over which an electron of given

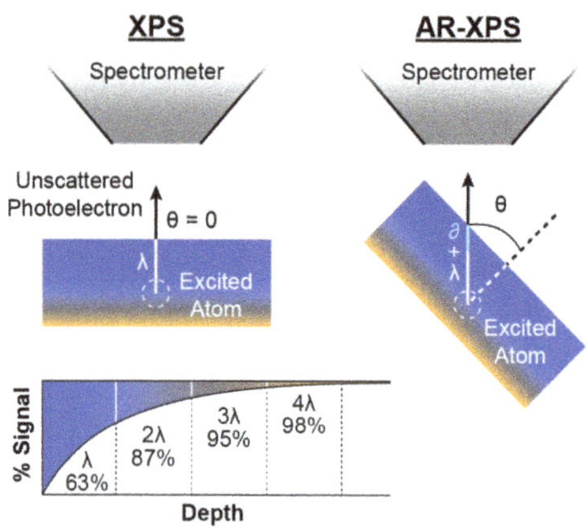

Figure 8.2. Relationship between IMFP and sampling depth, and impact of sample tilting on photoelectron path through the sample. δ indicates the additional distance a photoelectron emitted from the same depth as the untilted sample must travel.

energy can travel between successive inelastic collisions. Scattering can take place in the solid state (i.e. the sample) as well as in the liquid or gas phase. For this reason, XPS measurements are almost always performed under high or ultrahigh vacuum conditions (HV or UHV, $\leqslant 1 \times 10^{-6}$ or 1×10^{-8} mbar, respectively) in order to minimize the amount of additional scattering that takes place after photoelectrons exit the sample. Ambient pressure XPS (AP-XPS, section 8.2.3) is an emerging technique that makes it possible to circumvent this requirement in order to enable the analysis of, e.g., liquid samples, but requires specialized chambers, vacuum pumping arrangements and x-ray sources with very high energy and/or brightness in order to enable the measurement.

There are a variety of terms that, based on IMFP, are used to describe the analysis depth of an XPS measurement, but perhaps the most straightforward is the sampling depth, which is defined as the depth of material from which a specified percentage of the photoelectron signal originates [5]. As 95% of XPS signal originates from a depth of 3λ (figure 8.2), this will serve as the working definition of sampling depth used below. As the IMFP depends on the KE of the photoelectron, it is also dependent on the energy of the x-ray excitation. As a result, the use of more energetic x-ray sources (such as those at synchrotrons) will result in a larger IMFP and therefore a larger sampling depth. Furthermore, IMFP will vary as a function of the specific core level excitation, with photoelectrons from higher orbitals having larger IMFPs due to their lower BE and therefore higher KE. Finally, the $\cos\theta$ term in equation (8.2) implies that the sampling depth is tunable as a function of the angle between the sample normal and the electron energy analyzer, as increasing angle increases the effective depth (i.e. the path length) which photoelectrons must travel through in order to be detected (figure 8.2). This is the basis of angle-resolved XPS (AR-XPS), a non-destructive depth-profiling technique that will be discussed briefly in section 8.2.3.

Values of IMFP for pure elements have been measured directly and are summarized in figure 8.3(a) [6, 7]. Various calculators also exist for determining IMFP as a function of kinetic energy for elemental as well as some simple inorganic and organic compounds [8]. For most samples relevant to battery applications, however, such calculators can only be used as a rough estimation. As a general rule of thumb, the sampling depth of XPS measurements is ca. 5–10 nm; however, this can be much larger for materials comprised of very light elements. For example, the sampling depth for Li 1s photoelectrons in Li metal (BE ~54 eV) excited by Al Kα x-rays is ~16 nm. In contrast, the sampling depth for the same Li 1s photoelectrons emitted from a lithiated graphite anode is half that of Li metal (~8 nm). Care must also be taken in estimating sampling depth for heterogeneous samples, as the presence of surface layers (e.g. solid electrolyte interphases) or regions of variable composition (such as those present in laminates) will result in different IMFPs depending on the specific location in the sample from which photoemission takes place. In principle, it is possible to estimate the thickness of an overlayer if its IMFP is known [9, 10]; however, such analyses also require that the layer thickness and composition is uniform across the analyzed area. This requirement holds even for samples with non-planar geometries [11, 12], and for most samples relevant to

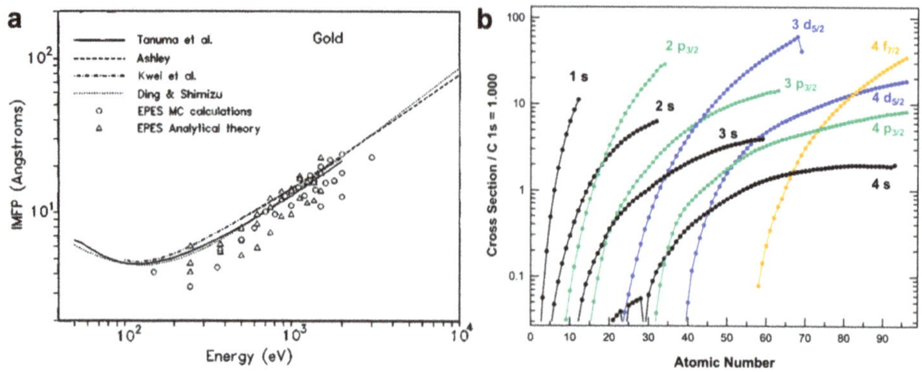

Figure 8.3. (a) Measured and calculated values of IMFP in gold as a function of KE. Reprinted from [7], with the permission of AIP Publishing. (b) Calculated elemental absorption cross-section for various core levels excited by Al Kα x-rays relative to the C 1s cross-section. After Scofield [4].

battery applications this is not the case. As a result, the estimation of layer thickness is often not possible for battery samples, and any such analyses encountered in the literature should be met with skepticism.

8.1.3 X-ray photoelectron spectra

As noted above, the XPS technique is based upon the measurement of the intensity of photoelectron emission as a function of BE (or KE) in order to obtain information about a sample's elemental composition and chemical makeup. An XPS spectrum is generally plotted in terms of BE as a function of spectral intensity, with BE decreasing from left to right (i.e. with increasing KE, figure 8.4). As absolute intensity (e.g., counts or counts per second, c.p.s.) can vary significantly from sample to sample, intensity is generally plotted as arbitrary units (a.u.) and sample-to-sample comparisons are then normalized to the maximum intensity of features of interest. For appropriately prepared samples it is also possible to directly compare changes in spectral intensity (e.g. to monitor the deposition of a successively thick overlayer); however, the same caveats related to layer thickness/uniformity noted in the previous section also hold for such analyses.

Collection of two types of XPS spectra are generally performed when analyzing a given sample. First, a scan over a large energy range with relatively low energy resolution but correspondingly high count rates is performed (i.e., high pass energy), referred to as a wide scan or survey spectrum (figure 8.4(a)). Survey spectra reveal which elements are present at the sample surface and provide a means of performing quantitative analysis of the overall sample composition. Quantification is generally performed by weighting spectral contributions from individual core levels by Scofield relative sensitivity factors (RSFs, figure 8.3(b)) [4], which account for relative differences in absorption cross-section between different core level excitations, as well as for the different kinetic energies (and therefore information depths) of the different core levels.

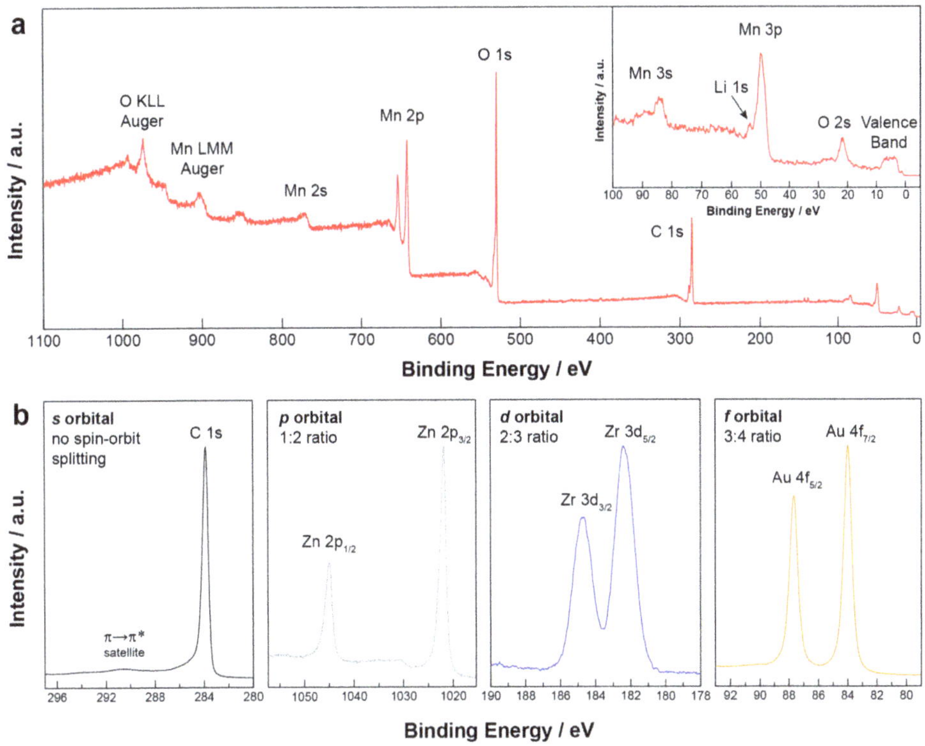

Figure 8.4. (a) Survey spectrum from $LiMn_2O_4$ cathode particles collected with a 40 eV pass energy. Inset shows detail below 100 eV. (b) Core level spectra from a variety of surfaces collected with a 20 eV pass energy. From left to right: C 1s from highly ordered pyrolytic graphite, Zn 2p from a Zn metal anode, Zr 4d from a $Li_{6.25}Al_{0.25}La_3Zr_2O_{12}$ solid electrolyte, Au 4f from a gold electrode. With the exception of C 1s, all other core levels exhibit spin–orbit splitting with fixed peak area ratios related to the specific orbital, as indicated. Individual peaks are labeled with their spin multiplicity. (Author's own work, unpublished.)

Second, a scan (or series of scans) over a narrower energy range with higher energy resolution but correspondingly lower count rates (i.e., low pass energy) is performed, referred to as a core level spectrum. As the name indicates, these spectra are used to analyze specific core levels of the elements of interest in a sample. Most core levels in XPS intrinsically exhibit two peaks due to a phenomenon called spin–orbit splitting, which is the result of differences in orbital energy levels arising from electron spin (± ½). With the exception of *s* orbitals (figure 8.4(b)), which only have spin ½ electrons, all other orbitals exhibit spin–orbit splitting (figures 8.4(c) and (d)). spin–orbit peak doublets exhibit characteristic peak ratios, which are the same for all orbitals of a given type, and characteristic energy separation, which increases as a function of *Z* and decreases with increasing valence level (i.e., 2p > 3p). Shifts in core level BE provide information on the chemical environment of the specific element (e.g., oxidation state, speciation, etc…), and can be used to determine, e.g., the extent of charge/discharge in an electrode material or whether specific compounds have formed at the sample surface. It is the analysis of chemical information

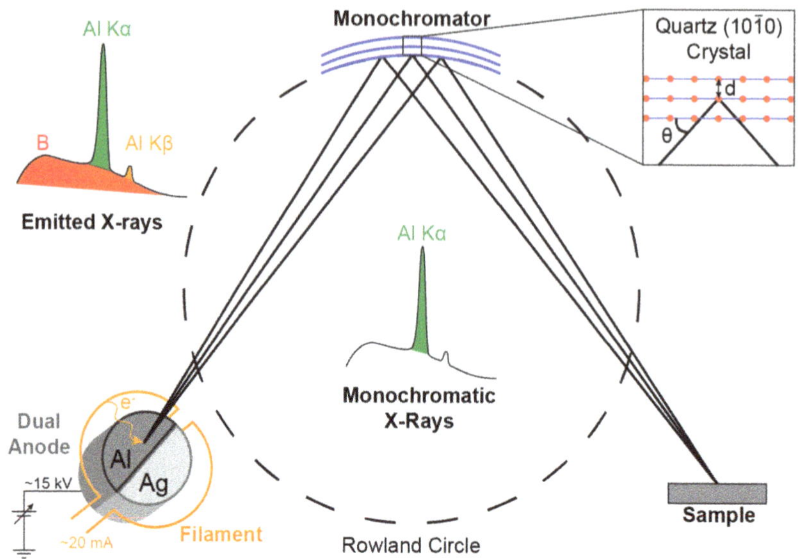

Figure 8.5. Schematic of a dual anode x-ray source with a monochromator, illustrating the energy distribution of emitted x-ray photons, the function of the monochromator, and the resulting energy distribution of x-rays after diffraction from the monochromator. 'B' in the emitted x-ray energy spectrum indicates the contribution of bremsstrahlung radiation.

provided by core level BE shifts that is the greatest power of the XPS technique and is also the greatest source of confusion and misinterpretation of XPS data—an issue that is common across disciplines and not limited to the battery literature. A detailed discussion of the interpretation of BE shifts and associated analysis is given in section 8.3.

8.1.4 Instrumentation

Although there are almost as many different UHV chamber and sample manipulator designs as there are XPS instruments, the core hardware for performing for the actual XPS measurement has remained relatively unchanged for quite some time.

X-ray sources

Lab-based x-ray sources rely on the bombardment of a high voltage metallic anode with electrons emitted from a hot filament cathode in order to produce x-rays via fluorescence (figure 8.5). A variety of anode materials with varying energies can be used (table 8.1), and the most common sources are Mg and Al. Higher energy sources are useful for increasing the sampling depth as well as for changing the position of Auger features in the event of overlap with core levels of interest; however, the broader linewidths of these sources (~2 eV) also result in poorer energy resolution.

Fluorescence-based x-ray generation also results in the production of significant bremsstrahlung (background) and satellite (e.g. $K\alpha_2/K\beta$) x-rays, as well as high

Table 8.1. Characteristic x-ray energies and linewidths of several common x-ray sources.

X-ray line	Energy (eV)	Width (eV)
Mg Kα	1253.6	0.70
Al Kα	1486.6	0.85
Ag Lα	2984.3	2.6
Ti Kα	4510.0	2.0
Cr Kα	5417.0	2.1

energy electrons and heat, all of which degrade the quality of a measured XPS spectrum if they reach the sample (figure 8.5). As a result, most modern XPS instruments also include a monochromator, generally consisting of one or more quartz crystals that are used to diffract the generated x-rays according to Bragg's law:

$$n\lambda = 2d\sin\theta \tag{8.3}$$

where n is the diffraction order, λ is the wavelength of the radiation, d is the crystal lattice spacing and θ is the angle between the crystal plane and the incident light. By controlling the angle of the quartz crystal relative to the source, the energy of a single x-ray line can be selected and prevent any additional x-rays emitted from the source from reaching the sample (figure 8.5). The use of monochromators significantly improves spectral quality, but at the expense of some of the brightness of the source. Per equation (8.3), if the relative energies of two x-ray lines are integer multiples of one another (e.g., Ag Lα and Cr Kα at ~2x and ~4x, respectively, relative to Al Kα), the same monochromator can be used for multiple anodes by only slight adjustments of the monochromator angle.

Energy analyzers

Detection of x-ray photoelectrons is most commonly performed using a hemispherical energy analyzer (figure 8.6(a)). Hemispherical analyzers rely on the principle of electrostatic deflection, with concentric inner and outer hemispherical plates held at positive and negative potentials, respectively, used to deflect incoming electrons and only allow those electrons within a specific range of KE (as defined by the equipotential surface between the two sections) to pass through the analyzer and be detected. Photoelectrons are first focused by a series of lenses before entering the hemispherical section. The lenses also retard the kinetic energy of the electrons to a fixed energy, called the pass energy (E_p). The pass energy defines the energy resolution (ΔE) of the spectrum to be collected via the ratio $\Delta E/E_p$, which is a constant value determined by the analyzer radius (R_0), acceptance angle (α) and entrance slit width (w), all of which are fixed during spectrum acquisition [2] (figure 8.6(a)). Lower values of E_p result in higher resolution (i.e., peaks with small full width at half maximum, FWHM) but lower signal, whereas higher E_p results in increased signal but lower resolution (figures 8.6(b) and (c)). Spectra collected using

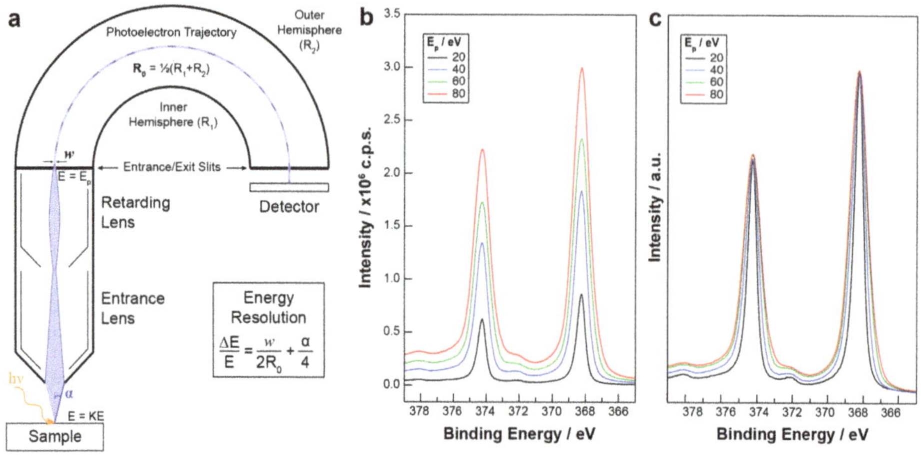

Figure 8.6. (a) Schematic of a hemispherical analyzer and representative paths traveled by photoelectrons from sample to detector. (b) Ag 3d core level spectra demonstrating effect of E_p on signal and (c) same data normalized to show the impact on energy resolution. (Author's own work, unpublished.)

a fixed E_p require scanning the lens voltages to allow for measurement across the desired range of BE.

Charge neutralization

Charge neutralization systems are particularly important when analyzing battery materials, as samples often contain poorly conductive and/or insulating layers or are themselves poorly conductive. The XPS measurement results in constant removal of electrons from the sample during x-ray irradiation, which, if not replaced, result in the development of a positive charge at the sample surface. This results in a modification to equation (8.1), where the measured KE (KE_{meas}) of photoelectron peaks depend not only on the x-ray energy and core level BE, but also on the analyzer work function (ϕ_a, a constant) and the sample surface charge (E_s):

$$KE_{meas} = h\nu - BE - \phi_a - E_s \tag{8.4}$$

For highly conductive samples, compensating electrons are easily supplied from the bulk of the material, resulting in $E_s = 0$ (figure 8.7(a)). In this case KE_{meas} is only modified by ϕ_a, as the sample is grounded to the analyzer. For poorly conductive or insulating samples, however, there is very little or no compensating supply of electrons to replace the exiting photoelectrons, resulting in a positive E_s (figure 8.7(b)). This positive E_s lowers the KE of exiting photoelectrons, resulting in an apparent shift of the spectrum to higher BE and, in extreme cases, the inability to collect an XPS spectrum at all. This condition can be corrected using an electron flood gun, which supplies an external source of low energy electrons (generally $\leqslant 5$ eV) to compensate this surface charge. Rather than attempting to exactly match E_s, which is impractical to measure experimentally, charge compensation via flood gun is usually performed with slight overcompensation, resulting in $E_s < 0$ that is

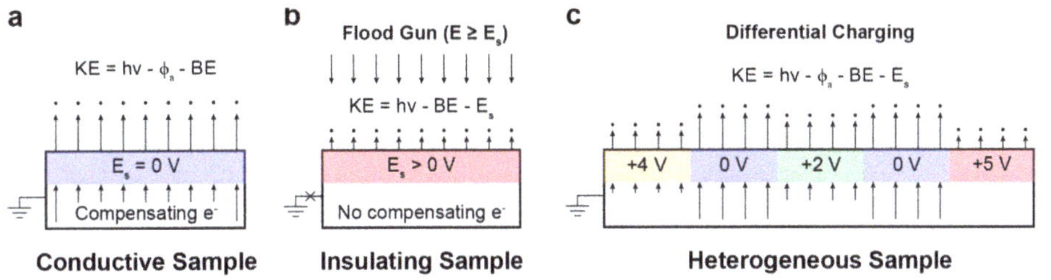

Figure 8.7. (a) Charge compensation for a conductive sample. (b) Lack of charge compensation for an insulating sample resulting in surface charging. An external supply of low energy electrons from a flood gun can compensate this surface charge. (c) Differential charging on a heterogeneous sample with regions of varying conductivity resulting in non-uniform surface charge.

controlled by the flood gun rather than the sample. Conditions are then optimized to provide the best peak FWHM for analysis. Although conceptually simple, charge neutralization via flood gun is not a catch-all solution, particularly when regions of differing conductivity are present on the sample surface. In such samples, regions with differing values of E_S develop (figure 8.7(c)), resulting in non-uniform charge compensation and correspondingly non-uniform BE shifts that must be accounted for when analyzing spectra (see section 8.3.2 for more detail).

Sample manipulation

Although not integral to the XPS instrumentation *per se*, chamber design, sample holders and corresponding sample mounting requirements are important to consider when analyzing battery materials. All systems can easily handle solid materials such as pellets, thin films, foils and laminates. Depending on sample holder geometry it is also usually possible to analyze powders by supporting on a vacuum-compatible adhesive (e.g., conductive carbon tape or insulating Kapton tape) or by embedding into a compliant metal support such as indium foil. Analysis of unsupported powders is also possible in some cases, e.g., by packing into a shallow plate; however, care must be taken when loading such samples so that they do not accidentally disperse inside the chamber either by too rapidly pumping down or by spilling during sample transfer inside the system. AP-XPS tools have further enabled the analysis of liquid electrolytes and liquid-coated electrodes, allowing for *operando* measurements of electrode and electrolyte materials.

Most systems are equipped with introductory/preparation chambers that enable loading of samples without breaking vacuum inside the spectrometer. Systems with introduction chambers that have inlets located inside of controlled atmosphere glove boxes or vacuum transfer systems that enable 'air-free' transfer of samples from a glove box are also becoming more common, which is a critical capability when analyzing air-sensitive materials. If such transfer systems are not available, great care must be taken when interpreting the results of XPS measurements, as the analyzed volume of an air-sensitive sample will almost certainly have been modified from its initial state during transfer through air into the XPS instrument. Note, however, that 'air-free' transfer systems do not guarantee that chemical modification of a surface has not taken place—it only makes such modifications less likely. Even the sub-part-per-million O_2 and H_2O levels in a common glove box are sufficient to induce significant chemical modifications to reactive surfaces such as Li metal, which can oxidize even under UHV conditions [13].

8.1.5 Complementary surface analysis techniques

Although XPS can provide a broad spectrum of information about materials surface chemistry, it is only one tool among many surface-sensitive analysis methods available at the lab scale. Below, a brief description is given of several complementary surface-sensitive tools to contextualize results that will be described in section 8.2. Complementary synchrotron-based capabilities are not considered in this chapter (with the exception of HAXPES, discussed in section 8.2.3), and the interested reader is referred to chapter 3 for a detailed discussion of synchrotron methods. Fourier Transform Infrared (FTIR) and Raman spectroscopy can also provide complementary, surface-specific information, and a detailed discussion of these techniques can be found in chapter 7.

Auger electron spectroscopy (AES)

AES uses high energy electrons, rather than photons, to excite core level electrons from the surface of the sample in an otherwise identical process to Auger electron emission in XPS (figure 8.1(b)). AES is generally more sensitive to lighter elements, as the Auger emission probability decreases with increasing Z (figure 8.8(a)). As-measured Auger spectra yield a series of adsorption edges specific to the elements present in the sample, and the ability to more readily focus electron beams relative to x-rays make spatial mapping more straightforward than for XPS (figure 8.8(b)). Auger spectra are more commonly viewed in their differentiated form (figure 8.8(c)), which makes it easier to distinguish relative shifts and changes in line shape, both of which can be used to extract information about the chemical state of the elements present (figure 8.8(c)). Although more commonly used for qualitative or semi-quantitative analysis of sample composition, quantitative analysis of AES spectra is also possible, with detection limits as low as ~100 ppm depending on the sample and specific element of interest [1].

Ultraviolet photoelectron spectroscopy (UPS)

Similar to XPS, UPS relies on the photoelectric effect to analyze photoemitted electrons from a sample surface. Instead of high energy x-rays, however, UPS utilizes relatively low photon energies (usually He(I) or He(II) at 21.2 or 40.8 eV, respectively) that result in emission of valence photoelectrons. Although UPS has not been widely utilized to study battery materials to date, it can provide useful information about the work function and band structure of electrode materials when used in combination with XPS, other spectroscopic techniques and theory [16] (figure 8.9(a)). It has also been used to monitor changes in the electronic structure of organic molecular electrodes in response to charge/discharge [17] (figure 8.9(b)).

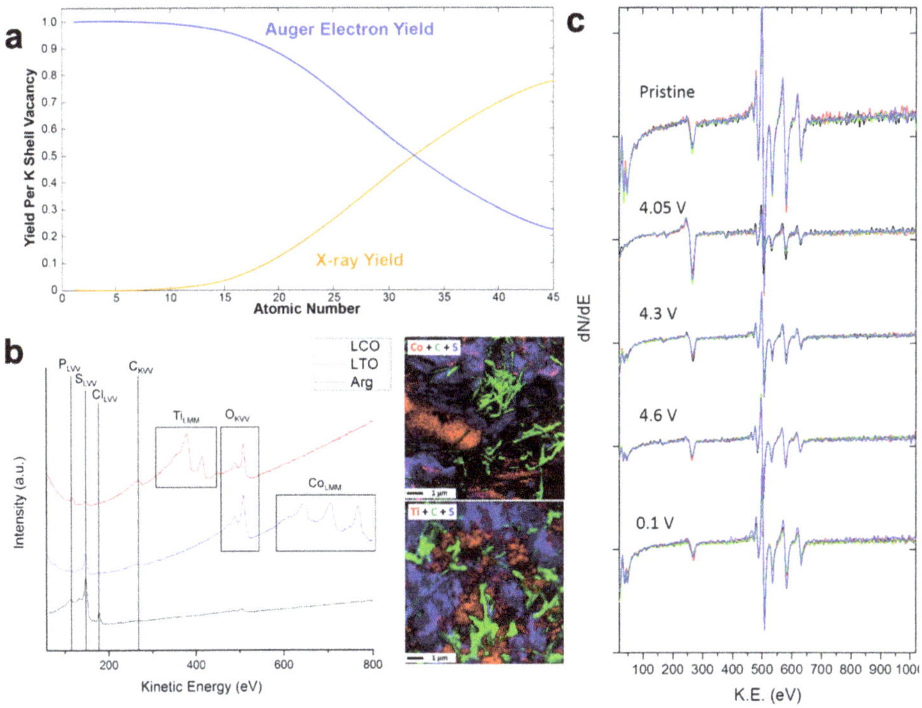

Figure 8.8. (a) Auger electron and x-ray fluorescence yield as a function of Z. (b) As-measured Auger spectrum of $LiCoO_2$ (LCO) cathode, $Li_4Ti_5O_{12}$ (LTO) anode and Li_6PS_5Cl (Arg) solid electrolyte and corresponding AES mapping of cathode (top) and anode (bottom) composite electrodes. Reprinted from [14], Copyright (2017), with permission from Elsevier. (c) Differentiated *in situ* AES spectrum of $LiMn_2O_4$ cathode showing changes in line shape as a function of applied potential. Reprinted from [15]. Copyright (2017) American Chemical Society.

Low energy ion scattering (LEIS)

Also known as Ion Scattering Spectroscopy (ISS), LEIS relies on the bombardment of a sample surface with ionized noble gases (e.g., He^+, Ne^+, Ar^+) to obtain information about the composition of the surface. The ions are backscattered from the surface, with some energy loss that is proportional to the momentum transfer, which is directly related to the mass of the ion and, importantly, the atomic mass of the species from which the ion is backscattered (figure 8.10(a)), providing elemental sensitivity. LEIS is unique in that it is sensitive only to the topmost monolayer of the sample, as only species that remain ionized after backscattering can be detected and this is highly unlikely after backscattering from species deeper in the sample. Typical ion energies used are 0.5–5.0 keV. Given the extreme surface sensitivity of this technique, its primary utility is the analysis of battery materials with relatively uniform, well-defined surfaces. In this case it is possible to identify properties such as preferential surface segregation of species in multicomponent oxides [18], as well as other surface-specific chemical information [19].

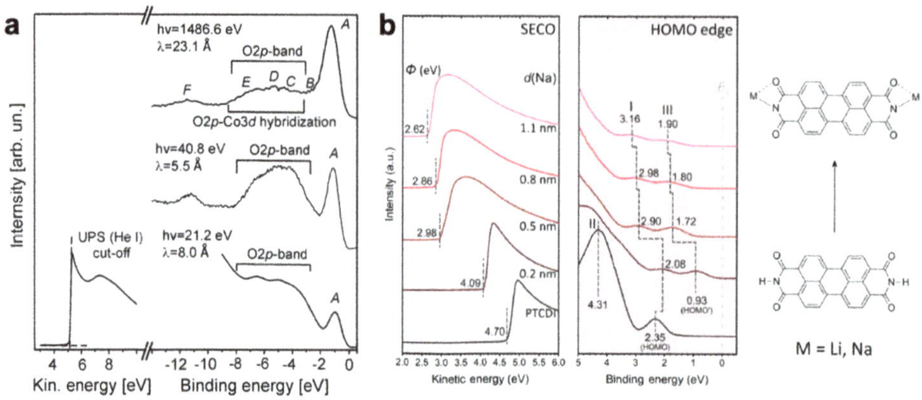

Figure 8.9. (a) XPS (top) He(II) (middle) and He(I) (bottom) UPS spectra detailing the valence band structure of a $LiCoO_2$ cathode. Reprinted from [16]. Copyright (2014) American Chemical Society. (b) *In situ* He(I) UPS measurements of the secondary electron cutoff (SECO, left) and highest occupied molecular orbital (HOMO, right) of an organic anode material, showing the impact of Na metal addition on the electronic structure of the molecule. A proposed reaction scheme for the ionic association is given in the scheme on the right. Reprinted by permission from Springer: Springer Nature *Nano Research* [17] Copyright (2020).

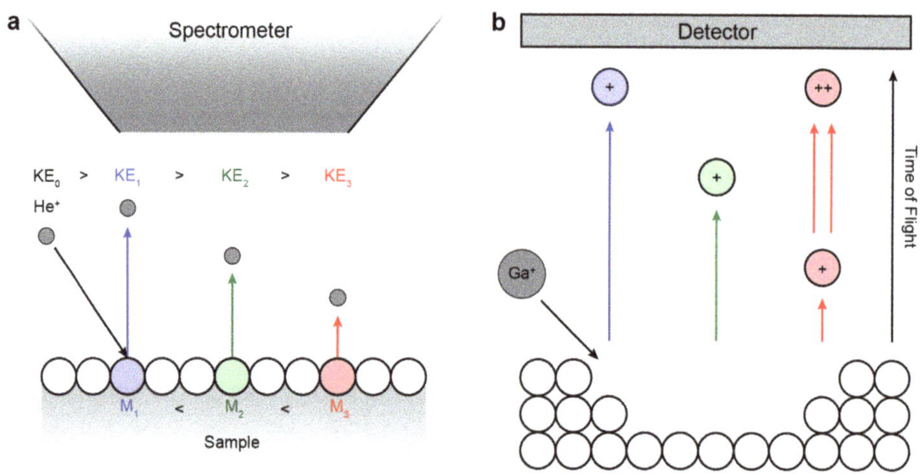

Figure 8.10. (a) Schematic of the LEIS measurement. Inelastic scattering of He^+ from atoms of differing mass ($M_1 > M_2 > M_3$) lowers the KE proportionally ($KE_1 < KE_2 < KE_3$). Only the top monolayer is sampled. (b) Schematic of the ToF-SIMS measurement. Heavier ions (e.g., Ga^+) are used to remove material, which is ionized (i.e., secondary ions) and accelerated through a time-of-flight sector towards a detector. Lighter ions travel faster through the ToF sector than heavier ions of the same charge and are detected earlier. Ions with higher charge ($z > 1$) are also detected earlier than ions of the same mass with $z = 1$.

Time of flight secondary ion mass spectrometry (ToF-SIMS)

ToF-SIMS is in some sense a complementary technique to LEIS in that it also relies on the bombardment of a sample with energetic ions in order to obtain compositional information about a sample. Rather than analyzing scattering of the primary

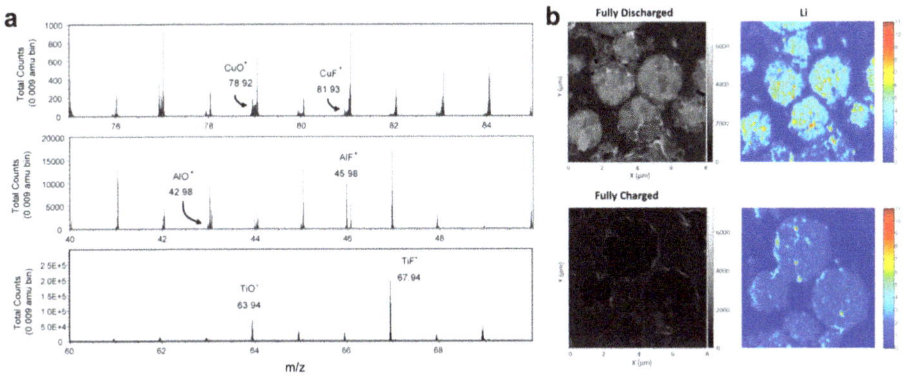

Figure 8.11. (a) ToF-SIMS mass spectra collected from Cu (top), Al (middle) and Ti (bottom) current collectors. Reprinted from [20], copyright (2009), with permission from Elsevier. (b) 2D ToF-SIMS map showing Li distribution in a fully discharged and charged $Li_{1.0}Mn_{0.54}Ni_{0.13}Co_{0.13}O_2$ cathode. Reprinted from [21], copyright (2015), with permission from Elsevier.

ion beam from the topmost monolayer, however, ToF-SIMS instead utilizes heavy primary ions to continuously remove atoms from the sample, which are ionized in the process (called secondary ions) and analyzed directly (figure 8.10(b)). Secondary ions removed from the surface are then passed through a long drift sector, where they are spatially separated by ToF according to their KE. The KE of secondary ions is dictated by the degree of momentum transfer from the primary ion beam, with heavier elements receiving less kinetic energy than lighter elements, as well as by the charge of the secondary ion ($z = \pm 1$, ± 2, etc…). For this reason, ToF-SIMS measurements are plotted as a function of mass to charge state ratio (m/z), rather than mass (figure 8.11(a)), as an ion with $z = +2$ will travel twice as fast through the ToF sector as a +1 ion with identical mass, yielding an additional peak at half the m/z of the +1 ion. In addition to depth profiling and 2D/3D imaging capabilities (figure 8.11(b)), ToF-SIMS measurements can detect extremely low concentrations ($\geqslant$1 ppb), making the technique ideal for measuring trace impurities in a sample as well.

8.2 XPS analysis of battery materials

As described in section 8.1, the XPS measurement is a powerful tool for understanding the surface (electro)chemistry of materials for batteries. In this section, specific examples will be given for the types of questions XPS is well-suited to answering—either as a standalone technique or in concert with the other analytical tools discussed in section 8.1.5.

8.2.1 Electrode charge/discharge

Due to its sensitivity to changes in BE as a result of local electronic environment, XPS is able to track changes in the oxidation state of electrode materials in response to charge/discharge and monitor their stability during extended cycling. In most cases it is possible to monitor qualitative changes in oxidation state by observing changes in peak position and/or line shape as a function of state of charge [22–24] (SOC, figure 8.12). In general, peak shifts to higher BE correspond to an increase in oxidation state (e.g. increasing SOC) while peak shifts to lower BE correspond to a decrease in oxidation state (e.g., decreasing SOC).

Quantitative analysis of mixed charge states is more complicated, however, and depends on the specific elements involved. Many of the first and second row transition metals are well-suited to quantitative analyses of charge/discharge behavior as they exhibit simple peak shapes with BEs that are well-separated as a function of oxidation state [25–29] (figure 8.13(a)). However, electrode materials containing Cu, Fe, Ni, Mn and Co require more rigorous analysis to quantify mixed valence, as these species can adopt oxidation states with complex peak shapes when unpaired electrons are present [30–32] (figures 8.13(b) and (c)). This phenomenon is referred to as multiplet splitting and will be discussed further in section 8.3.3. If suitable peak models are unavailable, mixed oxidation states can also be resolved by measuring primary peak BE shifts as a function of known material stoichiometry [22], or by analyzing higher valence core levels where multiplet splitting is less pronounced [33] (e.g., 3p vs 2p, figure 8.13(d)). Note that although such higher valence core levels do afford simpler peak shapes, they also suffer from the disadvantage of low signal-to-noise and overlap with other higher valence core levels, making them more difficult to detect and deconvolute—in particular when significant interphase layers are present. For Mn specifically, it has also been shown that multiplet splitting of the Mn 3s core level can be used as an indicator of Mn oxidation state [34–36]; however, other work has cast doubt on this

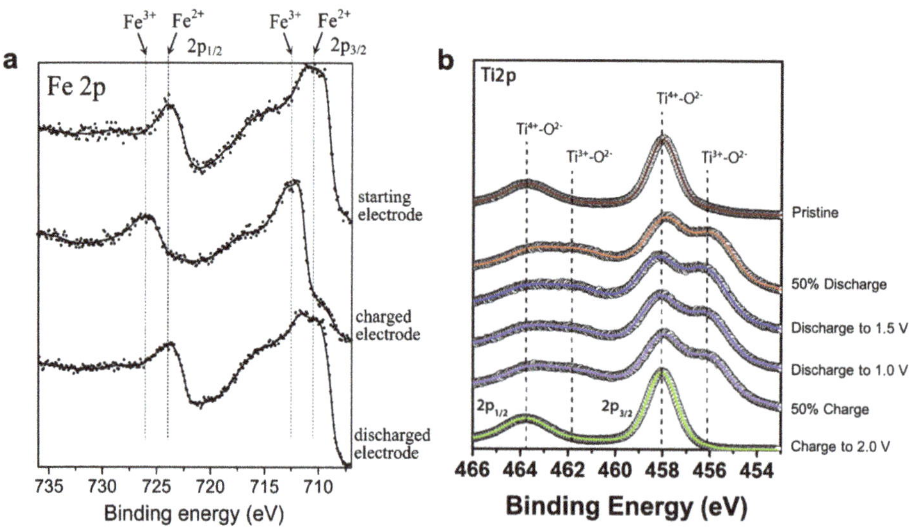

Figure 8.12. (a) Fe 2p core level spectra demonstrating effects of charge/discharge on Fe oxidation state in a $LiFePO_4$ cathode. Reprinted with permission from [23]. Copyright (2010) American Chemical Society. (b) Ti 2p core level spectra demonstrating effects of charge/discharge on Ti oxidation state in a $Li_4Ti_5O_{12}$ anode. Voltages are vs. Li/Li^+. Reprinted with permission from [24]. Copyright (2016) American Chemical Society.

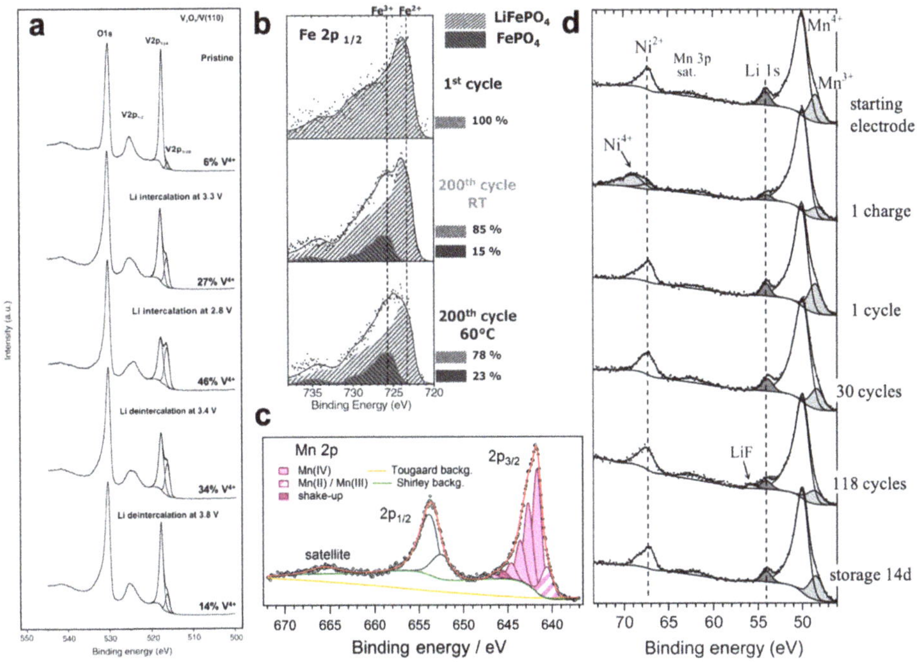

Figure 8.13. (a) Quantitative analysis of $V^{4+/5+}$ redox during Li intercalation/deintercalation into a V_2O_5 thin film. Reprinted from [26], Copyright (2007), with permission from Elsevier. (b) Quantification of Fe^{2+} and Fe^{3+} content in cycled $LiFePO_4$ cathodes taking into account the complex peak envelopes due to multiplet splitting of Fe 2p photoelectrons. Reproduced from [30]. Copyright IOP Publishing Ltd All rights reserved. (c) Deconvolution of Mn oxidation states including Mn^{4+} multiplet splitting envelope in $LiNi_{0.5}Mn_{1.5}O_4$. Reproduced from [31] with permission of The Royal Chemical Society. (d) Quantification of Mn and Ni redox as a function of cycling using higher valence core levels (Ni and Mn 3p) that do not exhibit strong multiplet splitting. Reprinted with permission from [33]. Copyright (2010) American Chemical Society.

methodology [37], and the same trends do not necessarily hold for ligands other than oxygen (e.g. hydroxides, fluorides, sulfides, etc…), which is relevant to consider when interphase layers are present. Reference [37] does note that peak fitting of the Mn 3s remains a viable method for characterizing mixed valence provided suitable standard peak models are available.

XPS is also well-suited to track intermediate species as a function of SOC for conversion chemistries such as metal oxides [38], lithium–sulfur [39] and lithium–air [40] (figure 8.14(a)). In Li–S systems, for example, accurate determination of electrode speciation requires careful consideration of reaction products that may also produce overlapping signals in the core level regions of interest (e.g. S 2p). Complementary techniques such as FTIR and Raman spectroscopy can be used to aid in understanding cathode chemistry and the nature of any reaction products that are formed during charge/discharge [41, 42]. *In situ* methodologies have also been developed using low vapor pressure ionic liquids to prevent electrolyte evaporation under UHV conditions and enable XPS measurements of changes in electrode

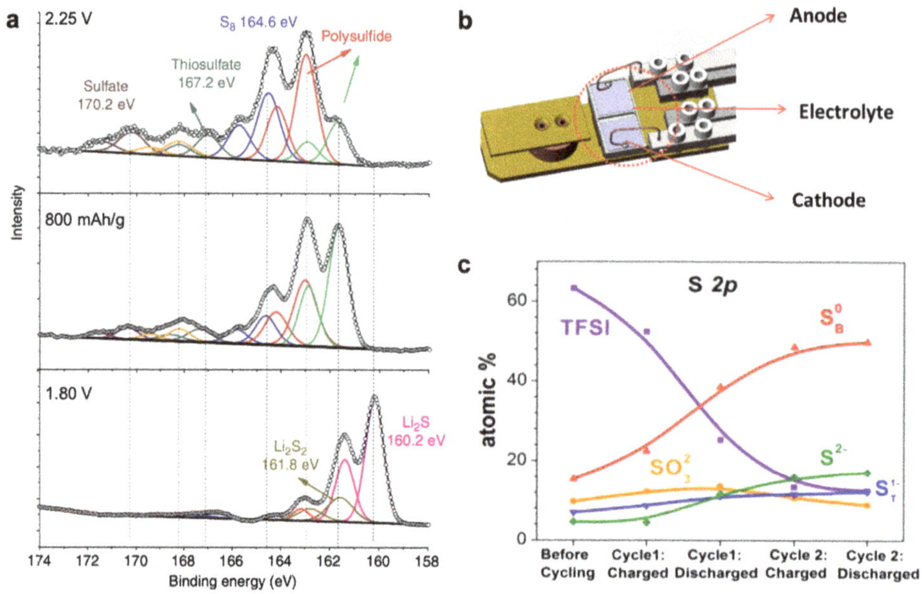

Figure 8.14. (a) Changes in sulfur speciation on a MnO_2 cathode support a function of discharge. Reprinted by permission from Macmillan Publishers Ltd: *Nature Communications* [39], copyright (2015). (b) *In situ* XPS measurement setup and (c) corresponding measurement of changes in sulfur speciation as a function of charge state. Reprinted with permission from [43]. Copyright (2017) American Chemical Society.

speciation as well as spatial mapping of reaction products during charge/discharge [43] (figure 8.14(b)).

8.2.2 Interphase chemistry and speciation

Perhaps the most common use of XPS for battery analysis is to understand the composition of reaction layers that form in response to both cycling of the battery and changes in electrolyte/electrode chemistry. Such layers are commonly referred to as the solid–electrolyte interphase (SEI) or, more generally, as interphases. In most cases these interphases are much thicker (~100–1000 nm) than the analytical depth of the XPS measurement (~5–10 nm), and care must therefore be taken in interpreting XPS data measured from battery electrodes to ensure the chemical information obtained is relevant to the process being studied. Nevertheless, XPS information about near-surface speciation can be of significant value to understanding changes in electrode chemistry as a function of electrolyte composition and/or electrochemical treatment. Furthermore, destructive and non-destructive XPS methodologies can also be used to understand how interphase composition and speciation change as a function of depth (see sections 8.2.3 and 8.2.4). In all cases extreme care must be taken when assigning speciation in interphases, as there is a vast literature proposing a wide range of binding energies for the same compounds [44]. Even very careful work can contain errors in assignment, and complementary

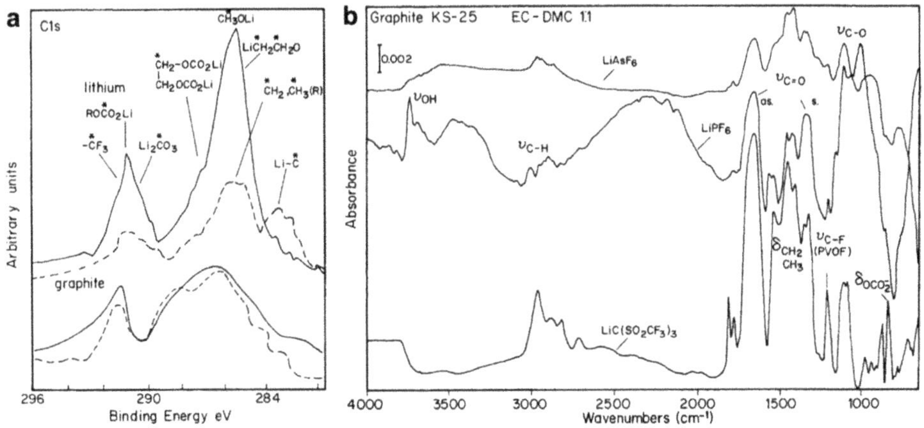

Figure 8.15. (a) C 1s core level spectra showing differences in interphase composition between graphite and Li metal electrodes exposed (Li) or cycled (graphite) in 1 M $LiAsF_6$ in 1:1 EC:DMC. (b) FTIR spectra for graphite electrodes cycled in various electrolytes used to corroborate interphase speciation from XPS of cycled electrodes. Reprinted from [50], Copyright (1999), with permission from Elsevier.

measurements to confirm species assignment are often critical to accurate characterization of interphase chemistry.

Anode interphases

Among the many possible anode materials, there is particularly intense interest within the battery community to understand interphase chemistry on graphite [44, 45] and lithium metal [46, 47] due to their technological relevance and (in the case of graphite) widespread use in many important battery systems. Graphite represents a particularly challenging surface on which to analyze interphase components via XPS, as signal from graphite overlaps with the organic interphase components in C 1s core level spectra. As a result, there are often no other species with which to reliably charge reference data, significantly complicating the analysis. XPS analysis of Li metal is also challenging due to the fact that the cross-section for x-ray absorption for Li 1s electrons is very low, resulting in a very weak XPS signal [4]. Furthermore, different species in Li 1s core level spectra are not as well-separated in BE, making it challenging to generate unique, quantitatively accurate peak fits. Nevertheless, XPS can provide uniquely useful information about interphase composition on both surfaces.

Graphite and Li, along with most electrodes used in batteries, tend to contain a mixture of both inorganic and organic components that depends on the electrode and specific electrolyte used [48–50] (figure 8.15). Careful correlation of species proposed from peak deconvolution of the C 1s and Li 1s core levels with spectral signatures from other, complementary core levels (e.g., O 1s, F 1s, P 2p, N 1s, S 2p, etc) can help confirm the presence or absence of a variety of decomposition products. Controlled gas-phase dosing experiments onto UHV-cleaned Li surfaces also provide useful information about the spectral signatures of simple Li compounds [13, 51], and other recent work has attempted to provide clearer BE referencing guidance for more uniform identification of Li reaction products relevant to Li anodes as well as Li-ion electrode surfaces in general [52] (see section 8.3.2). No matter how careful the analysis, however, it is sometimes not possible to unambiguously determine the SEI composition from XPS measurements alone. Complementary techniques such as FTIR and Raman spectroscopies [48, 50] and ToF-SIMS [53] can provide more detailed bonding and compositional information, yielding a more complete understanding of interphase structure (figure 8.15b). Note that most other spectroscopic techniques are more bulk sensitive, so care must be taken to understand any deviations from the near-surface chemistry measured by XPS.

The possibility of multiple chemical species with more or less identical spectral makeup makes precise identification of organic interphase species a particularly challenging process. Although not a widespread practice, there has been some effort to utilize valence band spectra of model interphase components as a means of distinguishing between chemically similar interphase components [54, 55] (figure 8.16). As valence band photoelectrons are the most sensitive to the chemical bonding environment of a given species, they can produce meaningful differences even for small changes in chemical identity (i.e., methyl versus ethyl groups on an organolithium compound, figure 8.16(a)). Although this methodology is likely not suitable for highly complex interphases, it is an interesting route by which more detailed chemical information about interphase speciation can be obtained relative

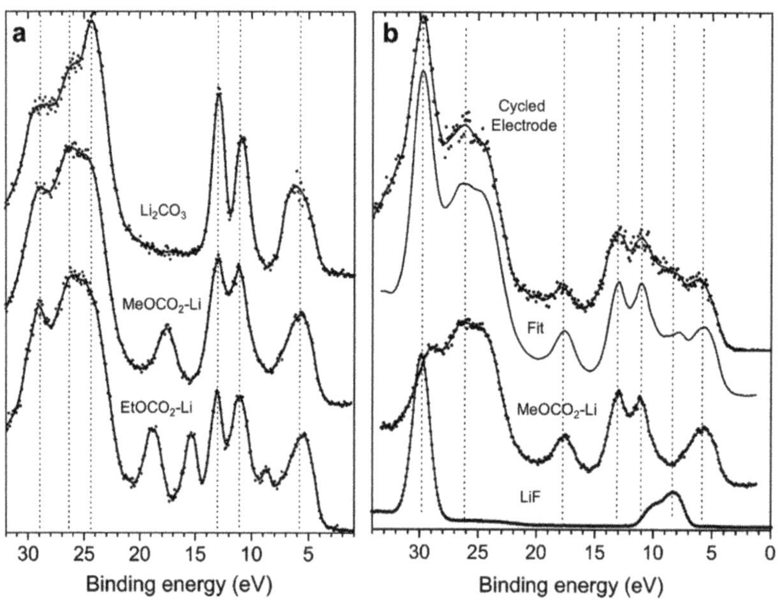

Figure 8.16. (a) Valence band spectra of different reference organolithium compounds. From top to bottom: Li_2CO_3, $MeOCO_2$-Li and $EtOCO_2$-Li. (b) Use of reference valence band spectra to determine interphase speciation on a cycled stainless steel electrode. From top to bottom: valence band spectrum of cycled electrode, fit derived from reference spectrum component analysis, reference spectrum for $MeOCO_2$-Li and reference spectrum for LiF. Reprinted with permission from [54]. Copyright (2005) American Chemical Society.

to traditional XPS analyses. Complementary UPS measurements, coupled with theory, would be useful additions to this methodology as well.

In contrast to graphite and lithium metal, interphase speciation on other electrode surfaces can be more straightforward to determine, particularly in cases where reaction products form with the inorganic components of the electrode. On silicon anodes, for example, the presence of lithiated phases and electrolyte reaction products can readily be distinguished and compared as a function of electrolyte chemistry and discharge potential [56] (figure 8.17). XPS has also been utilized to understand changes in surface chemistry in response to phase changes during charge/discharge in other inorganic anode materials such as $Li_4Ti_5O_{12}$ [24]. As with graphite and Li anodes, the presence of any proposed inorganic reaction products must be corroborated with the presence of similar species in other relevant core levels to ensure accurate characterization of interphase chemistry.

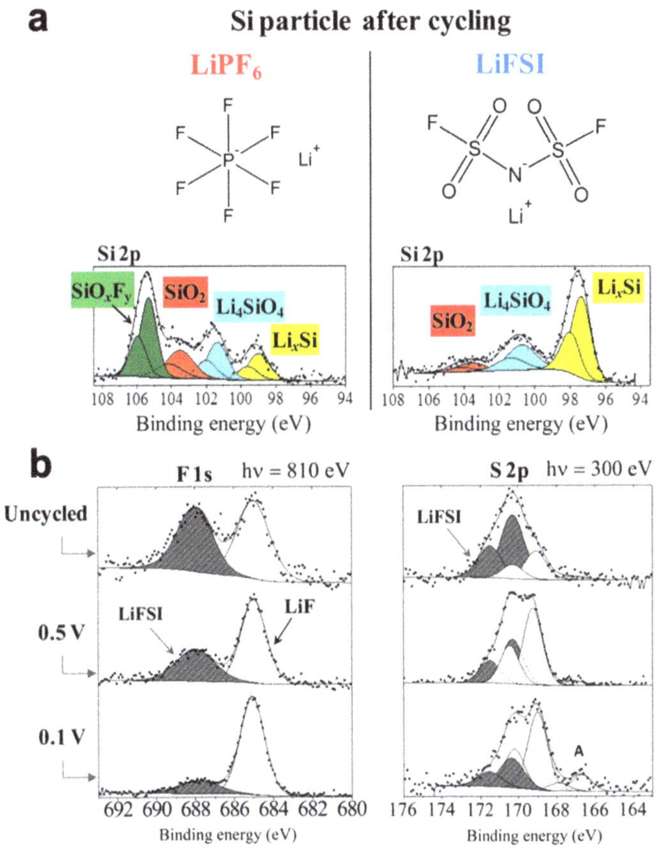

Figure 8.17. (a) Si 2p core level spectra comparing interphase composition on a silicon anode as a function of electrolyte chemistry ($LiPF_6$ versus LiFSI in 2:1 EC:DEC). (b) Evolution of the interphase composition in F 1s and S 2p core level spectra as a function of potential during discharge in LiFSI electrolyte. Additional sulfur species are unidentified but related to interphase formation or x-ray degradation (peak A). Reprinted with permission from [56]. Copyright (2013) American Chemical Society.

Cathode interphases

Interphases on cathode surfaces are also possible to analyze, although comparatively less literature exists for cathode interphases relative to anodes. As with anodes, XPS can be used to understand the influence of electrolyte chemistry [15, 57], cathode composition [58] and passivating coatings [59] on cathode interphase composition. As the drive to higher voltage cathode materials strains the anodic stability of existing electrolytes, even with optimized additives, mechanistic understanding of cathode interphase chemistry is becoming increasingly important. For example, choice of anode has been shown to impact cathode interphase composition [60] (figure 8.18(a)). *In situ* scanning Auger microscopy measurements can enable single-particle level analysis of interphase composition [15], which, coupled with complementary XPS measurements, enables the correlation of electrode oxidation state with interphase chemistry (figure 8.18(b)). XPS has also been used to demonstrate the existence of cathode corrosion and crossover via detection of cathode elements on cycled anode surfaces [61, 62] (figure 8.18(c)).

All-solid-state systems

Battery systems that replace traditional liquid electrolytes with solid-state ionic conductors have attracted significant interest in the battery community in recent years and have helped spur renewed interest in lithium metal anodes as well. On the one hand, all-solid-state systems represent an ideal architecture for XPS analysis, in particular *in situ* and *operando* measurements, as the lack of high vapor pressure

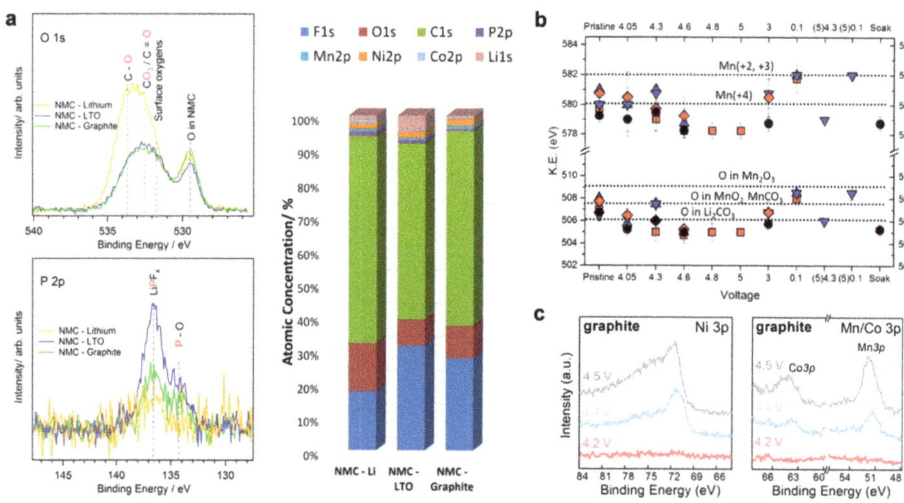

Figure 8.18. (a) Effect of anode choice on NMC 111 cathode interphase composition. Reproduced from [60]. Copyright IOP Publishing Ltd CC BY 4.0. (b) *In situ* AES measurements of Mn and O chemistry as a function of potential for $LiMn_2O_4$ particles. Red and blue data represent *in situ* cycled samples in EMI-BF_4 and Py_{13}TFSI ionic liquids, while black data represent samples cycled *ex situ* in 1 M $LiPF_6$ in 1:1 EC:DMC. Reprinted with permission from [15]. Copyright (2017) American Chemical Society. (c) XPS of graphite anodes revealing crossover of dissolved Ni, Mn and Co species from NMC 811 cathode as a function of cutoff potential. Reprinted with permission from [62]. Copyright (2020) American Chemical Society.

components enable the direct insertion of functional devices into a UHV environment. However, the shallow information depth and relatively poor spatial resolution of XPS measurements also puts a significant constraint on electrode thicknesses and battery geometry in order to measure the evolution of interfacial chemistry at solid-state interfaces.

Several techniques have been developed in order to circumvent these limitations, including vacuum deposition of thin layers of electrode [27, 64, 65] or solid electrolyte [66] materials (figure 8.19(a)), as well as *operando* plating of metal using a remote electron source [63, 67] (figure 8.19(b)). These techniques have the advantage of providing clearer information about interphase formation relative to the *ex situ* measurements discussed above, as there is no environmental exposure whatsoever prior to analysis. By taking further advantage of the UHV environment to generate surfaces of varying cleanliness, it is also possible to understand how intrinsic materials chemistry (e.g. composition, dopants, etc…) and extrinsic reaction layers formed from environmental exposure can influence the extent of interphase formation [29] (figure 8.20). Of course, *ex situ* XPS studies of solid electrolytes can provide valuable information as well. For example, it can be used to understand the influence of material preparation on performance [68] and monitor materials stability after extended cycling [69].

8.2.3 Concentration gradients and depth profiling

In addition to measurements of surface oxidation state and interphase chemistry, XPS can also be used to understand gradients in composition at electrode surfaces. This can be achieved both by non-destructive and destructive methods. Per equation (8.2), information depth is tunable as a function of the angle of photoemission from

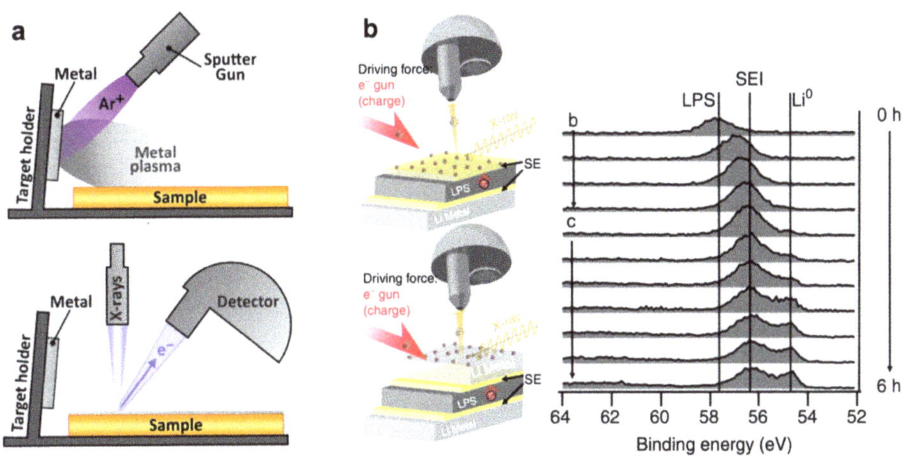

Figure 8.19. (a) Vacuum deposition of Li metal enabling *in situ* XPS measurement of interphases formed with solid electrolytes. Reprinted from [27], Copyright (2015), with permission from Elsevier. (b) Deposition of Li metal using a remote electron source to enable *operando* XPS measurement of interphase formation on a Li_2S-P_2S_5 solid electrolyte. Reproduced from [63]. CC BY 4.0.

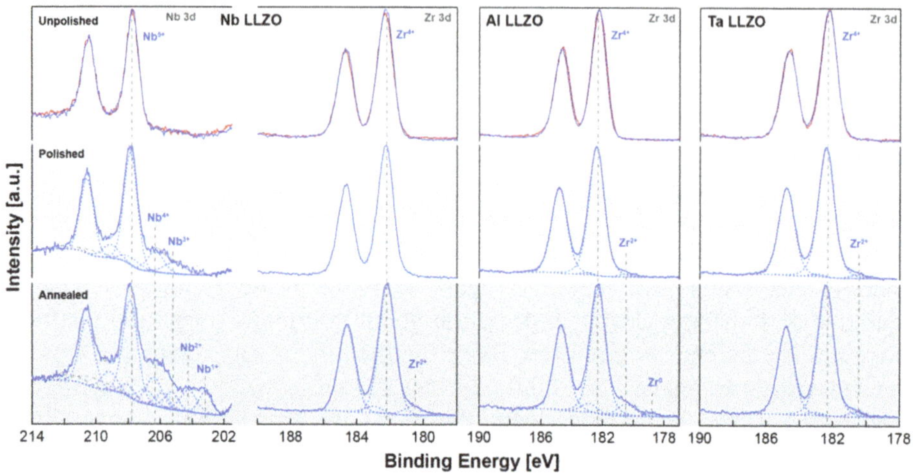

Figure 8.20. Influence of surface chemistry and dopant type on the extent of reactivity of $Li_7La_3Zr_2O_{12}$ (LLZO) solid electrolytes in contact with Li metal. Reproduced from [29]. Copyright (2019) WILEY-VCH Verlag GmbH & Co. KGaA, Weinheim.

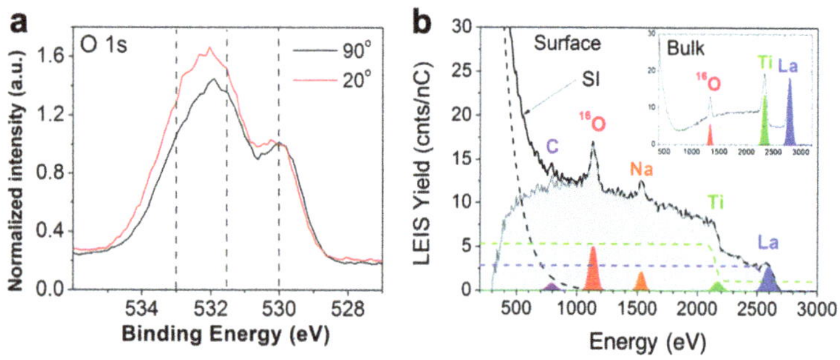

Figure 8.21. (a) AR-XPS measurement of Mg-intercalated MnO_2 showing the higher concentration of OH^- species relative to O^{2-} from the MnO_2 lattice, indicating a surface conversion reaction. Reproduced from [71] with permission of The Royal Society of Chemistry. (b) $^4He^+$ LEIS measurements of a $Li_{0.33}La_{0.55}TiO_3$ solid electrolyte revealing surface termination by La. Inset compares the bulk composition after Ar^+ sputtering. Reproduced from [18] with permission of The Royal Society of Chemistry.

the sample relative to the analyzer, meaning that as the sample is tilted, the information depth becomes smaller as the effective path length electrons must travel to escape the surface and be detected increases (figure 8.21(a)). This is the basis of angle-resolved XPS (AR-XPS), which is a non-destructive means of gaining information about how sample composition changes from the 'bulk' to the surface. Although this technique so far has not been widely applied for battery samples, AR-XPS has been shown to be able to detect the presence of near-surface insertion [70] and conversion [71] reactions at electrode surfaces (figure 8.21(a)). Although not a completely non-destructive technique, LEIS can also be used to gain information about the composition of the first 1–2 monolayers of a sample surface and understand, e.g., the presence of surface reconstructions [18] or the surface distribution of specific elemental species on interface-engineered materials [19] (figure 8.21(b)).

More commonly, destructive depth profiling is performed with XPS to measure gradients in surface composition. This is generally achieved by sputtering with noble gas ions (e.g., Ar^+) to remove controlled amounts of material per sputter cycle. Using this method, it is possible to monitor changes in elemental composition as a function of sputter time [72] (figure 8.22), and if sputter rates are known (e.g., via calibration from a standard of known thickness), can be correlated to layer thickness [73]. Note, however, that different elements sputter at different rates, and that generally organic materials sputter much faster than inorganic materials [74], so care must be taken not to over-interpret changes in material composition with depth, as these artefacts can result in depth profiles that are not truly representative of the material being studied. Compositional heterogeneity across the sample surface (e.g., cathode particles in a laminate) can also limit the utility of XPS depth profile information. Finally, sputter rates can be quite slow for common ion guns used in XPS instruments, and the limit of detection for XPS measurements is generally ⩾0.1–1.0 at%, so if extensive depth profiling or very high sensitivity is needed, ToF-SIMS is a preferrable method and can also provide more detailed spatial information than is generally possible with XPS [75, 76].

In some cases it is also possible to extract meaningful information about changes in the chemical speciation of interphases as a function of depth; however, doing so requires corroboration by complementary information from other techniques, in particular for organic species [77]. This is because high energy ion bombardment

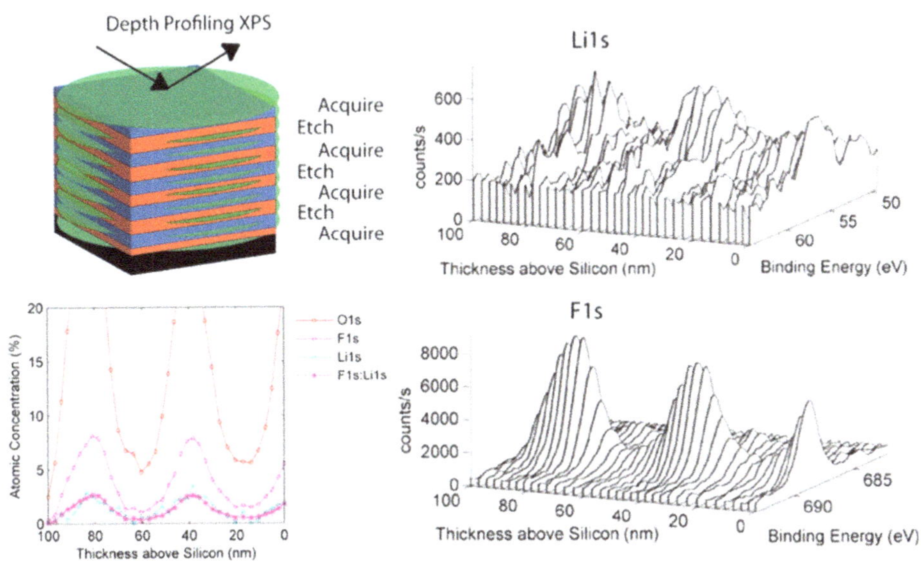

Figure 8.22. XPS depth profiling of a block copolymer solid electrolyte consisting of polystyrene (PS, red) poly (oligo(oxyethylene)methacrylate) (POEM, blue) lamellae, showing preferential segregation of both cation (Li 1s) and anion (F 1s) to the POEM layer. Reprinted with permission from [73]. Copyright (2015) American Chemical Society.

almost always drives the formation of additional reaction products that are not intrinsic to the sample being studied [74]. Cluster ions (e.g., C_{60}^{+}) have been shown to result in less spurious chemistry than noble gas ions [78]; however, even this approach has been shown to generate extrinsic reaction products on battery electrodes [79]. As a result, any attempt to use XPS to understand changes in electrode speciation as a function of depth must be approached with extreme caution.

8.2.4 Synchrotron methods

The application of high brightness, energy tunable x-ray excitations possible at synchrotron lights sources enable unique measurements that are not possible using fixed-energy x-ray sources in lab-based instruments. Two synchrotron-based XPS techniques that are particularly relevant to battery applications are highlighted below.

Synchrotron XPS and hard x-ray photoelectron spectroscopy (HAXPES)
As shown in equation (8.2), IMFP (and therefore information depth) is directly proportional to the KE of the x-ray excitation. Unlike lab-based x-ray sources, synchrotron x-ray light sources are able to access significantly higher x-ray energies (⩾5 keV), as well as introduce the ability to tune x-ray energy over a wide range of KE. These features of synchrotron light sources enable two different analytical capabilities. First, by utilizing high KE x-rays (the basis of HAXPES), XPS information is available from much deeper within the sample. This increase in information depth can enable measurements of electrode/interphase chemistry at buried interfaces [80] (e.g., when interphase thickness is greater than the information depth of a lab-based instrument) as well as *operando* measurements of cycling behavior in ASSBs with thicker (and therefore more realistic) electrodes [81] (figure 8.23).

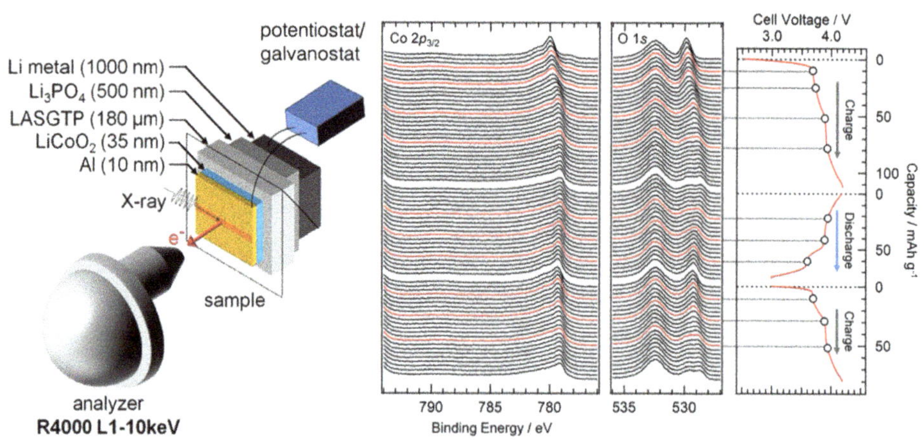

Figure 8.23. *Operando* HAXPES measuring changes in Co 2p and O 1s core levels in response to charge/discharge of a $LiCoO_2$ cathode in an all-solid-state battery configuration. Reproduced from [81]. CC BY 4.0.

Second, tuning the x-ray KE enables non-destructive depth profiling analyses, making it much more straightforward to extract information about changes in chemical speciation as a function of depth. This is because there is no physical perturbation of the interphase, and therefore there is a much lower probability of extrinsic chemical reactions taking place as compared to sputter depth profiling. Depending on the specific x-ray energies used, it is possible to obtain information from the first few nm to as much as 50 nm below the surface, enabling measurements of interphase composition gradients [82–84] as well as any changes in electrode composition/redox as a function of depth [31, 85] (figure 8.24). Note that similar depth profiles are also possible using angle-resolved HAXPES measurements. Changes in x-ray KE also yield changes in the cross-section for x-ray absorption for different core levels, potentially providing enhanced sensitivity for species (e.g. Li) that are not easily resolved with lab-based instruments [86].

Ambient pressure XPS (AP-XPS)

AP-XPS experiments provide the unique ability to measure *in situloperando* behavior of electrodes exposed to gases or liquid electrolytes at pressures ranging from a few mbar to near atmospheric pressure. Such measurements circumvent the normal XPS requirement for UHV conditions via the use of very small entrance apertures to the analyzer optics, small working distances and intricate differential pumping apparatus to ensure that all of the emitted photoelectrons are not inelastically scattered before they can be detected [87] (figure 8.25(a)). Lab-based instruments do exist, although they are usually optimized for gas-phase experiments and are generally not well-suited to working with high vapor pressure liquid

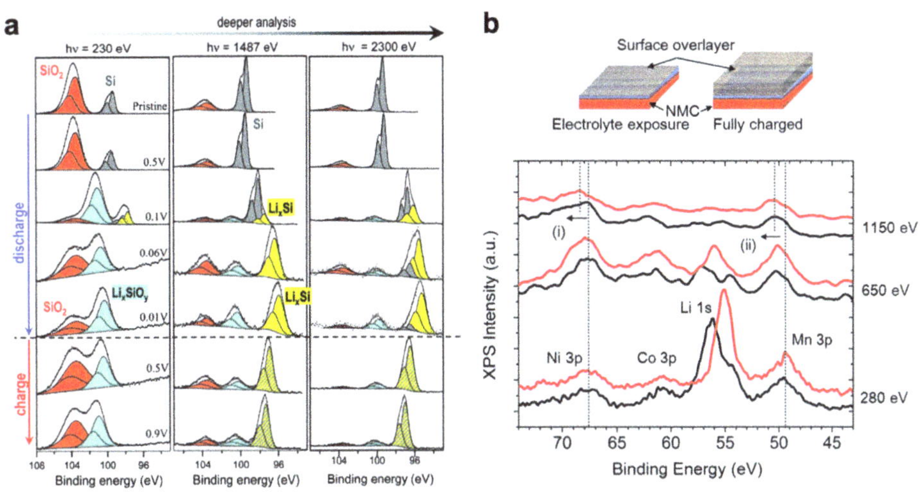

Figure 8.24. (a) Synchrotron XPS measurements of changes in interphase composition and extent of lithiation as a function of x-ray KE (depth) in a Si anode. Reprinted with permission from [82]. Copyright (2012) American Chemical Society. (b) Extent of electrode redox and interphase formation as a function of x-ray KE for an electrolyte-exposed (black) and charged (red) $Li_xNi_{0.4}Mn_{0.4}Co_{0.18}Ti_{0.02}O_2$ cathode. Reproduced from [85] with permission of The Royal Society of Chemistry.

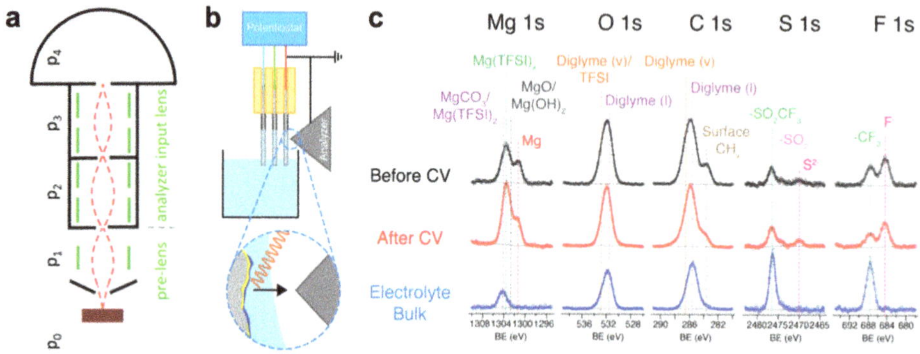

Figure 8.25. (a) Schematic of one type of differential pumping setup for AP-XPS, where p_0 indicates the operating pressure and p_1–p_4 represent differential pumping stages that sequentially lower the pressure to UHV conditions by the time photoelectrons reach the detector. Reproduced from [87]. CC BY 3.0. (b) *In situ* AP-XPS experimental setup to monitor interphase evolution of a working electrode in an electrolyte via the creation of a thin electrolyte meniscus and (c) corresponding experimental data showing changes in Mg anode interphase composition before and after Mg plating/stripping as well as the signal contribution from the bulk electrolyte. (b) and (c) Reprinted with permission from [90]. Copyright (2017) American Chemical Society.

samples. Synchrotron-based instruments provide the added benefit of higher x-ray intensity, yielding more photoelectrons and therefore providing significantly improved signal-to-noise. AP-XPS measurements have been used to understand changes in electrode composition *in situ* as a function of annealing conditions [88], as well as to investigate changes in electrode interphase speciation during *operando* cycling of electrodes [89, 90] (figures 8.25(b) and (c)).

8.3 Practical considerations and technique limitations

Although clearly much can be learned from XPS analysis of battery materials, careful sample preparation and rigorous analysis of XPS data is absolutely required to obtain meaningful results. Otherwise, misleading or even incorrect conclusions about the materials and/or electrochemical phenomena being studied can be reached. Many examples of such analyses exist in the literature, and in most cases are simply the result of incomplete understanding of the XPS technique and how to appropriately treat, analyze and, most importantly, qualify the conclusions reached from measured XPS spectra. This section explores common pitfalls to analyzing XPS data and some practical guidelines for how to derive reasonable conclusions from a given XPS measurement.

Note that in some cases it is simply not possible to draw a strong conclusion from XPS data alone! The reasons for this are numerous, including poor signal-to-noise, core level overlap, insufficient binding energy separation to accurately distinguish speciation and substantial differential charging, among many others. Attempts to over-interpret XPS data are one of the biggest drivers of misinformation related to the use of XPS in the battery literature. Collecting additional, corroborating data using other analytical techniques is often essential to developing the accurate, atomic- and molecular-level insight that is needed to advance the community's understanding of the interfacial chemistry of battery materials.

8.3.1 Sample considerations

Sample preparation can have a significant impact on the results obtained by XPS measurements. This is because even a few nanometers of contamination can completely dominate the resulting XPS signal. In almost all cases it is necessary to rinse cycled samples in a clean solvent, as residual salt left behind after electrolyte evaporation can easily leave a coating that is too thick to see beneath via XPS. Care must be taken to utilize a complementary solvent—ideally the same one used for the electrolyte, but at least one that should not induce additional chemical reactions to take place with the electrode. Such rinsing can lead to changes in interphase composition [91, 92], however, and care must be taken to understand whether such changes could meaningfully impact the results presented in a given study.

Reactive metal anodes (e.g., Li Na, Mg, etc...) are particularly sensitive to preparation and any exposure to ambient conditions, even those in a controlled-atmosphere glove box with <1 ppm H_2O and O_2 content, creating significant ambiguity as to the origin of any observed reaction products on the metal surface. Indeed, oxidation of cleaned Li surfaces can even be seen within minutes under UHV conditions [13]. Extreme care must be taken to avoid environmental exposure as much as possible when analyzing any electrode surface and acknowledge that even handling in a controlled atmosphere may not guarantee that the interphase chemistry has not been modified at least somewhat [92]. In such cases comparative studies can be instructive [93], where identical handling can at least guarantee that differences in interphase chemistry are most likely driven by the electrochemistry being studied and not extrinsic environmental factors.

Sample format can also significantly impact the clarity of XPS results. For example, laminate electrodes (e.g., active material held together with conductive carbon by a binder) are often complicated to analyze due to the fact that the conductive additives and binders can easily coat the active material particles with a thick enough layer that results in little to no XPS signal from the active cathode or anode material of interest. Of course, interphase chemistry is still accessible on such samples as interphases will form on the top surface of the electrode regardless, but such structures can present a problem when trying to analyze active material oxidation state, necessitating other characterization methods to access electrode redox properties. This can sometimes be circumvented by working with binder-free electrodes, but sample handling and insertion into the XPS tool can become quite complicated. As noted in section 8.1.4, it is also possible to support active material particles on vacuum-compatible adhesives or by embedding in soft metal foils such as indium, however thick particle layers can lead to sample charging due to the poor electronic conductivity of most electrode materials, an issue that will be discussed further below. Materials with significant porosity must also be treated with care, as any absorbed/adsorbed solvent/gas species will result in long pump down times and can degrade the vacuum quality of the XPS instrument.

Finally, it is vitally important to understand whether the materials being analyzed are sensitive to x-ray exposure. It has been shown that both salts [94] and interphase components [95] can degrade after prolonged irradiation, in particular by the high intensity x-rays produced at a synchrotron light sources. As a result, it is always necessary to monitor peak shapes and relative intensities as a function of time when collecting spectra, and to always use the minimum x-ray exposure time possible to collect spectra of sufficient quality to perform the desired analysis. Collection of survey spectra before and after collection of core level spectra can reveal global information about any changes that took place during the course of the measurement. Parallel collection of core level spectra (i.e., collecting one scan of each core level of interest before collecting subsequent scans) has also been proposed as a preferrable method for spectrum acquisition relative to serial collection (i.e., collecting all scans of one core level before moving on to the next) as it makes it possible to more clearly identify and control for any x-ray exposure-related changes to peak shapes [96].

8.3.2 Binding energy referencing and differential charging

Charge referencing for accurate BE measurements is the most challenging and fraught process in all of XPS analysis, but it is also the most critically important to producing meaningful results. Traditional procedures, and even the ASTM/ISO standard for charge referencing [97], suggests setting the position of 'adventitious' carbon (i.e. C–C or C–H) to 284.8 eV. However, this procedure is only truly valid for homogeneous and conductive (i.e., metallic) samples, and even then, may not be truly accurate [96]. Furthermore, many examples exist where different assignments of the adventitious carbon peak have been made over an extremely wide BE range—from as low as 284.0 eV to as high as 285.6 eV—introducing significant error to subsequent peak assignments made in the work. As a result, one should always verify the charge reference used in a given experiment and ensure that it is consistent with the strategy utilized to analyze one's own data, otherwise erroneous peak assignments can be made independently of any sample-specific considerations.

In addition to systematic issues with adventitious carbon referencing, sample heterogeneity and the presence of insulating regions/layers, which are almost unavoidable on battery samples, introduce the additional possibility of differential charging, where, as discussed in section 8.1.4, insulating regions become positively charged relative to the rest of the sample, resulting in a shift to positive binding energy relative to conductive regions of the sample (figure 8.26(a)). In some cases, charge neutralization can resolve such differential charging; however, spatial differences in conductivity can result in differences in the effectiveness of the flood gun across the sample surface and may simply result in differential charging to lower BE instead. It is likely that the combination of inconsistent charge referencing strategies and unrecognized and/or incorrectly compensated differential charging are responsible for the extremely wide spread of reported BE values for many common species found on battery electrode surfaces [44], which for some species are $\geqslant$2 eV,

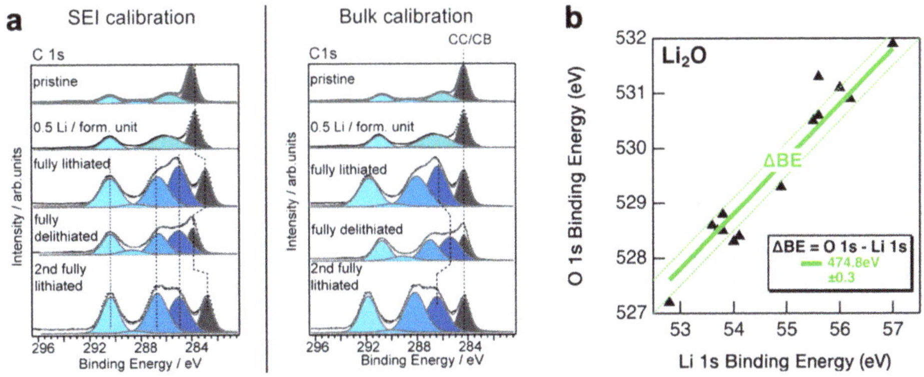

Figure 8.26. (a) Example of the effect of differential charging, where charge referencing to either the insulating components of the interphase (left) or the conductive carbon electrode (right) results in non-uniform peak shifts of the other species. Reprinted with permission from [98]. Copyright (2016) American Chemical Society. (b) Summary of reported literature values for the Li 1s and O 1s peak positions for Li_2O, demonstrating that regardless of the wide spread in absolute BE reported, there is a constant peak separation characteristic of this species. Reprinted with permission from [52]. Copyright (2018) American Chemical Society.

an order of magnitude larger than the experimental error of ± 0.2 eV for most XPS measurements.

This begs the question: how can one reliably charge reference inhomogeneous battery samples? Unfortunately, there is no straightforward answer to this question, and indeed the answer depends heavily on the specific sample being analyzed. What is clear is that *utilizing the position of adventitious carbon should not be relied upon as the sole means of charge correction.* While adventitious carbon charge referencing can certainly be used as a starting point, other methodologies should be employed to confirm its accuracy and to identify whether differential charging may be present. Some suggestions are given below for additional analyses to perform in order to produce a more accurate BE reference for battery samples:

1. **Alternative charge reference.** If other known species are present (e.g., graphite, metals, oxides, fluorides), these can be used as direct charge references with defined BE positions, although care must be taken to verify that the selected BE value is consistent with standard values reported in the literature, and not an outlier. Conductive components are not reliable for analyzing interphase components if differential charging is present.
2. **Characteristic peak separation.** Another means of more reliably performing speciation analysis is to utilize the peak separation between different compounds commonly present on battery electrodes [52]. If a surface species exhibits peaks in two core level spectra with a characteristic peak separation, it can be used as a fingerprint for identification of particular species in the absence of a fixed charge reference (figure 8.26(b)). If Li_2O specifically is found to be present, the position of either peak can be used as an absolute BE reference.

3. **Charge reference for species of interest only.** In some cases differential charging will make it impossible to use a global charge reference for analyzing all components in the sample [98]. In this case, a tradeoff must be made to charge reference the data such that it is possible to analyze the species of interest (e.g., the interphase components) while ignoring the chemistry of other components of the sample (e.g. active material or conductive additives). This approach is illustrated in figure 8.26(a).
4. **Sample biasing.** Although not a data treatment method, it is also possible to apply a sample bias to directly shift peaks related to conducting components of the sample relative to insulating components, and in doing so understand the origins of different peaks within a given sample and how best to charge reference the resulting data [99].

8.3.3 Peak fitting

Once a reasonable charge reference has been determined, it is often desirable to perform speciation analysis of the sample by deconvolution of core level spectra into component peaks. This procedure is another source of significant error in the XPS literature, but many common issues are straightforward to address, as they relate to the intrinsic physics of the XPS measurement. For example, the presence of spin–orbit splitting (section 8.1.3) introduces additional requirements to developing physically meaningful peak models. Although the spin–orbit splitting for some low Z elements is very small and can be ignored when fitting (e.g., Al, Mg 2p), or is large enough that there is no possibility for peak overlap, allowing peaks to be fit independently of one another (e.g., Mn 2p), in most cases peak separations are small enough that it is possible to have overlap between the spin–orbit doublets. For such core levels, *the peak model must account for spin–orbit splitting for every species proposed.* Excellent examples of this are given in figures 8.14(a) and 8.17 for the S 2p and Si 2p core levels, respectively, where each species is fit by two peaks with appropriate, fixed peak ratios and separation. Another example is given in figure 8.27 for the Ti 2p and Nb 3d core levels, where even though both spin–orbit peaks are reasonably well-separated, the presence of multiple oxidation states yields overlapping contributions that require a peak model that accounts for both spin–orbit components. For such core levels, any fits that only utilize single peaks are fundamentally incorrect, and any conclusions drawn from such peak models are therefore also incorrect.

Another physical effect that is sometimes relevant to the analysis of battery materials is multiplet splitting. This effect arises from coupling between the unpaired core electron that is generated by photoionization and unpaired valence electrons, which results in multiple final states and therefore multiple additional peaks within a single spin–orbit peak that also depend on the chemical makeup of the material (figure 8.28). Species most relevant to battery researchers that exhibit multiplet splitting are Mn^{4+}, Mn^{3+}, Mn^{2+}, Co^{3+}, Co^{2+}, Ni^{3+} and Fe^{3+}, as well as high spin Ni^{+2} and Fe^{+2} materials [100]. It is important to note that, as with spin–orbit splitting, multiplet splitting is present *regardless of concentration or signal-to-noise*

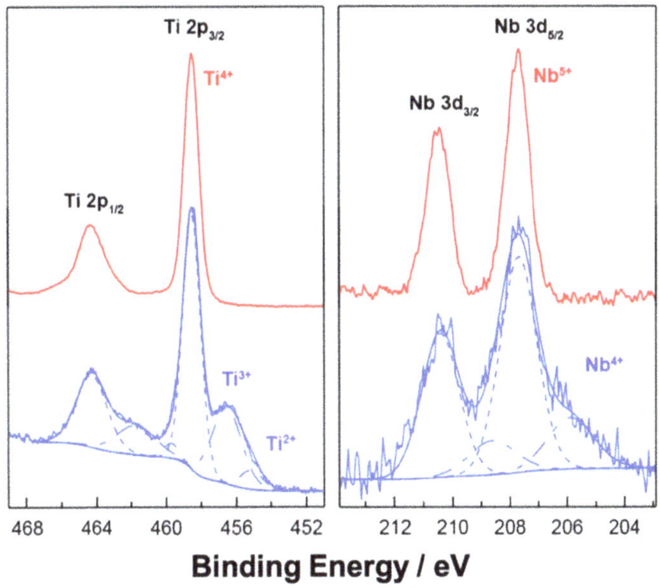

Figure 8.27. Ti 2p and Nb 3d core level spectra of $SrTiO_3$ as-prepared (red) and after sputter deposition of Li metal (blue), highlighting the need for peak models that account for both spin–orbit components when multiple, overlapping oxidation states are present. Reprinted with permission from [65]. Copyright (2018) American Chemical Society.

ratio and therefore must be taken into account whenever analyzing the oxidation state of such species. Peak models have been developed to fit these complex envelopes for many compounds with multiplet splitting [32, 101, 102]; however, robust models have yet to be generated for many battery-relevant materials. Nevertheless, existing peak models can be used as a starting point and, with appropriate constraints, a reasonably accurate analysis of mixed oxidation states may be possible.

Beyond physical effects, several practical considerations must be taken into account when developing accurate peak models. First, appropriate choices must also be made relating to the peak shape and background envelope used to fit a given species and core level. In general, simple convolutions of Gaussian–Lorentzian peak envelopes are sufficient to fit most species in XPS analysis, with a 70:30 Gaussian–Lorentzian mixture serving as the most commonly employed peak shape. A notable exception is for metals and metal-like species (e.g., graphite), which exhibit significant asymmetries to higher BE as well as satellite features that must be accounted for (figure 8.4(b)). The Shirley model is the most commonly applied background envelope [103]; however, in some cases where there is low signal-to-noise or peak overlap, linear backgrounds are more appropriate. The Tougaard model can also be applied for background subtraction [104], but it is not as widely used and is generally less straightforward to apply than the Shirley model.

When developing a peak model, one must always attempt to use the fewest peaks possible to fit a given core level spectrum. Additional peaks will always improve the

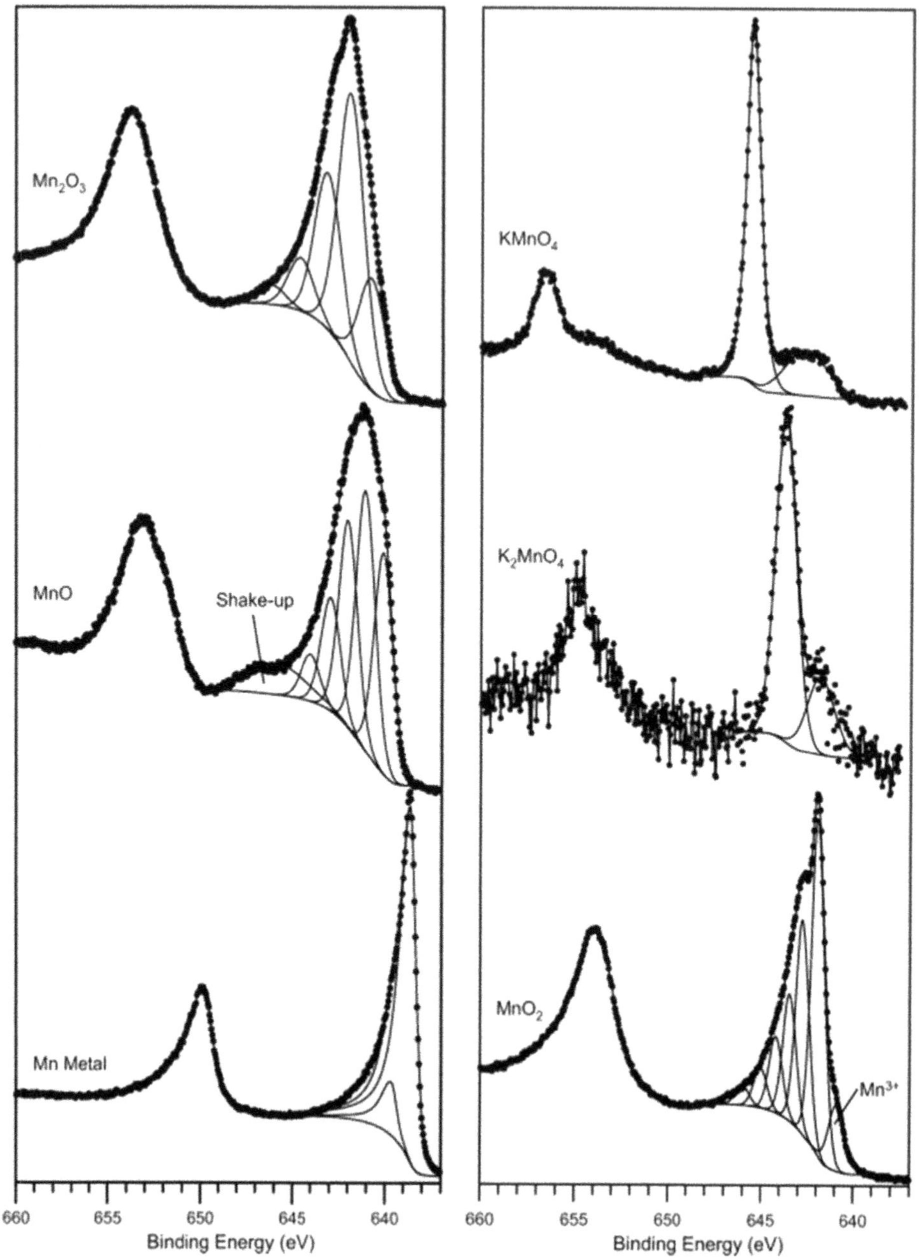

Figure 8.28. Multiplet splitting for a variety of different Mn compounds. Reprinted from [102], Copyright (2011), with permission from Elsevier.

fit but may ultimately result in a meaningless peak model. Appropriate constraints on peak position and FWHM must also be applied. Generally, it is preferrable to develop a peak model without first constraining peak positions and finding the best fit possible with the fewest number of peaks. Doing so requires constraining the

FWHM first, which varies depending on the specific core level being fit. Unfortunately, there are not particularly well-developed best practices for constraining FWHM; however, FWHM $\geqslant$ 3–4 eV are generally not physically meaningful for most core levels of interest. One exception is satellite features, which can be quite broad (FWHM $\geqslant$ 4 eV). When setting constraints, FWHM should be the same for all peaks fit within the same core level (within $\pm$ 0.2 eV), with some notable exceptions. For example, metal oxide peaks in the O 1s core level generally have lower FWHM than other oxygen-containing species (~1–1.5 eV versus 1.5–2.5 eV, respectively), and metallic species usually exhibit very low FWHM ($\leqslant$ 1 eV) relative to other oxidized species.

Once a reasonable peak model has been developed, peak positions should be fixed to within ±0.2 eV, and *the exact same peak model should be used to fit core level spectra for all samples of interest*, whether or not all of the same peaks are present on every sample. For this reason, it is also advisable to start with a sample exhibiting the most complex peak envelope in order to develop a robust model. Iterative fitting can then be used to fine-tune peak positions and thereby arrive at a self-consistent peak model that is able to fit all of the data. At this point it is appropriate to assign speciation based on the BE of constituent peaks, keeping in mind that additional constraints on peak area may be necessary based on stoichiometry or other speciation concerns. In particular, any peaks used to fit a given core level (e.g., C–O species in the C 1s) must exhibit a corresponding peak in all other relevant core levels (e.g., C–O species in the O 1s). Finally, these species should be constrained by stoichiometry, i.e., if Li_2O is proposed, Li and O signals corresponding to this species should be present in a roughly 2:1 ratio, or manually constrained so that extra, overlapping signal can be assigned to other species present on the surface [52, 105].

Abbreviations

α	Acceptance Angle
ϕ_a	Analyzer Work Function
ΔE	Energy Resolution
ν	Frequency
a.u.	Arbitrary Units
AES	Auger Electron Spectroscopy
AP-XPS	Ambient Pressure X-Ray Photoelectron Spectroscopy
AR-XPS	Angle-Resolved X-Ray Photoelectron Spectroscopy
ASSB	All-Solid-State Battery
ASTM	American Society for Testing and Materials
BE	Binding Energy
c.p.s.	Counts Per Second
DEC	Diethyl Carbonate *(electrolyte solvent)*
DMC	Dimethyl Carbonate *(electrolyte solvent)*
EC	Ethylene Carbonate (*electrolyte solvent*)
EMI-BF_4	Ethylmethylimidazolium Tetrafluoroborate *(electrolyte salt)*
E_p	Pass Energy
E_s	Surface Charge

ESCA	Electron Spectroscopy for Chemical Analysis
eV	Electron Volt
FTIR	Fourier Transform Infrared Spectroscopy
FWHM	Full Width at Half Maximum
h	Planck's Constant
HAXPES	Hard X-Ray Photoelectron Spectroscopy
HOMO	Highest Occupied Molecular Orbital
HV	High Vacuum
I_0	Incident Intensity
I_k	Measured Intensity
IMFP, λ	Inelastic Mean Free Path
ISO	International Organization for Standardization
ISS	Ion Scattering Spectroscopy
KE	Absolute Kinetic Energy
KE_{meas}	Measured Kinetic Energy
LCO	Lithium Cobalt Oxide, $LiCoO_2$
LEIS	Low Energy Ion Scattering
LiFSI	Lithium Fluorosulfonyl Imide
LLZO	Lithium Lanthanum Zirconium Oxide—$Li_7La_3Zr_2O_{12}$
NMC	Lithium Nickel-Manganese-Cobalt Oxide - $LiNi_xMn_yCo_zO_2$, $(x + y + x = 1)$
POEM	Poly(oligo(oxyethylene)methacrylate)
ppm	Part-Per-Million
PS	Polystyrene
$Py_{13}TFSI$	1-methyl-3-propylpyrrolidinium bis(trifluoromethane)sulfonyl imide *(electrolyte salt)*
R_0	Hemispherical Analyzer Radius
RSF	Relative Sensitivity Factor
SECO	Secondary Electron Cutoff
SEI	Solid Electrolyte Interphase
SOC	State of Charge
TFSI	Bis(trifluoromethane)sulfonyl imide *(electrolyte anion)*
ToF-SIMS	Time of Flight Secondary Ion Mass Spectrometry
UHV	Ultrahigh Vacuum
UPS	Ultraviolet Photoelectron Spectroscopy
w	Entrance Slit Width
XPS	X-Ray Photoelectron Spectroscopy

References

[1] Vickerman J C 1997 *Surface Analysis: The Principal Techniques* (Chichester: Wiley)

[2] Briggs D and Seah M P 1983 *Practical Surface Analysis by Auger and X-ray Photoelectron Spectroscopy* (Chichester; Aarau: Wiley; Salle + Sauerländer)

[3] Hüfner S 2003 *Photoelectron Spectroscopy: Principles and Applications* (Berlin: Springer)

[4] Scofield J H 1976 Hartree–Slater subshell photoionization cross-sections at 1254 and 1487 eV *J. Electron. Spectrosc. Relat. Phenom.* **8** 129–37

[5] Powell C J 1988 The quest for universal curves to describe the surface sensitivity of electron spectroscopies *J. Electron. Spectrosc. Relat. Phenom.* **47** 197–214

[6] Seah M P and Dench W A 1979 Quantitative electron spectroscopy of surfaces: a standard data base for electron inelastic mean free paths in solids *Surf. Interface Anal.* **1** 2–11
[7] Powell C J and Jablonski A 1999 Evaluation of calculated and measured electron inelastic mean free paths near solid surfaces *J. Phys. Chem. Ref. Data* **28** 19–62
[8] Powell C J and Jablonski A 2010 *NIST Electron Inelastic-Mean-Free-Path Database* (Gaithersburg, MD: National Institute of Standards and Technology)
[9] Hill J M, Royce D G, Fadley C S, Wagner L F and Grunthaner F J 1976 Properties of oxidized silicon as determined by angular-dependent x-ray photoelectron spectroscopy *Chem. Phys. Lett.* **44** 225–31
[10] Cumpson P J 2000 The thickogram: a method for easy film thickness measurement in XPS *Surf. Interface Anal.* **29** 403–6
[11] Shard A G, Wang J and Spencer S J 2009 XPS topofactors: determining overlayer thickness on particles and fibres *Surf. Interface Anal.* **41** 541–8
[12] Shard A G 2012 A straightforward method for interpreting XPS data from core–shell nanoparticles *The Journal of Physical Chemistry* C **116** 16806–13
[13] Zavadil K R and Armstrong N R 1990 Surface chemistries of lithium: detailed characterization of the reactions of H_2S, SO_2, and SO_2C_{12} using XPS and EELS *Surf. Sci.* **230** 61–73
[14] Auvergniot J, Cassel A, Foix D, Viallet V, Seznec V and Dedryvère R 2017 Redox activity of argyrodite Li_6PS5Cl electrolyte in all-solid-state Li-ion battery: an XPS study *Solid State Ionics* **300** 78–85
[15] Tang C-Y, Leung K, Haasch R T and Dillon S J 2017 $LiMn_2O_4$ surface chemistry evolution during cycling revealed by in situ auger electron spectroscopy and x-ray photoelectron spectroscopy *ACS Appl. Mater. Interfaces* **9** 33968–78
[16] Ensling D, Cherkashinin G, Schmid S, Bhuvaneswari S, Thissen A and Jaegermann W 2014 Nonrigid band behavior of the electronic structure of $LiCoO_2$ thin film during electrochemical Li deintercalation *Chem. Mater.* **26** 3948–56
[17] Lian X *et al* 2020 An *in situ* spectroscopy investigation of alkali metal interaction mechanism with the imide functional group *Nano Res.* **13** 3224–9
[18] Téllez H, Aguadero A, Druce J, Burriel M, Fearn S, Ishihara T, McPhail D S and Kilner J A 2014 New perspectives in the surface analysis of energy materials by combined time-of-flight secondary ion mass spectrometry (ToF-SIMS) and high sensitivity low-energy ion scattering (HS-LEIS) *J. Anal. At. Spectrom.* **29** 1361–70
[19] Hu L, Brüner P, Grehl T, Brongersma H H and Cabana J 2017 Control of chemical structure in core–shell nanocrystals for the stabilization of battery electrode/electrolyte interfaces *Chem. Mater.* **29** 5896–905
[20] Myung S-T, Sasaki Y, Sakurada S, Sun Y-K and Yashiro H 2009 Electrochemical behavior of current collectors for lithium batteries in non-aqueous alkyl carbonate solution and surface analysis by ToF-SIMS *Electrochim. Acta* **55** 288–97
[21] Sui T, Song B, Dluhos J, Lu L and Korsunsky A M 2015 Nanoscale chemical mapping of Li-ion battery cathode material by FIB-SEM and TOF-SIMS multi-modal microscopy *Nano Energy* **17** 254–60
[22] Ramana C V, Massot M and Julien C M 2005 XPS and Raman spectroscopic characterization of $LiMn_2O_4$ spinels *Surf. Interface Anal.* **37** 412–6
[23] Castro L, Dedryvère R, El Khalifi M, Lippens P-E, Bréger J, Tessier C and Gonbeau D 2010 The spin-polarized electronic structure of $LiFePO_4$ and $FePO_4$ evidenced by in-lab XPS *J. Phys. Chem.* C **114** 17995–8000

[24] Verde M G, Baggetto L, Balke N, Veith G M, Seo J K, Wang Z and Meng Y S 2016 Elucidating the phase transformation of $Li_4Ti_5O_{12}$ lithiation at the nanoscale *ACS Nano* **10** 4312–21

[25] Nordlinder S, Augustsson A, Schmitt T, Guo J, Duda L C, Nordgren J, Gustafsson T and Edström K 2003 Redox behavior of vanadium oxide nanotubes as studied by X-ray photoelectron spectroscopy and soft x-ray absorption spectroscopy *Chem. Mater.* **15** 3227–32

[26] Światowska-Mrowiecka J, Maurice V, Zanna S, Klein L and Marcus P 2007 XPS study of Li ion intercalation in V_2O_5 thin films prepared by thermal oxidation of vanadium metal *Electrochim. Acta* **52** 5644–53

[27] Wenzel S, Leichtweiss T, Krüger D, Sann J and Janek J 2015 Interphase formation on lithium solid electrolytes—an *in situ* approach to study interfacial reactions by photoelectron spectroscopy *Solid State Ionics* **278** 98–105

[28] Matsuyama T, Deguchi M, Mitsuhara K, Ohta T, Mori T, Orikasa Y, Uchimoto Y, Kowada Y, Hayashi A and Tatsumisago M 2016 Structure analyses using X-ray photoelectron spectroscopy and x-ray absorption near edge structure for amorphous MS_3 (M: Ti, Mo) electrodes in all-solid-state lithium batteries *J. Power Sources* **313** 104–11

[29] Zhu Y *et al* 2019 Dopant-dependent stability of garnet solid electrolyte interfaces with lithium metal *Adv. Energy Mater.* **9** 1803440

[30] Castro L, Dedryvère R, Ledeuil J-B, Bréger J, Tessier C and Gonbeau D 2012 Aging Mechanisms of $LiFePO_4$//graphite cells studied by xps: redox reaction and electrode/electrolyte interfaces *J. Electrochem. Soc.* **159** A357

[31] Sachs M, Gellert M, Chen M, Drescher H-J, Kachel S R, Zhou H, Zugermeier M, Gorgoi M, Roling B and Gottfried J M 2015 $LiNi_{0.5}Mn_{1.5}O_4$ high-voltage cathode coated with $Li_4Ti_5O_{12}$: a hard X-ray photoelectron spectroscopy (HAXPES) study *Phys. Chem. Chem. Phys.* **17** 31790–800

[32] Bondarchuk O, LaGrow A P, Kvasha A, Thieu T, Ayerbe E and Urdampilleta I 2021 On the x-ray photoelectron spectroscopy analysis of $LiNi_xMn_yCo_zO_2$ material and electrodes *Appl. Surf. Sci.* **535** 147699

[33] Dedryvère R, Foix D, Franger S, Patoux S, Daniel L and Gonbeau D 2010 Electrode/electrolyte interface reactivity in high-voltage spinel $LiMn_{1.6}Ni_{0.4}O_4/Li_4Ti_5O_{12}$ lithium-ion battery *J. Phys. Chem.* C **114** 10999–1008

[34] Junta J L and Hochella M F 1994 Manganese (II) oxidation at mineral surfaces: a microscopic and spectroscopic study *Geochim. Cosmochim. Acta* **58** 4985–99

[35] Nelson A J, Reynolds J G and Roos J W 2000 Core-level satellites and outer core-level multiplet splitting in Mn model compounds *J. Vacuum Sci. Technol.* A **18** 1072–6

[36] Galakhov V R, Demeter M, Bartkowski S, Neumann M, Ovechkina N A, Kurmaev E Z, Lobachevskaya N I, Mukovskii Y M, Mitchell J and Ederer D L 2002 Mn 3s exchange splitting in mixed-valence manganites *Phys. Rev.* B **65** 113102

[37] Ilton E S, Post J E, Heaney P J, Ling F T and Kerisit S N 2016 XPS determination of Mn oxidation states in Mn (hydr)oxides *Appl. Surf. Sci.* **366** 475–85

[38] Klein F, Pinedo R, Hering P, Polity A, Janek J and Adelhelm P 2016 Reaction mechanism and surface film formation of conversion materials for lithium- and sodium-ion batteries: an XPS case study on sputtered copper oxide (CuO) thin film model electrodes *J. Phys. Chem.* C **120** 1400–14

[39] Liang X, Hart C, Pang Q, Garsuch A, Weiss T and Nazar L F 2015 A highly efficient polysulfide mediator for lithium–sulfur batteries *Nat. Commun.* **6** 5682

[40] Younesi R, Hahlin M, Björefors F, Johansson P and Edström K 2013 Li–O_2 battery degradation by lithium peroxide (Li_2O_2): a model study *Chem. Mater.* **25** 77–84

[41] Diao Y, Xie K, Xiong S and Hong X 2012 Insights into Li–S battery cathode capacity fading mechanisms: irreversible oxidation of active mass during cycling *J. Electrochem. Soc.* **159** A1816

[42] Xiao L, Cao Y, Xiao J, Schwenzer B, Engelhard M H, Saraf L V, Nie Z, Exarhos G J and Liu J 2013 Molecular structures of polymer/sulfur composites for lithium–sulfur batteries with long cycle life *J. Mater. Chem.* A **1** 9517–26

[43] Nandasiri M I, Camacho-Forero L E, Schwarz A M, Shutthanandan V, Thevuthasan S, Balbuena P B, Mueller K T and Murugesan V 2017 *In situ* chemical imaging of solid-electrolyte interphase layer evolution in Li–S batteries *Chem. Mater.* **29** 4728–37

[44] Verma P, Maire P and Novák P 2010 A review of the features and analyses of the solid electrolyte interphase in Li-ion batteries *Electrochim. Acta* **55** 6332–41

[45] Peled E and Menkin S 2017 Review—SEI: past, present and future *J. Electrochem. Soc.* **164** A1703

[46] Lin D, Liu Y and Cui Y 2017 Reviving the lithium metal anode for high-energy batteries *Nat. Nanotechnol.* **12** 194–206

[47] Krauskopf T, Richter F H, Zeier W G and Janek J 2020 Physicochemical concepts of the lithium metal anode in solid-state batteries *Chem. Rev.* **120** 7745–94

[48] Aurbach D, Markovsky B, Shechter A, Ein-Eli Y and Cohen H 1996 A comparative study of synthetic graphite and Li electrodes in electrolyte solutions based on ethylene carbonate-dimethyl carbonate mixtures *J. Electrochem. Soc.* **143** 3809

[49] Aurbach D, Weissman I, Schechter A and Cohen H 1996 X-ray photoelectron spectroscopy studies of lithium surfaces prepared in several important electrolyte solutions. A comparison with previous studies by Fourier transform infrared spectroscopy *Langmuir* **12** 3991–4007

[50] Aurbach D, Markovsky B, Weissman I, Levi E and Ein-Eli Y 1999 On the correlation between surface chemistry and performance of graphite negative electrodes for Li ion batteries *Electrochim. Acta* **45** 67–86

[51] Etxebarria A, Koch S L, Bondarchuk O, Passerini S, Teobaldi G and Muñoz-Márquez M Á 2020 Work function evolution in Li anode processing *Adv. Energy Mater.* **10** 2000520

[52] Wood K N and Teeter G 2018 XPS on Li-battery-related compounds: analysis of inorganic SEI phases and a methodology for charge correction *ACS Appl. Energy Mater.* **1** 4493–504

[53] Schroder K W, Dylla A G, Harris S J, Webb L J and Stevenson K J 2014 Role of surface oxides in the formation of solid–electrolyte interphases at silicon electrodes for lithium-ion batteries *ACS Appl. Mater. Interfaces* **6** 21510–24

[54] Dedryvère R, Gireaud L, Grugeon S, Laruelle S, Tarascon J-M and Gonbeau D 2005 Characterization of lithium alkyl carbonates by x-ray photoelectron spectroscopy: experimental and theoretical study *J. Phys. Chem.* B **109** 15868–75

[55] Dedryvère R, Leroy S, Martinez H, Blanchard F, Lemordant D and Gonbeau D 2006 XPS valence characterization of lithium salts as a tool to study electrode/electrolyte interfaces of Li-ion batteries *J. Phys. Chem.* B **110** 12986–92

[56] Philippe B, Dedryvère R, Gorgoi M, Rensmo H, Gonbeau D and Edström K 2013 Improved performances of nanosilicon electrodes using the salt LiFSI: a photoelectron spectroscopy study *JACS* **135** 9829–42

[57] Alvarado J *et al* 2018 A carbonate-free, sulfone-based electrolyte for high-voltage Li-ion batteries *Mater. Today* **21** 341–53

[58] Erickson E M, Li W, Dolocan A and Manthiram A 2020 Insights into the cathode–electrolyte interphases of high-energy-density cathodes in lithium-ion batteries *ACS Appl. Mater. Interfaces* **12** 16451–61

[59] Baggetto L, Dudney N J and Veith G M 2013 Surface chemistry of metal oxide coated lithium manganese nickel oxide thin film cathodes studied by XPS *Electrochim. Acta* **90** 135–47

[60] Björklund E, Brandell D, Hahlin M, Edström K and Younesi R 2017 How the negative electrode influences interfacial and electrochemical properties of $LiNi_{1/3}Co_{1/3}Mn_{1/3}O_2$ cathodes in Li-ion batteries *J. Electrochem. Soc.* **164** A3054

[61] Yang L, Takahashi M and Wang B 2006 A study on capacity fading of lithium-ion battery with manganese spinel positive electrode during cycling *Electrochim. Acta* **51** 3228–34

[62] Li W, Liu X, Xie Q, You Y, Chi M and Manthiram A 2020 Long-term cyclability of NCM-811 at high voltages in lithium-ion batteries: an in-depth diagnostic study *Chem. Mater.* **32** 7796–804

[63] Wood K N, Steirer K X, Hafner S E, Ban C, Santhanagopalan S, Lee S-H and Teeter G 2018 Operando x-ray photoelectron spectroscopy of solid electrolyte interphase formation and evolution in Li_2S–P_2S_5 solid-state electrolytes *Nat. Commun.* **9** 2490

[64] Schwöbel A, Jaegermann W and Hausbrand R 2016 Interfacial energy level alignment and energy level diagrams for all-solid Li-ion cells: impact of Li-ion transfer and double layer formation *Solid State Ionics* **288** 224–8

[65] Connell J G, Zhu Y, Zapol P, Tepavcevic S, Sharafi A, Sakamoto J, Curtiss L A, Fong D D, Freeland J W and Markovic N M 2018 Crystal orientation-dependent reactivity of oxide surfaces in contact with lithium metal *ACS Appl. Mater. Interfaces* **10** 17471–9

[66] Hausbrand R, Fingerle M, Späth T and Guhl C 2017 Energy level offsets and space charge layer formation at electrode-electrolyte interfaces: X-ray photoelectron spectroscopy analysis of Li-ion model electrodes *Thin Solid Films* **643** 43–52

[67] Connell J G, Fuchs T, Hartmann H, Krauskopf T, Zhu Y, Sann J, Garcia-Mendez R, Sakamoto J, Tepavcevic S and Janek J 2020 Kinetic versus thermodynamic stability of LLZO in contact with lithium metal *Chem. Mater.* **32** 10207–15

[68] Sharafi A, Yu S, Naguib M, Lee M, Ma C, Meyer H M, Nanda J, Chi M, Siegel D J and Sakamoto J 2017 Impact of air exposure and surface chemistry on Li–$Li_7La_3Zr_2O_{12}$ interfacial resistance *J. Mater. Chem.* A **5** 13475–87

[69] Zhang W, Richter F H, Culver S P, Leichtweiss T, Lozano J G, Dietrich C, Bruce P G, Zeier W G and Janek J 2018 Degradation mechanisms at the $Li_{10}GeP_2S_{12}/LiCoO_2$ cathode interface in an all-solid-state lithium-ion battery *ACS Appl. Mater. Interfaces* **10** 22226–36

[70] Kim J, Buchner F and Behm R J 2018 Temperature-dependent insertion and adsorption of lithium on spinel $Li_4Ti_5O_{12}$(111) thin films—an angle-resolved XPS study *Phys. Chem. Chem. Phys.* **20** 18319–27

[71] Sahadeo E, Song J, Gaskell K, Kim N, Rubloff G and Lee S B 2018 Investigation of the water-stimulated Mg^{2+} insertion mechanism in an electrodeposited MnO_2 cathode using x-ray photoelectron spectroscopy *Phys. Chem. Chem. Phys.* **20** 2517–26

[72] Edström K, Gustafsson T and Thomas J O 2004 The cathode–electrolyte interface in the Li-ion battery *Electrochim. Acta* **50** 397–403

[73] Gilbert J B, Luo M, Shelton C K, Rubner M F, Cohen R E and Epps T H 2015 Determination of lithium-ion distributions in nanostructured block polymer electrolyte thin films by x-ray photoelectron spectroscopy depth profiling *ACS Nano* **9** 512–20

[74] Andersson A M, Henningson A, Siegbahn H, Jansson U and Edström K 2003 Electrochemically lithiated graphite characterised by photoelectron spectroscopy *J. Power Sources* **119–121** 522–7

[75] Dupré N, Moreau P, De Vito E, Quazuguel L, Boniface M, Bordes A, Rudisch C, Bayle-Guillemaud P and Guyomard D 2016 Multiprobe study of the solid electrolyte interphase on silicon-based electrodes in full-cell configuration *Chem. Mater.* **28** 2557–72

[76] Li J and Manthiram A 2019 A comprehensive analysis of the interphasial and structural evolution over long-term cycling of ultrahigh-nickel cathodes in lithium-ion batteries *Adv. Energy Mater.* **9** 1902731

[77] Gu Y *et al* 2018 Designable ultra-smooth ultra-thin solid-electrolyte interphases of three alkali metal anodes *Nat. Commun.* **9** 1339

[78] Chang L-S, Lin Y-C, Su C-Y, Wu H-C and Pan J-P 2011 Effect of C60 ion sputtering on the compositional depth profiling in XPS for $Li(Ni,Co,Mn)O_2$ electrodes *Appl. Surf. Sci.* **258** 1279–81

[79] Niehoff P, Passerini S and Winter M 2013 Interface investigations of a commercial lithium ion battery graphite anode material by sputter depth profile x-ray photoelectron spectroscopy *Langmuir* **29** 5806–16

[80] Ciosek Högström K, Malmgren S, Hahlin M, Gorgoi M, Nyholm L, Rensmo H and Edström K 2014 The buried carbon/solid electrolyte interphase in Li-ion batteries studied by hard x-ray photoelectron spectroscopy *Electrochim. Acta* **138** 430–6

[81] Kiuchi H, Hikima K, Shimizu K, Kanno R, Toshiharu F and Matsubara E 2020 Operando hard x-ray photoelectron spectroscopy of $LiCoO_2$ thin film in an all-solid-state lithium ion battery *Electrochem. Commun.* **118** 106790

[82] Philippe B, Dedryvère R, Allouche J, Lindgren F, Gorgoi M, Rensmo H, Gonbeau D and Edström K 2012 Nanosilicon electrodes for lithium-ion batteries: interfacial mechanisms studied by hard and soft x-ray photoelectron spectroscopy *Chem. Mater.* **24** 1107–15

[83] Malmgren S, Ciosek K, Hahlin M, Gustafsson T, Gorgoi M, Rensmo H and Edström K 2013 Comparing anode and cathode electrode/electrolyte interface composition and morphology using soft and hard x-ray photoelectron spectroscopy *Electrochim. Acta* **97** 23–32

[84] Dong Y, Young B T, Zhang Y, Yoon T, Heskett D R, Hu Y and Lucht B L 2017 Effect of lithium borate additives on cathode film formation in $LiNi_{0.5}Mn_{1.5}O_4$/Li cells *ACS Appl. Mater. Interfaces* **9** 20467–75

[85] Lin F, Nordlund D, Markus I M, Weng T-C, Xin H L and Doeff M M 2014 Profiling the nanoscale gradient in stoichiometric layered cathode particles for lithium-ion batteries *Energy Environ. Sci.* **7** 3077–85

[86] Tsuji J, Fujita M, Haruyama Y, Kanda K, Matsui S, Ozawa N, Yao T and Taniguchi K 2005 Excitation energy dependence for the Li 1s x-ray photoelectron spectra of $LiMn_2O_4$ *Anal. Sci.* **21** 779–81

[87] Starr D E, Liu Z, Hävecker M, Knop-Gericke A and Bluhm H 2013 Investigation of solid/vapor interfaces using ambient pressure x-ray photoelectron spectroscopy *Chem. Soc. Rev.* **42** 5833–57

[88] Tepavcevic S, Connell J G, Lopes P P, Bachhav M, Key B, Valero-Vidal C, Crumlin E J, Stamenkovic V R and Markovic N M 2018 Role of structural hydroxyl groups in enhancing performance of electrochemically-synthesized bilayer V_2O_5 *Nano Energy* **53** 449–57

[89] Lu Y-C, Crumlin E J, Veith G M, Harding J R, Mutoro E, Baggetto L, Dudney N J, Liu Z and Shao-Horn Y 2012 *In situ* ambient pressure x-ray photoelectron spectroscopy studies of lithium-oxygen redox reactions *Sci. Rep.* **2** 715

[90] Yu Y, Baskin A, Valero-Vidal C, Hahn N T, Liu Q, Zavadil K R, Eichhorn B W, Prendergast D and Crumlin E J 2017 Instability at the electrode/electrolyte interface induced by hard cation chelation and nucleophilic attack *Chem. Mater.* **29** 8504–12

[91] Herstedt M, Abraham D P, Kerr J B and Edström K 2004 X-ray photoelectron spectroscopy of negative electrodes from high-power lithium-ion cells showing various levels of power fade *Electrochim. Acta* **49** 5097–110

[92] Edström K, Herstedt M and Abraham D P 2006 A new look at the solid electrolyte interphase on graphite anodes in Li-ion batteries *J. Power Sources* **153** 380–4

[93] Connell J G, Genorio B, Lopes P P, Strmcnik D, Stamenkovic V R and Markovic N M 2016 Tuning the reversibility of Mg anodes via controlled surface passivation by H_2O/Cl^- in organic electrolytes *Chem. Mater.* **28** 8268–77

[94] Shterenberg I, Salama M, Gofer Y and Aurbach D 2017 X-ray photodecomposition of Bis (trifluoromethanesulfonyl)imide, Bis(fluorosulfonyl)imide, and hexafluorophosphate *J. Phys. Chem.* C **121** 3744–51

[95] Nordh T, Younesi R, Brandell D and Edström K 2015 Depth profiling the solid electrolyte interphase on lithium titanate ($Li_4Ti_5O_{12}$) using synchrotron-based photoelectron spectroscopy *J. Power Sources* **294** 173–9

[96] Greczynski G and Hultman L 2020 X-ray photoelectron spectroscopy: towards reliable binding energy referencing *Prog. Mater Sci.* **107** 100591

[97] ASTM International 2015 *E1523-15 Standard Guide to Charge Control and Charge Referencing Techniques in X-Ray Photoelectron Spectroscopy* (West Conshohocken, PA: ASTM International)

[98] Maibach J, Lindgren F, Eriksson H, Edström K and Hahlin M 2016 Electric potential gradient at the buried interface between lithium-ion battery electrodes and the sei observed using photoelectron spectroscopy *J. Phys. Chem. Lett.* **7** 1775–80

[99] Bergman S L, Sahasrabudhe G S, Ji H, Cava R J and Bernasek S L 2017 Useful x-ray photoelectron spectroscopy-based chemical tool: differential charging studies of complex composite materials *Chem. Mater.* **29** 4162–6

[100] Biesinger M C 2008 Multiplet splitting *X-ray Photoelectron Spectroscopy (XPS) Reference Pages* http://xpsfitting.com/2008/09/multiplet-splitting.html

[101] Biesinger M C, Lau L W M, Gerson A R and St C R 2010 Smart resolving surface chemical states in XPS analysis of first row transition metals, oxides and hydroxides: Sc, Ti, V, Cu and Zn *Appl. Surf. Sci.* **257** 887–98

[102] Biesinger M C, Payne B P, Grosvenor A P, Lau L W M, Gerson A R and St C R 2011 Smart Resolving surface chemical states in XPS analysis of first row transition metals, oxides and hydroxides: Cr, Mn, Fe, Co and Ni *Appl. Surf. Sci.* **257** 2717–30

[103] Shirley D A 1972 High-resolution x-ray photoemission spectrum of the valence bands of gold *Phys. Rev.* B **5** 4709–14

[104] Tougaard S 1988 Quantitative analysis of the inelastic background in surface electron spectroscopy *Surf. Interface Anal.* **11** 453–72

[105] Payne B P, Biesinger M C and McIntyre N S 2012 Use of oxygen/nickel ratios in the XPS characterisation of oxide phases on nickel metal and nickel alloy surfaces *J. Electron. Spectrosc. Relat. Phenom.* **185** 159–66

IOP Publishing

Batteries
Materials principles and characterization methods
Chen Liao

Chapter 9

Scanning electrochemical microscopy: a versatile tool for inspecting the reactivity of battery electrodes

Dipobrato Sarbapalli, Abhiroop Mishra, Kendrich O. Hatfield, Zachary T. Gossage and Joaquín Rodríguez-López

9.1 Introduction

Since Volta's first batteries in the early 1800s, energy storage technologies such as Li-ion batteries have developed into reliable and essential devices for modern society. Demand for the Li-ion battery has increased in the past 30 years [1–3] and today it ranks among the most disruptive technologies in human history [4]. Producing high-performance, safe, and long-lasting batteries requires a better understanding of the electrochemical processes, mass and charge transport mechanisms at various length scales, and aging and materials degradation during battery operation [5–8]. Additionally, since lithium is scarce [9, 10] there is motivation within the battery community to develop and understand batteries based on other chemistries such as potassium-ion [11, 12] zinc-ion [13–15] and sodium-ion [16], among others [17–21]. Emerging technologies like redox flow batteries [22–24] and supercapacitors [25–27] also create the need to develop versatile approaches for understanding manifold processes occurring at battery interfaces at-large.

In these efforts, scanning electrochemical microscopy (SECM) has emerged as a powerful tool with versatile analytical modes for understanding the complex processes and performance of materials within batteries [28, 29]. SECM is a scanning probe technique developed by the Bard group in the late 1980s [30], using a small electrode probe (i.e. an SECM tip) to map and quantify localized electrochemical phenomena over a sample surface. SECM can provide *in situ* information on each major component in a typical battery—anode, cathode, electrolyte, and the separator, examples of which are depicted in figure 9.1. SECM has been applied to systems such as Li-ion batteries, Li–air batteries, and redox flow batteries to develop a better understanding of the operational mechanisms and degradation processes

doi:10.1088/978-0-7503-2682-7ch9

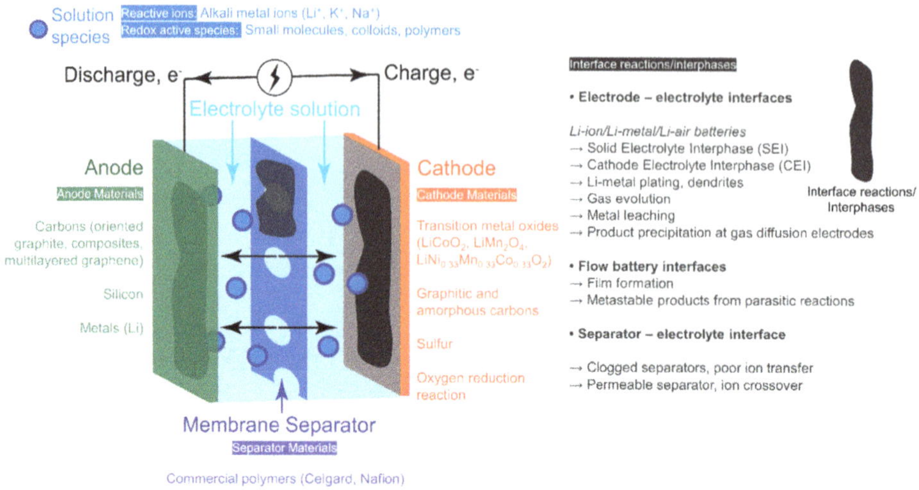

Figure 9.1. Schematic of battery components studied via SECM, including examples of electrode and separator materials discussed in this chapter and their respective interface reactions/interphases.

involved [31–36]. The power of SECM lies in its ability to study ion and electron transfer—both of which are the fundamental processes driving all batteries. Further, SECM provides insight on the complex interfacial electrochemistry that occurs within battery systems at length scales ranging from μm to nm [37–41]. For instance, this dynamic range is particularly useful in exploring the formation and progression of the solid–electrolyte interphase (SEI), which is critical to safe operation of Li-ion batteries [42]. While SECM is regarded as a mature technology [43, 44] and related techniques with new capabilities are arising, the advantages of SECM in the evaluation of battery chemistries are only starting to be realized as evidenced by the growing literature in the field [29, 45, 46].

The aim of this chapter is to establish a reference platform for understanding SECM operation and its application in battery electrochemistry research. To accomplish this, we first describe the working principles of SECM and the unique information that can be acquired with this technique. Subsequently, we present literature examples highlighting the use of SECM to systematically study various battery systems, while providing context within SECM theory. Next, we describe how SECM can be coupled to other characterization techniques, allowing multi-modal studies. For example, Raman spectroscopy and atomic force microscopy (AFM) have been coupled with SECM [47–50] enabling extraction of chemical and topographical information, respectively, with simultaneous electrochemical measurements. While this chapter primarily focuses on SECM, we briefly comment on other emerging scanning probe electrochemical methods related to it. Finally, we conclude the chapter with a perspective of some state-of-the-art advances in SECM, along with future avenues of battery research that may be pursued with SECM characterization. We hope that this chapter inspires readers to develop unique experiments using SECM in battery research.

9.2 Principles of scanning electrochemical microscopy

The SECM probe is often, but not necessarily, a flat disc ultramicroelectrode (UME) that is positioned in close proximity to a surface of interest. Henceforth, we will refer to the SECM probe as the 'tip' or 'probe' and the analyte material as the 'substrate.' We often refer to redox-active species in solution as the 'redox mediator,' or simply, the 'mediator.' A typical SECM experiment involves rastering the SECM tip across the substrate while measuring the tip and/or substrate currents as a result of *electrochemical reactions*. For simplicity, we might consider one such reaction as the reduction of an oxidized species (O) gaining an electron to generate the reduced species (R), as shown in equation (9.1):

$$O + e \underset{k_b}{\overset{k_f}{\rightleftharpoons}} R \tag{9.1}$$

Here the subscripts k_f and k_b refer to the rate constants for the forward (reduction) and backward (oxidation) steps of an electrochemical reaction at equilibrium. Throughout this chapter, we will exemplify SECM modes using this simplified notation. For instance, in the feedback mode of SECM, a small probe performs the transformation R → O (i.e. oxidizes species R), while the substrate performs the opposite (reduction) reaction, O → R. In such an SECM experiment one could extract the value of k_f of the sample by analyzing the feedback response and applying the formalisms shown in table 9.1. Sub-sections below provide a brief summary of SECM theory and working principles. The scope is limited to provide enough detail about this technique enabling a reader to comprehend the remainder of this chapter. For further detail and description, we direct the reader toward an SECM textbook [51], and several reviews [33, 52–55].

9.2.1 Ultramicroelectrodes and their operation

Ultramicroelectrodes are typically electrodes with their largest dimension less than 25 μm [56]. Key to the operation of UMEs is their ability to attain a mass-transfer limited steady-state current, as shown in figure 9.2 for the transformation R → O. Such a response results in a time-invariant current due to the formation of a radial diffusion layer of electrogenerated species, unlike electrodes of larger dimensions (» 25 μm) which display a transient response [56]. The UME steady-state response allows SECM to measure electrochemical properties as a function of tip–substrate interactions without convolving time. While UMEs can be made in many geometries (such as disc, hemisphere, band, cylindrical etc) [56], SECM typically involves tips with disc geometries that have relatively simple analytical expressions. A more detailed review of UMEs and their preparation can be found in the literature [56–60].

9.2.2 Approach curves

Approach curves are used to position the SECM tip at a known distance from the substrate via the use of a redox mediator such as the O/R redox pair described in

Table 9.1. Quantitative relationships in SECM measurements.

Equation number	Principles or modes in SECM	Equations	References
2	Steady state current at UME	$i_{T,\infty} = 4\text{nFaDC}$ (for $R_g > 10$) $i_{T,\infty} = 4\text{nFaDC}\beta(RG)$ (for $R_g < 10$)	[57, 61]
3	β(RG)	$\beta(\text{RG}) = 1 + \frac{0.23}{\left(RG^3 - 0.81\right)^{0.36}}$	[61]
4	α(RG)	$\ln 2 + \ln 2\left(1 - \frac{2}{\pi}\text{ArcCos}\frac{1}{RG}\right) - \ln 2\left[1 - \left(\frac{2}{\pi}\text{ArcCos}\frac{1}{RG}\right)^2\right]$	[61]
5	κ	$\frac{k_f a}{D}$	[61]
6	Negative feedback approach curve	$\frac{i_{T,neg}}{i_{T,\infty}} \approx \left[\frac{2.08}{RG^{0.358}}\left(L - \frac{0.145}{RG}\right) + 1.585\right] * \left[\frac{2.08}{RG^{0.358}}(L + 0.0023RG) + 1.57 + \frac{\ln R_g}{L} + \frac{2}{\pi RG}\ln\left(1 + \frac{\pi RG}{2L}\right)\right]$	[61]
7	Mass transfer limited positive feedback approach curves	$\frac{i_{T,pos}}{i_{T,\infty}} \approx \alpha(RG) + \frac{\pi}{4\beta(RG)\text{ArcTan}(L)} + \left(1 - \alpha(RG) - \frac{1}{2\beta(RG)}\right)\frac{2}{\pi}\text{ArcTan}(L)$	[61]
8	Feedback curve for finite kinetics, κ	$\frac{i_T}{i_{T,\infty}} \approx \frac{i_{T,pos}}{i_{T,\infty}}\left(L + \frac{1}{\kappa}, RG\right) + \frac{\frac{i_{T,neg}}{i_{T,\infty}}(L, RG) - 1}{\left(1+2.47RG^{0.31} L\kappa\right)\left(1 + L^{0.006RG+0.113_\kappa - 0.0236RG+0.91}\right)}$	[61]
9	AC-SECM	$\text{R}(\infty) = \frac{\rho}{4a}$ $\frac{R(\infty)}{R(L)} = \frac{i_d(L)}{i_T(\infty)}$	[65, 68] [83]
10	Butler–Volmer relation	$lnk_f = lnk^0 - \frac{\alpha F}{RT}(E - E^0)$	[56]

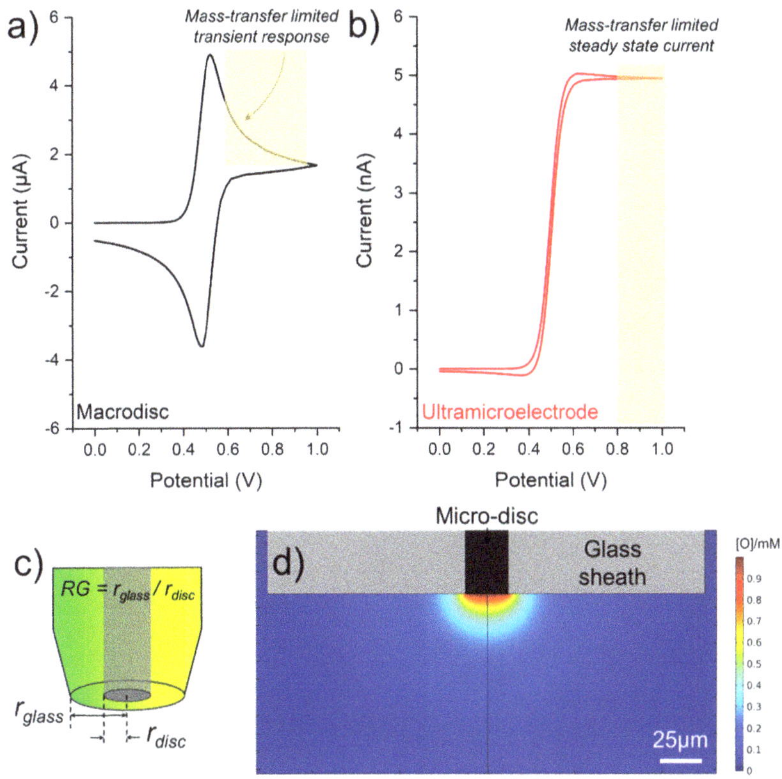

Figure 9.2. COMSOL finite element simulations of cyclic voltammograms (CVs) with (a) 2.5 mm diameter macrodisc electrode and (b) 25 μm diameter UME for a single electron oxidation of R → O with standard potential (E°) of 0.5 V. Mass transfer limited steady state current is illustrated in the CV. (c) RG, a geometric parameter, defined as the ratio of UME disc radius to the radius of the insulating glass sheath. (d) A COMSOL simulation of the radial diffusion layer generated under mass transfer-limited conditions. Color variation represents concentration of the generated O species when the initial concentration of R is 1 mM.

figure 9.2 (examples of mediators used with SECM are highlighted in reference [33]). In this mode, the tip is biased to a potential such that oxidation/reduction of the mediator is limited only by mass transfer (i.e. diffusion) of the mediator to the electrode surface. As the tip is brought near the substrate (shown in figure 9.3(a)), the tip current (i_T) deviates from the value obtained in bulk solution ($i_{T,\infty}$). The tip current can increase or decrease depending on the nature of the substrate. An insulating substrate hinders diffusion of the mediator to the tip, causing the tip current to decrease from $i_{T,\infty}$ (figure 9.3(b)), a conducting substrate can donate/accept electrons to/from the mediator (figure 9.3(c)), forming the original charge state and leading to a 'feedback loop' which results in a current higher than $i_{T,\infty}$. This operation is termed 'feedback mode,' with 'negative feedback' occurring over ideal insulators and 'positive feedback' occurring over ideal conductors.

Approach curves are often plotted as normalized tip current ($i_T/i_{T,\infty}$) versus normalized tip-substrate distance, L (= d/a, where d is the distance from tip to

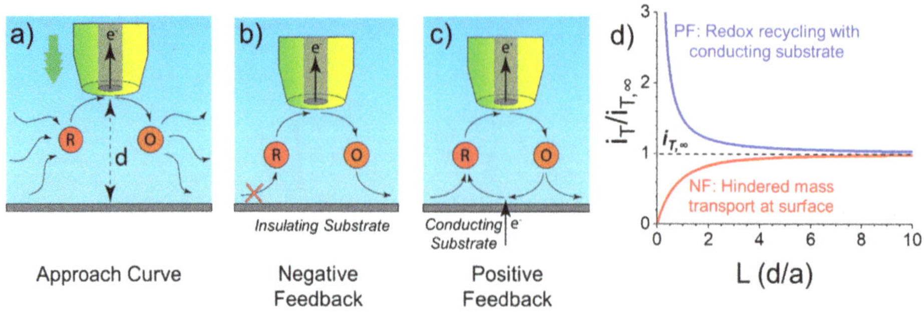

Figure 9.3. Operation of the feedback mode. (a) Schematic of an approach curve experiment, (b) negative feedback (NF) due to hindered diffusion of species R, (c) Positive feedback (PF) due to recycling of species O→R by the substrate, (d) NF and PF approach curves showing deviation of tip currents i_T from $i_{T,\infty}$ as the tip is brought near the substrate.

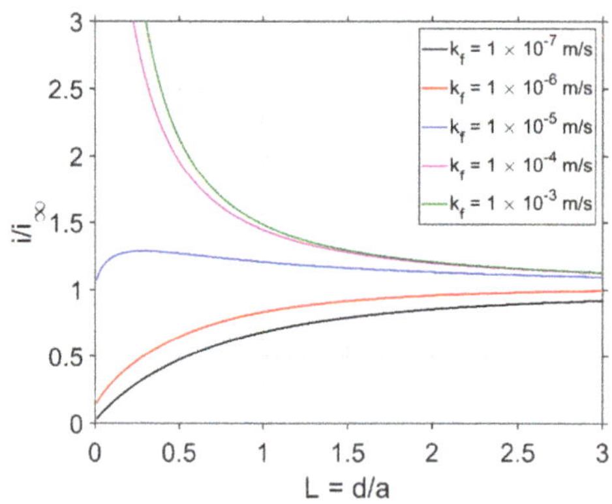

Figure 9.4. Analytical approach curves [61] for different electrochemical kinetics k_f between substrate and mediator. Diffusion coefficient of mediator, $D = 1 \times 10^{-10}$ m^2 s^{-1}, tip radius, a = 12.5 μm, RG = 3.

substrate and a is the tip radius) as shown in figure 9.3(d). While an absolute d value is unknown at the beginning of an experiment, the data can be fitted to previously developed analytical expressions [61], yielding final tip-substrate distance. The approach speed should be low enough to ensure there is no convection in solution so that diffusion is the only form of mass transfer present [62]. In a positive feedback approach curve, the tip current is a function of the substrate's ability to transfer an electron to/from the mediator in solution and ultimately back to the UME. Therefore, fitting positive feedback approach curves can yield the heterogeneous electron transfer rates between the mediator and substrate. Taking as an example the transformation R → O at the SECM probe and the reciprocal O → R at the substrate, one could extract the value of k_f at a sample of interest. k_f is defined within the Buttler–Volmer model [56] for electron transfer (refer to equation (10), table 9.1). Figure 9.4 shows how approach curves change with different k_f values. Standard rate constants, k^o, can also be determined by collecting and fitting multiple approach curves while applying different substrate potentials and using equation (10) [63, 64].

It should be noted that $k°$ is characteristic between an electrode, a redox mediator and the solution composition.

Good practice before any SECM experiment is to perform negative feedback approach curves over flat insulated substrates to determine an accurate *RG* of the tip. While figure 9.2 defines *RG* explicitly, practical *RG* values will deviate from the definition by some extent due to the microdisc being off-centered, glass sheath not perfectly shaped, among other reasons. Fitting these approach curves ($i_T/i_{T,\infty}$ versus L) to the analytical expressions, which are a function of L and *RG* [61], yields *RG* for that particular SECM tip. Prior knowledge of *RG* subsequently leads to an accurate measurement of k_f from positive feedback curves, since k_f is the only unknown parameter.

In certain cases, the use of mediators for the approach curves is undesirable as it may perturb the chemistry of the substrate and lead to parasitic reactions [65]. Alternatives include alternating current SECM (AC-SECM), ionic measurements, and laser and shear-force positioning [66, 67]. AC-SECM, which operates based on electrochemical impedance concepts, facilitates positioning without the use of a mediator by applying an oscillating potential to the SECM tip and measuring the solution resistance between the tip and counter electrode. The tip–substrate distance can be found by fitting a plot of resistance normalized by the resistance in bulk solution versus L (figure 9.5) [68]. The normalized conductance approach curve ($R(\infty)/R(L)$) follows the well-established amperometric feedback theory [69] and can be fitted to obtain absolute tip–substrate distances. Some specialized SECM probes enable direct measurement of electrolyte ions or other molecules in solution, leading to analogous approach curves [70, 71]. Also, groups have introduced additional hardware to enable precise positioning through shear-force or a laser alike to AFM (see section 9.4.2.2).

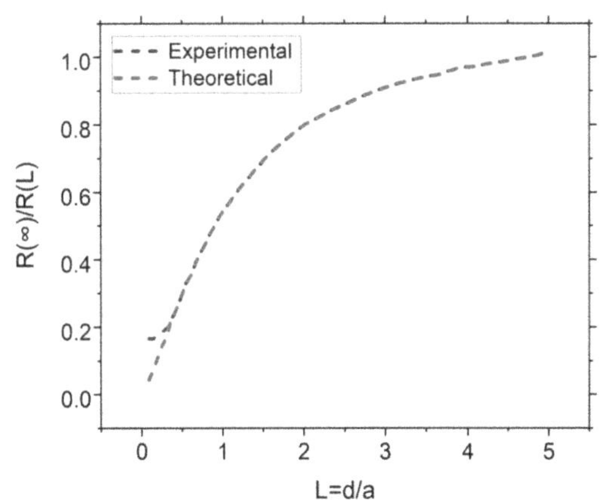

Figure 9.5. AC-SECM approach curve on an insulator (Teflon) with a 25 μm diameter UME in a 1 mM KCl solution. Reproduced and replotted for clarity from [68]. Copyright (1993) American Chemical Society.

9.2.3 SECM modes

In this section, we introduce the main modes of SECM operation, providing foreground for the experiments and analysis described in subsequent sections.

Feedback imaging: Feedback imaging is performed by measuring the tip current resulting from electrolysis of a redox mediator while the probe is rastered near a substrate. The tip–substrate distance must be on the order of the thickness of the diffusion layer generated at the tip to modify i_T. Feedback imaging provides a way to visualize electrochemically active and inactive regions (i.e. conductive versus insulating) on the substrate [72]. If the substrate is flat and tip–substrate distance is known through approach curves, feedback imaging can be used to map kinetics of electron transfer. This can be achieved by converting $i_T/i_{T,\infty}$ values in the feedback image into k_f by using analytical expressions [61]. In practice, this enables *in situ* mapping of passivating processes, e.g. interphase formation.

Substrate generation-tip collection (SG-TC): In this mode, the tip collects redox-active species generated by the substrate [68, 73] as depicted in figure 9.6(a). The tip current and collection efficiency (CE) depend heavily on the tip-substrate distance, and the activity along with applied potential of the substrate. These measurements are analogous to those performed in other generation/collection platforms such as the rotating ring-disk electrode (RRDE) [56], with the added benefit of obtaining spatially-resolved maps of sample reactivity. For identification, this mode requires some foreknowledge on the collected species, but it can be used to quantify known or unknown products generated from a substrate by virtue of their standard electrochemical potentials [33, 56].

Redox competition (RC) and ion-sensitive SECM: In this mode, the tip and substrate are held at potentials to perform the same redox reaction while measuring tip current [74]. Electrochemically active regions of the substrate will consume redox species, leaving the tip unable to perform redox reactions near that area, resulting in a decreased tip current. However, if the substrate region exhibits sluggish kinetics, the tip current will experience a lesser effect, thus giving rise to contrast in SECM images.

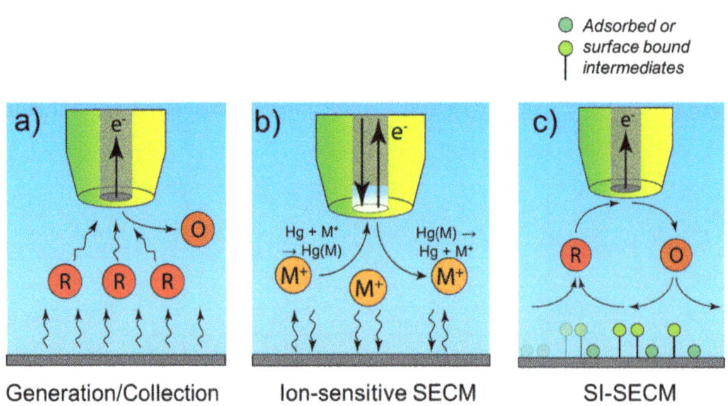

Figure 9.6. Additional modes of SECM. (a) Generation/collection, (b) ion-sensitive SECM, and (c) surface interrogation mode of SECM. Surface species are titrated by the electrogenerated mediator from SECM tip.

The concept of redox competition has been extended to ion-flux detection at battery interfaces. Here, a Hg SECM probe is used to quantify local ionic flux changes [75, 76]. The flux is monitored by amalgamation and stripping of alkali ions (Li^+, Na^+, K^+) at the tip (figure 9.6(b)). The measured amalgamation and stripping currents are correlated to the concentration of ions near the probe. Fluxes can be detected by monitoring changes in the tip current; for example, if a substrate is actively consuming ions, the tip current would decrease. Therefore, competition for ions between tip and substrate can inform the user about the activation of ion-consuming or ion-releasing processes at the substrate [77, 78].

Substrate interrogation SECM (SI-SECM): This mode utilizes an electrogenerated mediator at the tip to titrate surface-bound species, e.g. surface oxides, hydrides, reactive radicals, etc [79]. Homogeneous electron transfer between the mediator and surface species results in a transient positive feedback (figure 9.6(c)) that lasts until the surface species are exhausted. By quantifying the charge required to consume the surface-bound species, it is possible to obtain the surface coverage of species and its isotherm with respect to electrode potential. Simulations with finite-element method software such as COMSOL Multiphysics can also be used to extract rate constants associated with charge transfer between surface species and mediators, along with the formation and decay dynamics of these surface species [79–82].

9.2.4 Analytical expressions

SECM is a quantitative technique that involves precise positioning and displacements of the UME tip while measuring current. The resulting data can be quantitatively interpreted using analytical expressions for different SECM modes [61]. Table 9.1 presents some of the commonly used analytical expressions while table 9.2 lists description of the notations used in these expressions.

9.3 Battery applications

In this section, we discuss the applications of SECM in three major battery technologies: Li-ion, Li–air, and redox flow batteries. The use of SECM and the information it provides on different components of these systems is illustrated with literature examples. This section also serves as a guide, helping the reader to design experiments based on different aspects of the system to be analyzed.

9.3.1 Li-ion batteries

Li-ion batteries (LIBs) rely on (de)intercalation of Li^+ inside host electrodes to store energy. LIBs typically consist of a graphite anode and a lithium transition metal oxide cathode immersed in a non-aqueous Li^+ electrolyte [84]. During the (dis) charge cycles in a LIB, electrolyte decomposition and other parasitic reactions result in the formation of thin-films on the anode and cathode surfaces, called the solid–electrolyte interphase (SEI) [85] and cathode–electrolyte interphase (CEI), respectively [86]. Once formed, these ionically conducting interphases enable Li^+ (de) intercalation between electrodes while preventing further parasitic electrochemical reactions from occurring. Consequently, the performance of any LIB is dependent

Table 9.2. Notations used in table 9.1.

Notation	Description
a	Electrode radius
d	Tip-to-substrate distance
n	Number of electrons transferred in the redox process
D	Diffusion coefficient of the redox species
C	Bulk concentration of the redox species
F	Faraday's constant
$i_{T,\infty}$	Mass transfer limited, steady state current at UME
RG	Radius of glass sheath to radius of electrode (figure 9.2(c))
L	Ratio of distance between UME and substrate to electrode radius (d/a)
ρ	Solution resistivity
$R(\infty)$, $R(d)$	Resistance with UME positioned in the bulk, resistance at a normalized UME-substrate separation of d
$i_T(\infty)$, $i_T(d)$	Current with UME positioned in the bulk, current at a normalized UME-substrate separation of d
k^0	Standard heterogeneous rate constant
k_f	Forward rate constant
E	Applied potential at electrode
E_{eq}	Equilibrium potential
E^0	Standard potential
η	Overpotential (E–E_{eq})
α	Transfer coefficient

on the following processes: (i) formation and stability of SEI and CEI [42, 86] (ii) dynamics during (de)lithiation such as ionic fluxes, metal leaching, and phase transformations [87, 88], and (iii) mass transport properties within porous electrode materials [89]. The versatility of SECM makes it an attractive technique to investigate these critical phenomena. A deeper understanding of these processes would help in improving the existing LIB technologies and in the development of novel materials for LIBs. Table 9.3 summarizes the applications of SECM discussed in this section along with key information obtained from different SECM modes applied to different phenomena in LIBs.

9.3.1.1 Formation and stability of the solid–electrolyte interphase

Feedback mode SECM has been used to investigate the *in situ* formation of SEI [90, 91]. As the anode surface is covered by an electronically insulating film, sharp contrast is detected in SECM images as the initial positive feedback transitions towards a negative feedback response. SECM can detect this transition to study the formation and stability of the SEI. Figure 9.7(a) illustrates this phenomenon and shows the effect of SEI formation on SECM feedback.

Bülter *et al* studied the SEI stability on multiple anodes with feedback mode SECM [42]. They investigated spatiotemporal changes of the SEI by repeatedly

Table 9.3. Summary of SECM applications in Li-ion batteries.

Concept addressed	SECM mode used	Key information obtained	Relevant equation (table 9.1)	References
Temporal evolution of SEI	Feedback at a single point	Dynamic nature of SEI, feedback changes with time	2, 8	[42, 90]
Spatial evolution of SEI	Feedback imaging	Inhomogeneous nature of SEI and the passivation property	2, 8	[42, 90]
Ion transfer kinetics	Competition mode	Local ion transfer rate at an operating battery electrode	10	[78]
Mass transport properties in porous electrodes	Feedback mode	Ratio of diffusion coefficient of Li^+ between bulk solution and inside a porous electrode	See reference	[89]

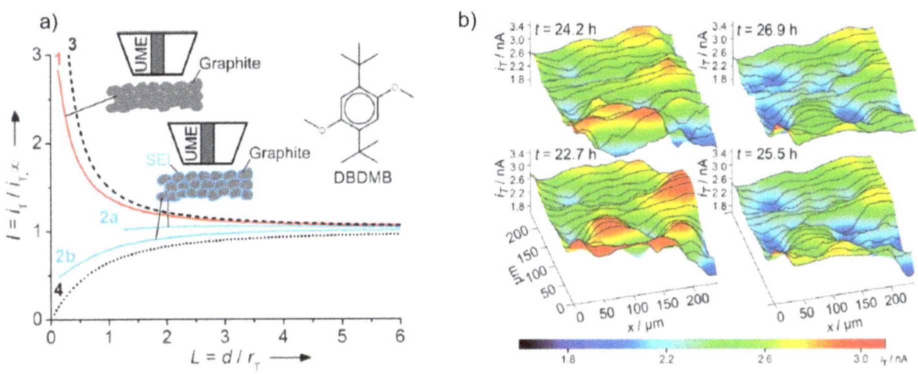

Figure 9.7. Effect of SEI formation on SECM feedback. (a) Feedback approach curve on fresh graphite electrode (1) before and (2) after SEI formation. The presence of the SEI decreases the feedback current measured at the tip. Curves 2a and 2b are from different regions of SEI covered electrode. Curves (3) and (4) represent ideal positive and negative feedback, respectively [92]. (b) Temporal evolution of SEI on an identical region of a graphite electrode using a 2,5-di-tertbutyl-1,4-dimethoxy benzene mediator recorded after different intervals of time [42], John Wiley & Sons, copyright (2014).

imaging the same region over different intervals of time. Figure 9.7(b) shows the change in feedback on a graphite substrate, clearly illustrating the differing surface evolution at various regions of the graphite. In similar studies, SECM revealed SEI passivity on highly oriented pyrolytic graphite (HOPG) and composite graphite electrodes using the feedback mode [90]. The SEI on the composite electrode was comparatively unstable due to solvent uptake and destabilizing particle–particle interactions. These studies highlight the dynamic nature of SEI on different electrode materials and illustrate the ability of SECM to provide critical information fundamental to battery performance.

Further understanding of the fundamental mechanisms of Li^+ (de)intercalation and SEI formation on different electrode materials was developed using SECM. In one such study, SEI formation was probed via the feedback mode using samples with different number of graphene layers as substrate [91]. Patterned substrates established that SEI still formed and feedback decreased, even in the case of low Li^+ (de)intercalation rate over the basal plane [93]. In another similar study, formation of SEI and Li^+ (de)intercalation was studied on TiO_2 based paste electrodes [92]. Feedback mode imaging was used to visualize the formation of SEI on the porous substrate and helped in identifying the potentials relevant to the formation process. Recently, Krueger *et al* utilized SECM to investigate SEI evolution on Li metal electrodes under realistic battery cycling conditions [94]. The methodology developed could be extended to other electrode and electrolyte systems for the investigation of interface evolution.

Recently, silicon anodes have garnered attention due to their high charge storage capacity [95]. However, they undergo a significant volume change during (dis)charge cycles [96], leading to significant loss of energy storage capacity [97]. Silicon clathrates provide an attractive alternative as they allow Li^+ storage without such detrimental structural changes [98] SECM was used to study the effect of preprocessing on the SEI properties of $Ba_8Al_{16}Si_{30}$ silicon clathrates [98]. It was found that a ball-milled sample showed unstable SEI which lost the passivation within the first cycle, however, chemical etching before ball-milling led to a stable SEI layer. This study underlines the use of SECM to evaluate novel electrode systems and SEI stability for improved next generation LIBs.

9.3.1.2 Formation and stability of CEI

Similar to SEI formation on the anode, a cathode electrolyte interphase (CEI) forms during battery operation [86, 99]. This interphase has received much attention recently as the operating potentials of next generation cathodes extend beyond the thermodynamic stability window of electrolytes [5, 100]. Thus, irreversible reactions such as oxidative decomposition of the electrolytes leads to the formation of an interphase on the cathode surface; a better understanding of this phenomenon is necessary to improve battery performance. In 2019, Liu *et al* used SECM to study the *in situ* formation of a CEI on a $LiMn_2O_4$ (LMO) cathode material, for use in aqueous Li-ion batteries [101]. Feedback imaging of the cathode during cycling suggested an increase in electronic conductivity, contrasting the trend observed for SEI in non-aqueous systems. Combined with chemical composition obtained from XPS depth profiling, it was speculated that the CEI suppressed the dissolution of Mn from the electrode and prevented continuous capacity loss during cycling. In another SECM study, feedback revealed the electronic character of the CEI at high voltage cathode materials, $LiNi_{0.5}Mn_{1.5}O_4$ and $LiNi_{0.33}Mn_{0.33}Co_{0.33}O_2$ [102]. Approach curves indicated negligible changes in positive feedback at individual locations before and after cycling, implying no insulating structures were formed. These studies provided critical information about CEI evolution and paved the way for SECM analysis of the CEI at other materials.

Studies investigating the CEI are relatively scarce compared to the SEI. Care must be taken to avoid potential experimental flaws that can lead to erroneous or ambiguous results. One potential area for misinterpretation is the common use of conductive carbons in making LIB electrodes. SECM approach curves/feedback images over these regions will exhibit positive feedback characteristics which should not be interpreted as the CEI (or SEI) being conductive in nature. This combination of conductive carbons with less-conductive, active material further complicates approach curves that usually provide knowledge of the tip–substrate distance and electron transfer kinetics. One possible approach towards addressing this issue is using binderless cathodes (derived from methods such as electrodeposition [103] and pulsed layer deposition [104]) or single particle experiments [105–107]. Thus, prior characterization and sample preparation is required when designing experiments for investigation of both CEI and SEI.

9.3.1.3 Investigation of ionic flux at interfaces

Several energy storage technologies including LIBs operate via the reversible shuttling of metal cations between electrodes. Investigation of ionic flux over electrified surfaces yields a deeper understanding of surface modification and degradation processes, guiding rational electrode design for next generation batteries. For SECM measurements of Li^+ and other metal cations, Hg based UMEs are employed for their ability to form highly reversible amalgams upon reduction of the cations [108]. They provide an additional advantage over typical platinum and gold UMEs by circumventing interfering processes such as solvent decomposition at highly reducing potentials [109, 110]. They are stable in both aqueous [108, 111, 112] and non-aqueous media [76, 78] making them suitable for studying different battery systems. Barton and Rodríguez-López developed Hg capped Pt-UMEs to study local Li^+ concentrations at activated electrodes, including Au and graphene [76, 91]. These initial studies established the methodology to use such probes for quantification of alkali ion fluxes at substrates via ion competition between tip and substrate.

Subsequently, Gossage *et al* combined ionic flux measurements on HOPG using a Hg-disc well (HgDW) UME with COMSOL Multiphysics finite element modelling to extract kinetic parameters of the ionic processes occurring while charging and discharging the substrate, as depicted in figure 9.8 [78]. The peak currents at the UME were continuously monitored by collecting CVs at the Hg probe during SEI formation and (de)intercalation of Li^+. This study revealed reversible ionic flux during SEI formation and further enabled extraction of intercalation kinetics by fitting the probe response at different substrate potentials in the intercalation region, as shown in figure 9.8(b). The results indicated much faster ion transfer rates at the HOPG edge plane compared to bulk graphite, figures 9.8(c) and (d).

In 2020, the Rodríguez-López group further used a substrate pulsing method for mapping Li^+ flux and passivation during SEI formation on multi-layer graphene [77]. The use of the ion-sensitive HgDW probe [75] facilitated the investigation of two processes as the SEI formation progressed: (i) Li^+ flux via reversible amalgamation in Hg, and (ii) electronic passivation via use of a mediator in feedback mode.

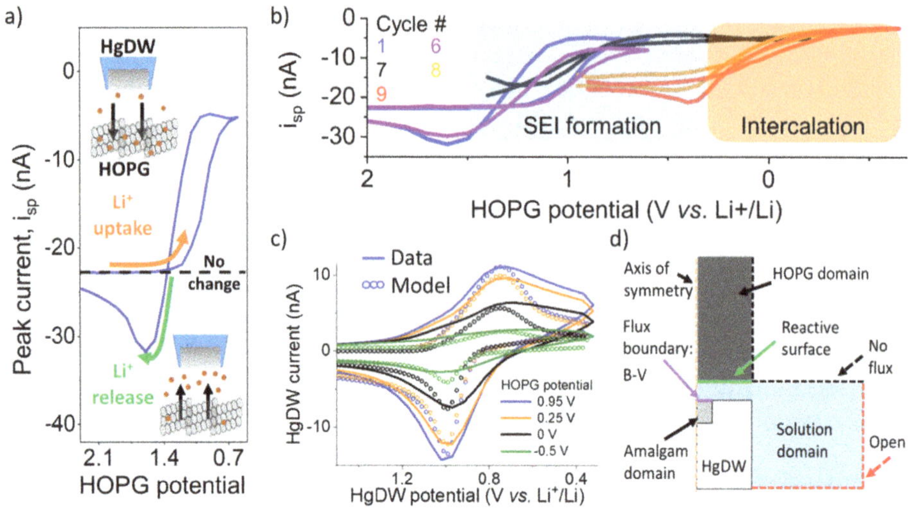

Figure 9.8. Use of HgDW probes and computational simulations for the investigation of ionic flux. (a) Illustration of Li^+ flux at HOPG measured by the HgDW UME, (b) Li^+ flux measurements via extracted stripping peak currents, i_{sp}, during SEI formation and (de)intercalation, (c) Experimental and modeled CVs for the HgDW at different HOPG potentials during intercalation, (d) 2D axisymmetric geometry for the COMSOL model to extract ion transfer kinetics. Reproduced from [78] with permission of The Royal Society of Chemisty.

Correlation between local ion and electron transfer seeds significant opportunity for improved insight on formation and stabilization of interphase structures [113].

The analytical methods described above can be extended to other emerging technologies. For instance, Barton *et al* demonstrated ionic flux measurements of K^+ during (de)intercalation at an HOPG electrode [114]. The results indicated higher flux of K^+ above HOPG edge planes compared with basal planes when biasing the HOPG to intercalate K^+. These types of fundamental information and analytical developments are widely applicable for understanding interfacial chemistry in next generation energy storage technologies, since the Hg probe can also measure concentrations of other alkali ions such as Na^+ [114], in addition to metals such as Zn^{2+} and Pb^{2+} [115].

9.3.1.4 Mass transport properties in electrodes

Aside from ion flux at the interface, another critical factor determining the rate of (dis)charge in LIBs is ionic transport within the pores and at grain boundaries on the electrodes. In a recent study, Li^+ transport on porous electrodes was approached by employing anthracene as a proxy [89]. SECM with anthracene as the mediator was used to obtain negative feedback response over $LiFePO_4$ films prepared on a composite carbon electrode. Negative feedback was obtained because anthracene undergoes electrochemically irreversible oxidation [116]. Subsequently, negative feedback currents were correlated to ionic diffusion of anthracene in porous electrodes, as shown in figure 9.9 [89]. Their experimental setup and methods with

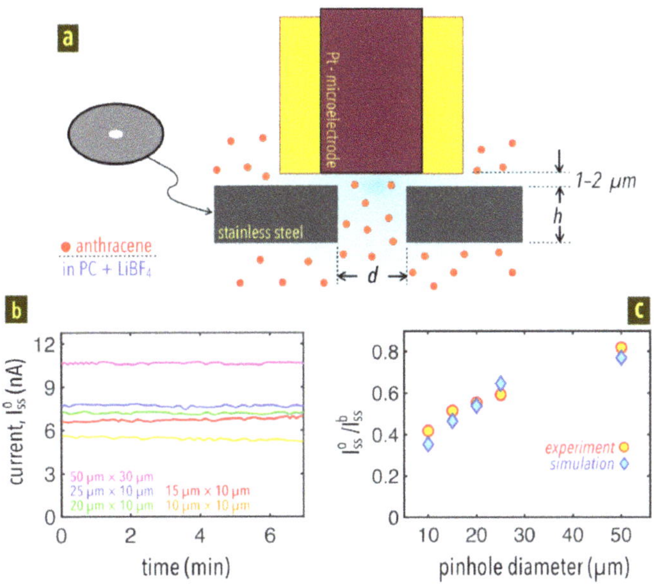

Figure 9.9. Investigation of mass transport properties in porous electrodes using SECM. (a) Schematic of the experimental setup depicting the proof of concept SECM measurement for determining porosity (2 mM anthracene in 0.25 M $LiBF_4$ in propylene carbonate). (b) Chronoamperometric response for the steady-state current at five different pinhole sizes. (c) Comparison of average of experimental and simulated current at each pinhole size with the corresponding bulk chronoamperometric response. Reproduced from [89]. Copyright (2020) American Chemical Society.

micropinholes clearly validated SECM applications for understanding ion diffusion not only at material surfaces, but also within materials. Measurements of internal ion diffusion can help direct engineering of faster ion transport in electrode composites and ionic conductors.

9.3.2 Redox flow batteries

Redox flow batteries (RFBs) rely on energy storage in the form of oxidized and reduced redox-active molecules dissolved in solution, with current collector electrodes being employed to charge and discharge these molecules. Healthy operation of a redox flow battery relies on processes such as (i) stability of charged molecules [117–119], and (ii) unhindered charge transfer between the molecules and electrodes [120], among other challenges such as molecule crossover across separator membranes [8]. Emerging flow batteries use non-aqueous electrolytes with organic redox molecules [121–124] as well as size-exclusion approaches that utilize particles and large polymers containing redox-active moieties where charge percolation on individual particles is crucial to efficient cycling [125–128]. Side effects of cycling include parasitic reactions and film formation. Table 9.4 summarizes the applications of SECM discussed in the following subsections along with key information obtained using different SECM modes to study redox flow batteries.

Table 9.4. Summary of SECM applications in redox flow batteries.

Concept addressed	SECM mode	Key information obtained	Relevant equations (table 9.1)	References
Lifetime/stability of charged complexes	Mass transfer limited steady state currents	Characterized decay as a function of time, empirical decay constants quantified	2	[129]
Electron transfer rates at the electrode-electrolyte interface	Approach curves	Comparative and/or quantitative characterization of heterogeneous electron transfer rate constants	8	[130–133]
Detecting interfacial structures which impede electron transfer	Feedback imaging	Feedback imaging and Tafel analysis on individual sites.	6, 7, 8	[120]
(Dis)Charge rates of redox-active particles	Feedback imaging, contact measurements, SI-SECM	Feedback imaging used to locate redox-active particles on a substrate Probe used to charge and discharge single particles via contact to single particles or SI-SECM	Specific equations in references	[47, 134, 135]

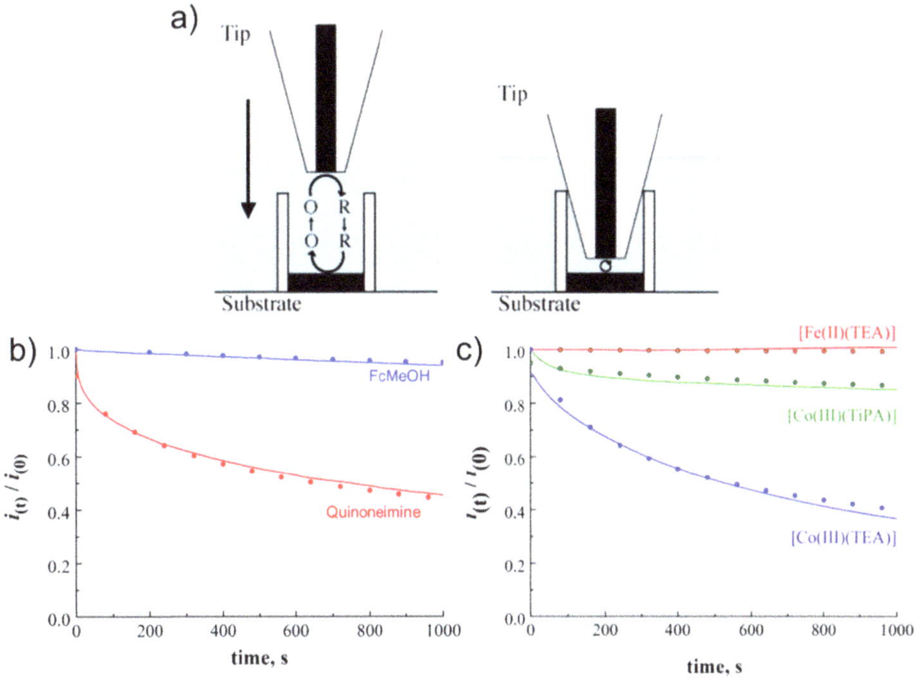

Figure 9.10. Schematics and results obtained from the e-SECM setup. (a) Thin layer cell design. (b) Proof of concept of e-SECM with the stable oxidized and neutral ferrocenemethanol (FcMeOH), and another redox species, quinoneimine which decreases over time due to hydrolysis. (c) Testing candidate molecules for an alkaline flow battery illustrates the stark differences in stability between Fe and Co triethanolamine (TEA) complexes. Ligand replacement with triisopropanolamine (TiPA) is shown to enhance the stability of Co complexes. Reprinted from [129], courtesy of Netzahualcóyotl Arroyo Currás.

9.3.2.1 Stability of charged molecules

The stability of electrogenerated molecules is typically studied through electrochemical techniques such as cyclic voltammetry, rotating ring disc electrode voltammetry, and bulk electrolysis, as well as spectroscopic and spectroelectrochemical approaches based on UV–vis absorbance and electron paramagnetic spectroscopy (EPR) to name a few [56, 117–119, 129, 130, 136, 137]. SECM can also be used to characterize degradation of charged molecules via a chemical pathway. Recently, Arroyo Currás developed an electrolysis-SECM mode (e-SECM) for rapidly characterizing the chemical stability of RFB molecules [129]. Typically, charged molecules (or mediators) can degrade via an irreversible chemical reaction, for example R → X, after the electrogeneration of R from O.

e-SECM utilizes a substrate to carry out electrolysis of the mediator in a thin-layer cell, as illustrated in figure 9.10(a). The use of a thin-layer cell drastically reduces the time to perform the bulk electrolysis. After the electrolysis, the presence of an irreversible chemical step can be detected by performing a chronoamperometric measurement at the tip to convert the electrogenerated mediator back to its original state. For a stable mediator, the tip current should remain constant over time due to a constant, reversible concentration (equation (2), table 9.1). However, a degradation mechanism would lead to the charged mediators depleting over time, leading to a drop in measured currents (figure 9.10(b)). Arroyo Currás modeled the current decrease over time with an empirical $t^{-1/2}$ function to extract decay rate constants of charged molecules used in alkaline-flow batteries, as illustrated in figure 9.10(c). This SECM approach offers a fast and simple quantitative characterization of degradation in charged mediators.

9.3.2.2 Charge transfer and film formation in redox-flow batteries

SECM is an excellent tool for evaluating electrochemical rate constants and charge transfer between redox-active species and current collectors for RFBs [64, 138]. For example, SECM approach curves have been used to measure electron transfer rate constants for redox-active polymers at platinum substrates [130, 131]. Similarly, approach curves were utilized to characterize the charge transfer kinetics between poorly conducting $LiFePO_4$ electrodes and ferrocene based mediators. This electrode–mediator combination is of practical use in redox flow lithium batteries, wherein charge from solid electrodes is shuttled using redox mediators [132]. SECM feedback mode was also used to quantify electron transfer at Cu current collectors utilized in semi-solid Li-ion flow batteries [133]. A drastic drop in feedback implied formation of insulating SEI films on the electrode surface. Consequently, the SEI, which is beneficial in LIB technology turned out to be a glaring problem in Li-ion based flow batteries. SECM can also indicate film development at the current collector during the cycling of a redox-flow battery. Watkins *et al* reported on film formation with dialkoxybenzene (DAB) molecules on a graphite current collector through open circuit feedback imaging and measurements of electron transfer kinetics as a function of substrate overpotential [120].

Open circuit feedback imaging is performed with the substrate at its open circuit potential (OCP), where no potential is applied to the substrate. Positive feedback can be obtained with conductive substrates at OCP if the substrate area is much larger than the tip [139, 140]. At OCP, an electron from another species in solution at a different location can be transferred via the conducting substrate and given to the electrogenerated species from the tip, leading to positive feedback [139, 140]. Any film which inhibits electron transfer from solution species to the electrode would lead to a drop in OCP feedback. On the other hand, if the substrate is biased, there might be enough driving force to generate feedback at the tip by transferring electrons across the passivating film. An example of this phenomenon is shown in figure 9.11(a), exhibiting the power of OCP imaging in detecting interfacial structures.

The example in figure 9.11(a) also examined the same film formation process by quantifying α (refer table 9.1, equation (10) and table 9.2) using Tafel plots from the SECM data illustrated in figure 9.11(b). SECM can be used to quantify the kinetic parameter k_f for different values of substrate overpotential η, which subsequently can be used to analyze electron transfer in the context of Butler–Volmer kinetics, using equation (10) in table 9.1. For uncomplicated one-electron redox reactions, α varies between 0.3 to 0.7 [56]. However, Watkins *et al*, reported very low 'apparent'

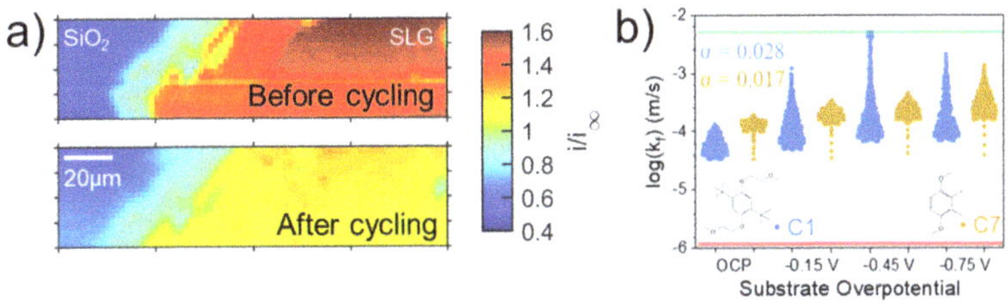

Figure 9.11. Characterizing film development on current collector graphene electrodes with SECM feedback images. (a) OCP feedback from single layer graphene (SLG) illustrating drastic decrease in feedback before and after cycling a DAB redox mediator. This decreased feedback is not observed when the substrate is biased. (b) Distribution of electron transfer kinetics (k_f) as a function of substrate overpotential and corresponding α for two different DAB molecules. The low α values indicate interfacial structures inhibiting electron transfer. Note green and red lines correspond to positive and negative feedback measurement limits, described by equations (5) and (6) in table 9.1 [120].

values of α <0.1. These artificial values imply that an increase in electrode overpotential does not manifest in higher k_f values, contrary to what is dictated by Butler–Volmer kinetics (equation (10), table 9.1). Such trends have been reported in literature, and arise due to porous insulating films where electron transfer proceeds through pinholes and through tunneling across the insulating film leading to a lesser dependence on substrate overpotentials than predicted by Butler–Volmer kinetics [42, 63, 141]. Likewise, redox-active films can show similar behavior with electron transfer taking place homogeneously between the charged film and electro-generated species from the SECM tip [142, 143]. Therefore, SECM may be used to locally probe deviations in α values as a strong indicator of surface phenomena on current collectors.

9.3.2.3 (Dis)charge of single colloidal redox-active particles

Redox-active colloids (RACs) are an emerging material in RFB technology, potentially reducing crossover between anolyte and catholyte via size-exclusion membrane separators [126] These colloidal particles consist of billions of redox--active moieties on an insulating polymeric backbone. Characterizing parameters such as (i) practical capacity of colloids, (ii) charge transfer diffusion coefficients, and (iii) charge transfer mechanisms are of vital importance to the practical understanding of such materials. While there has been progress made from a computational perspective in understanding microscopic charge transfer mechanisms [144], traditional electrochemical cycling methods interrogate multiple particles, yielding a bulk measurement indicative of the overall ensembled behavior. The use of SECM is advantageous in this scenario, since it can be used to isolate and characterize single particles.

Gossage *et al* utilized feedback SECM to locate dispersed viologen RAC nanoparticles using a 300 nm radius Pt tip [47, 134]. Once located, the SECM tip was brought over individual particles and contact was made between the RAC and the tip, enabling electrolysis of individual colloidal particles, as shown in figure 9.12.

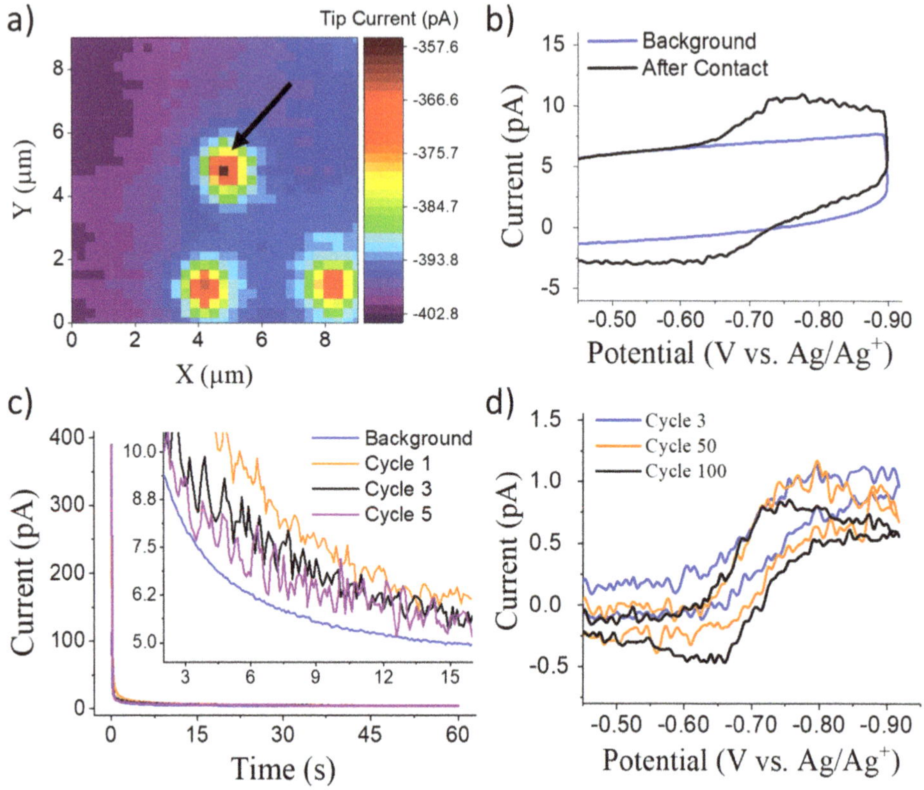

Figure 9.12. Characterizing (dis)charge of single colloidal particles using SECM. (a) Feedback imaging of RACs on a glass substrate, (b) CV in bulk solution (background) and response after contact with a single particle using a 300 nm radius SECM probe. (c) Chronoamperometric charge of single particles, which is further utilized to quantify parameters like coulombic efficiency and charge diffusion coefficients in the RACs. (d) Changes in the response at a single RAC while cycling with voltammetry [134] John Wiley & Sons, copyright (2018).

Charge transfer diffusion coefficients were quantified by measuring time dependent faradaic currents generated at the UME tip (shown in figure 9.12(b–c)) [145], which agreed with bulk results. Single particle electrolysis revealed a trend of improving coulombic efficiency with cycling, although full theoretical charge capacities were inaccessible. Cyclic voltammetry on individual particles revealed UME-type sigmoidal voltammetric curves. However, with time, the shape of the voltammogram (figure 9.13(d)) changed to resemble that of a macro-disc CV (see macro-disc CV in figure 9.2(a)). This suggests that cycling changed charge transport profiles within individual RACs. SECM thus illustrates its versatility to probe multiple parameters of practical importance in the development of RAC materials at a single-particle level. While powerful, this methodology is dependent on careful measurements to bring the tip in contact with single particles.

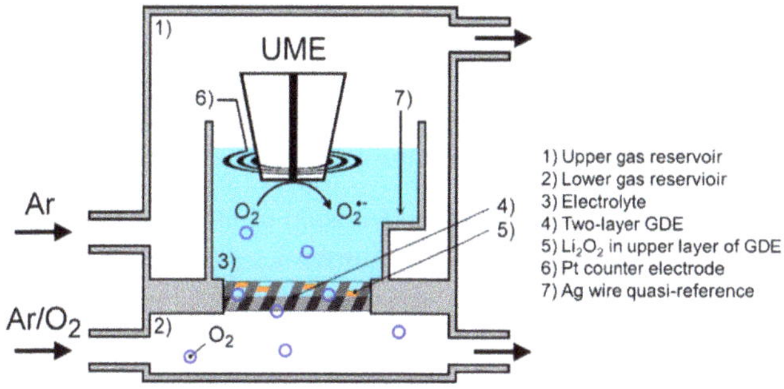

Figure 9.13. Cell schematic to study GDEs in Li-air batteries with an SECM tip. Such experimental configurations can be extended for studying redox species transport across membranes and other porous materials. Reproduced from [161], copyright (2015), with permission from Elsevier.

9.3.3 Li–air batteries

Li–air batteries supply electrical energy from Li metal oxidation by oxygen gas, which can theoretically supply 11 430 Wh kg^{-1} [146]. These batteries consist of a Li-metal anode and a gas diffusion (GDE) cathode. Li-air architectures can be classified by their electrolyte, including aprotic, aqueous, mixed aqueous-aprotic solutions, and solid-state electrolytes. The main processes governing the electrochemical performance of the Li–air battery include (i) oxygen transport and (ii) catalytic oxygen reduction/evolution reactions (ORR/OER) at the GDE cathode. ORR/OER reactions are particularly important, as their reaction rates depend strongly on the catalyst material. In addition, the Li–O_2 oxidation reaction leads to the formation of insulating Li-peroxides which can clog the GDE pores, preventing oxygen diffusion and blocking catalytically reactive sites. Comprehensive reviews about Li–air batteries, their mechanisms of operation, and challenges are highlighted in the literature. [147–151] SECM experiments have been developed to probe the two critical processes mentioned above, revealing design guidelines for future Li–air battery systems. The studies on Li–air batteries described henceforth utilize non-aqueous solvents. We do not discuss the challenges associated with Li-metal anodes, but other reviews highlight these issues and approaches for mitigating problems [152–154]. Table 9.5 summarizes the applications of SECM discussed in the following section along with key information obtained by using different modes to study phenomena in Li–air batteries.

9.3.3.1 Clogging, transport, and detection of transient electrogenerated species at GDE cathodes

As mentioned earlier, gas diffusion cathode clogging is a major issue in Li–air batteries. Lithium peroxide discharge products from reaction of Li^+ with superoxide radicals lead to blockage across the GDE [147, 148, 161]. SECM can characterize this clogging process *in situ*, as demonstrated by Schwager *et al* with a gas diffusion SECM cell [161]. Their cell design is shown in figure 9.13.

Table 9.5. Summary of SECM applications in Li–air batteries.

Concept addressed	SECM mode	Key information obtained	Relevant equations (table 9.1)	References
Diffusion through porous GDE cathodes	Mass transfer limited steady state currents, SG-TC modes	Characterization of blocking due to formation of insoluble Li_2O_2 products, detection of oxidizable intermediates upon ORR in Li^+ media at GDE cathodes	2	[155, 156]
Catalyst mediators for reducing insulating Li_2O_2	Steady state currents, Approach curves	Quantitative characterization of homogeneous redox reactivity between mediator and Li_2O_2	2, 8	[157, 158]
ORR catalysis	SECM collection: SG-TC or TG-SC, and redox competition	Qualitative characterization of efficient ORR/OER catalysts	N/A	[159, 160]

In these experiments, oxygen flux across the GDE was measured by a Pt SECM tip positioned near the surface via impedance approach curves. In 0.1 M tetrabutylammonium perchlorate in dimethyl sulfoxide (DMSO), the SECM tip allowed stable oxygen detection of gas diffusing through the GDE indicated by a steady state current from ORR. However, replacement of the electrolyte with lithium perchlorate led to a decreasing current at the SECM tip due to tip fouling by solid products, likely Li_2O_2. The authors developed a pulsing protocol to reduce oxygen (4 s at −0.7 V versus silver quasi-reference) and clean the tip electrochemically (6 s at 0.9 V versus silver quasi-reference) after ORR to overcome this problem.

Using their pulsing protocol, they characterized O_2 transport across GDEs of thicknesses ranging from 60–250 μm. These measurements enabled characterization of the time required to build up a steady flow of O_2 across the GDE during discharge (figure 9.14(a)). In addition to oxygen transport, this SECM setup allowed generation/collection measurements to detect oxidizable intermediates from the ORR reaction carried out at the GDE, as shown in figure 9.14(b). Their results indicate that not all reduction products at the GDE are insoluble as often described in the literature [147, 148]. Further, their pulsing protocol was combined with mapping to evaluate oxygen diffusion across the GDE while discharging. (figure 9.14(c, d))

9.3.3.2 Catalyst mediators

The oxidation of Li_2O_2 during Li–air battery charging requires large overpotentials since Li_2O_2 is electronically insulating. Therefore, catalysts which can homogeneously

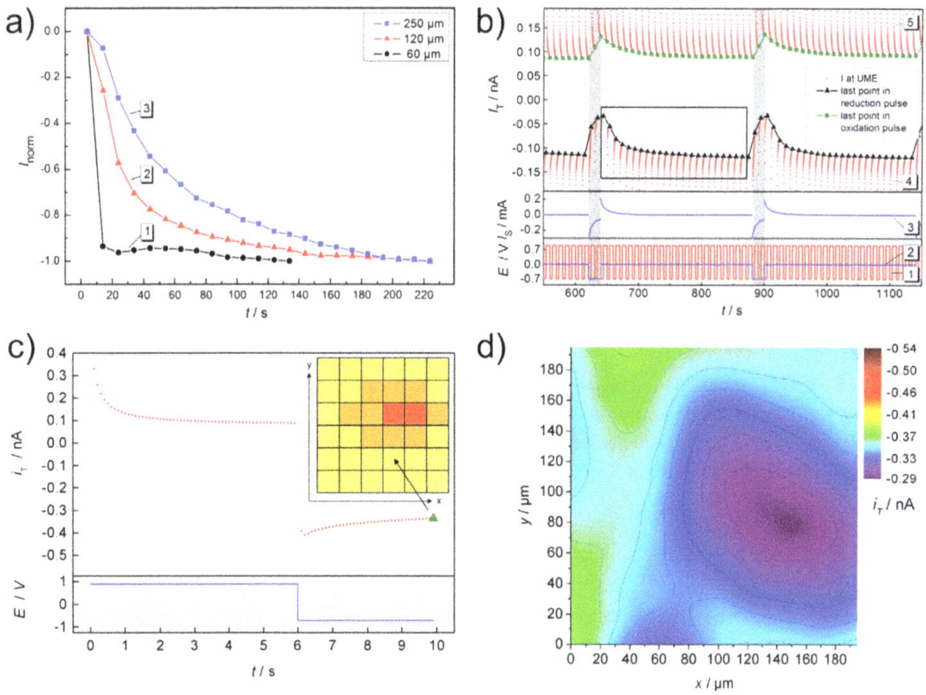

Figure 9.14. Experimental SECM data characterizing oxygen transport across GDE cathodes assembled within the cell illustrated in figure 9.14. (a) Characterizing transient oxygen flow across a GDE, (b) data collected in a measurement. Potential of the tip and GDE is shown in red and blue, respectively. I_S denotes the GDE current, while I_T denotes the tip current. When the GDE was biased to reducing potentials for ORR, the reducing UME currents during pulsing decreases owing to redox competition. (c) Schematic describing how the GDE was imaged. The tip current, i_T, was taken from the steady state value after the SECM tip was biased to reducing potentials to perform ORR (last data point, green triangle). (d) Resulting O_2 flux image from a cycled GDE electrode with lower magnitude of currents representing lower oxygen permeation. Reproduced from [156, 161], copyright (2016, 2015) with permission from Elsevier.

oxidize solid Li_2O_2 products and can be regenerated in a fast heterogeneous electrochemical reaction at the cathode are needed [157, 162]. Calvo *et al* [157] utilized a simple SECM substrate generation-tip collection (SG-TC) experiment to demonstrate the use of tetrathiafulvalene (TTF) as a catalyst for this charging process. The redox potential of TTF/TTF$^+$ (3.65 V vs. Li/Li$^+$) is more positive than the O_2/Li_2O_2 couple (2.96 V versus Li/Li$^+$), therefore TTF$^+$ undergoes homogeneous redox reaction with Li_2O_2, forming O_2 and TTF.

Prior to SECM SG-TC, the tip was biased to reduce oxygen, forming Li_2O_2 precipitates which blocked the tip. Then, the substrate was biased to generate TTF$^+$, and tip was biased to reduce the electrogenerated TTF$^+$. This protocol led to increasing tip current since Li_2O_2 precipitates are oxidized by excess TTF$^+$ generated by the substrate. This tip current increase allowed quantification of the performance of TTF/TTF$^+$ as a catalyst for improving Li–air battery charging. A similar

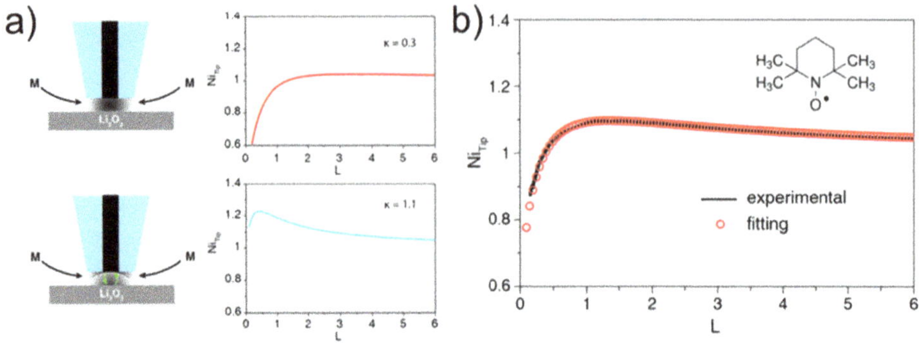

Figure 9.15. SECM approach curves on Li_2O_2 substrates using different mediators (a) yields different normalized electron transfer rate constants κ for different mediators. (b) Example of a least squares fit to extract κ between (2,2,6,6-Tetramethylpiperidin-1-yl)oxyl (TEMPO) and Li_2O_2. Reproduced from [158], copyright (2018) with permission of Springer.

screening approach can be utilized to quantify other catalysts for these battery systems.

Chen, *et al* demonstrated another elegant approach to measure reaction kinetics between electrogenerated catalysts and Li_2O_2 [158]. They used SECM with a Li_2O_2 pellet as a substrate, as shown in figure 9.15. The catalyst was electrogenerated at the tip, and approach curves were performed to quantify 'apparent' rate constants, k_{app} for the regeneration of catalysts by Li_2O_2. Comparison of k^0 (determined via Nicholson–Shain analysis from cyclic voltammetry [56]) and k_{app} of the catalysts revealed no correlations. This implied that Li_2O_2 oxidation requires intermediate steps such as adsorption of catalyst mediator on the peroxide surface; these multi-step electron transfers are known as inner-sphere reactions [163].

In addition to the charging process, discharging metal–air batteries requires ORR; a sluggish, inner-sphere, catalytic reaction [163]. The cathode reaction during discharge of a metal–air battery involves ORR to form peroxide radicals which react with metal ions to generate discharge products [148, 149, 164, 165]. Thus, ORR kinetics significantly dictate the discharge rates of metal–air batteries. Consequently, a major research thrust in metal–air batteries is to discover efficient ORR catalysts [149, 166]. SECM generation/collection [73, 167] and redox competition modes [74] are established tools able to address many electrocatalytic knowledge gaps, ranging from characterizing catalyst electrochemical reactivity to discovering new catalyst materials via combinatorial screening [29, 168–170]. However, SECM has seen limited use in the discovery and electrochemical characterization of metal–air battery catalysts [159, 160].

9.4 Related electrochemical scanning probe techniques and multi-modal characterization with SECM

The power of SECM to characterize localized electrochemical processes at interfaces is well established. However, this method has drawbacks which often make results difficult to interpret in the context of energy storage materials [171]. For example, the rough topography on many heterogeneous battery electrodes complicates interpretation of SECM feedback images. This is because changing tip–substrate distances are a convoluting factor in measuring the tip current. Also, SECM provides very limited local chemical and structural information, which is critical to understanding reactivity and designing better battery electrodes. Therefore, we describe alternative pipette-based electrochemical scanning probe methodologies which overcome some of these issues. We also discuss coupling SECM to spectroscopic and other scanning probe methods to create powerful multi-modal characterization [172] tools for battery analysis.

9.4.1 Other electrochemical scanning probe techniques

This section highlights the use of Scanning Ion Conductance Microscopy (SICM), Scanning Micropipette Contact Method (SMCM), and Scanning Electrochemical Cell Microscopy (SECCM) as alternatives to SECM measurements. We briefly describe their design and working principles along with a few examples illustrating their use in energy storage research.

9.4.1.1 Scanning ion conductance microscopy (SICM)

SICM utilizes a micro- or nanopipette as a probe to detect ionic currents in electrolyte solutions, as shown in figure 9.16(a) [173]. The ionic current is measured via reference-counter electrodes, with one electrode placed inside of the probe filled with electrolyte, and the other in the bulk solution. An external bias is applied to generate ionic current between the two electrodes.

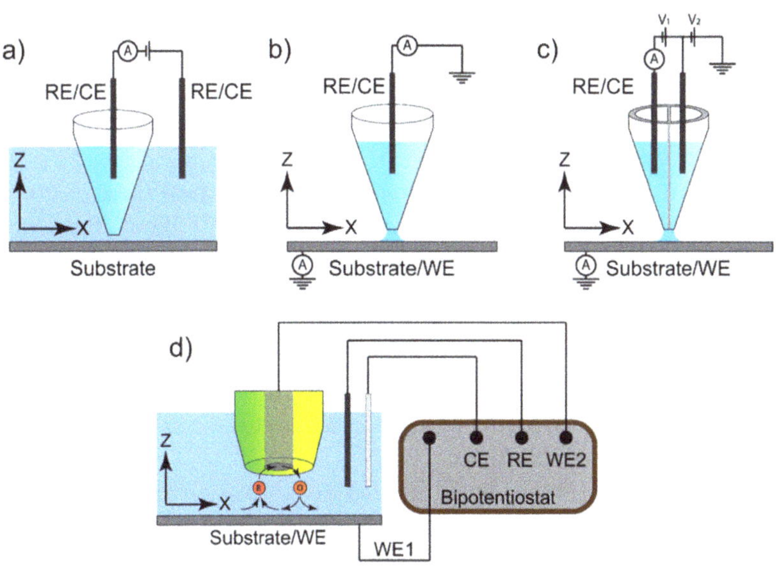

Figure 9.16. Schematic of (a) SICM, (b) SMCM, (c) SECCM compared to (d) SECM schematic. WE—working electrode, CE—counter electrode, RE—reference electrode.

The ionic current measured is a function of the pipette resistance and access resistance between the pipette tip and sample surface [173]. Therefore, by monitoring these resistances, one can estimate the tip–sample distance and construct topographical maps. The main advantage of SICM is the ability to make non-contact images of surface topography using inert ions while ignoring effects of local electrochemical reactivity.

9.4.1.2 Scanning micropipette contact method (SMCM) and scanning electrochemical cell microscopy (SECCM)

Contrasting to SECM and SICM approaches, SMCM and SECCM rely on micro- to nano-pipette measurements over a substrate exposed to a small liquid meniscus between the tip and substrate (figure 9.16(b)). When a substrate bias is applied, the tip response is influenced only by the region of the substrate enclosed within the liquid meniscus [32, 174, 175].

In the case of SMCM, a single-barreled pipette is used with a reference-counter electrode inside. The substrate potential is controlled to produce voltammograms which can be used to quantify local electron-transfer kinetics. Mapping substrate reactivity is possible by applying a potential step while measuring steady state currents at different locations [175].

SECCM is an advancement of SMCM wherein a double-barreled pipette (θ pipette, figure 9.16(c)) is utilized to probe surfaces. Instead of a single reference-counter electrode, two electrodes are maintained in each of the chambers. This enables precise positioning of a tip over the substrate by monitoring the ionic current between the two electrodes while the tip oscillates in a direction normal to the substrate [174]. Changes in ionic current are characteristic of the tip–substrate distance, similar to the concept of SICM measurements described previously [174]. Therefore, this enables the combined measurement of surface topography via ionic currents, along with local electrochemical activity by monitoring substrate currents [174].

9.4.1.3 Applications in energy storage

Reviews as recent as 2018 have highlighted the application of these electrochemical scanned probe techniques in energy storage [32, 176, 177]. Thus, the scope of this section is limited to literature published after 2018. While SICM can monitor topographical changes on battery surfaces *in operando* [178, 179]. the lack of electrochemical information from this technique is a potential drawback. Payne, *et al* recently illustrated an approach using SICM alongside COMSOL finite element simulations to track ionic conductivity in porous $LiFePO_4$ battery cathodes [180]. This advance in SICM experimentation extends to other permeable electrode materials, membranes, and coatings.

SMCM and SECCM have been applied to characterize the (dis)charge and topography of battery cathode materials at the nanoscale [181, 182]. Tao, *et al* [105] performed SECCM on individual $LiMn_2O_4$ cathode particles co-located with Scanning Electron Microscopy (SEM). These measurements revealed the unique electrochemical response from single particles which are lost in bulk measurements.

Takahashi *et al* [183] utilized SECCM to characterize topography and kinetics of Li^+ ion (de)intercalation at thin film $Li_4Ti_5O_{12}$ (LTO) and $LiFePO_4$ cathode materials prepared with no binders or additives. They also generated diffusion coefficient maps of Li^+ ion (de)intercalation in the LTO cathodes. Similarly, SECCM was utilized to characterize the interrelationships between ZrO_2 protective coating thicknesses on $LiCoO_2$ cathodes and Li^+ ionic current across the coating during (dis)charge [184]. SECCM cyclic voltammograms at the nanometer scale revealed that a 2–5 nm coating enables Li^+ de(intercalation) to be measured across the ZrO_2 layer, although at 5 nm thicknesses, currents decreased homogeneously across the entire scanned area. In another study, SMCM was shown to be compatible with ionic liquids for studying high energy cathode materials such as lithium nickel manganese cobalt oxide [185]. These emerging alternative techniques are promising complementary tools for battery studies requiring little or no additional hardware aside from an SECM.

9.4.2 Multimodal SECM techniques

The need for chemical and other forms of material characterization in SECM experiments has led to considerable effort in developing multi-modal tools with SECM. In these setups, SECM is coupled to spectroscopic and other scanning probe techniques as highlighted in this section.

9.4.2.1 Raman-SECM

Raman-SECM has been utilized only by a few groups for studying problems related to corrosion [186], catalysis [187] and surface modification [188]. The Rodríguez-López group at the University of Illinois has developed an in-house Raman SECM [47], which has been used to study energy storage systems alongside other applications such as characterizing reactivity at carbon defects [48]. A schematic of their Raman-SECM setup is shown in figure 9.17; this setup was used for a proof-of-concept study aimed at understanding film formation on model electrodes for flow batteries. The feedback behavior of the $[Fe(CN)_6]^{3-/4-}$ redox pair was probed over multi-layer graphene electrodes, showing the simultaneous quantification of the rate constant for electron transfer as the potential of the substrate was progressively made more positive in a potential window-opening experiment. Eventually, signatures consistent with exfoliation, defect formation, and the formation of a cyanide-containing film were detected via the Raman component and their relative weights on the reaction kinetics discerned in real time with SECM [48].

Experiments on redox-active colloids and films described in section 9.3.2.3 utilized co-localized Raman spectroscopy along with SI-SECM. As described previously, SI-SECM can be used to charge and discharge surface features, such as films, via a suitable electrogenerated mediator species at the SECM tip. Gossage, *et al* evaluated ethyl viologen RAC monolayers [47] with SI-SECM by reducing heptyl viologen at the tip, which reduced ethyl viologen in the film [189]. Once reduced by SI-SECM or the underlying electrode, the ethyl viologen could be oxidized by generating another mediator, such as the oxidized form of FcMeOH, to

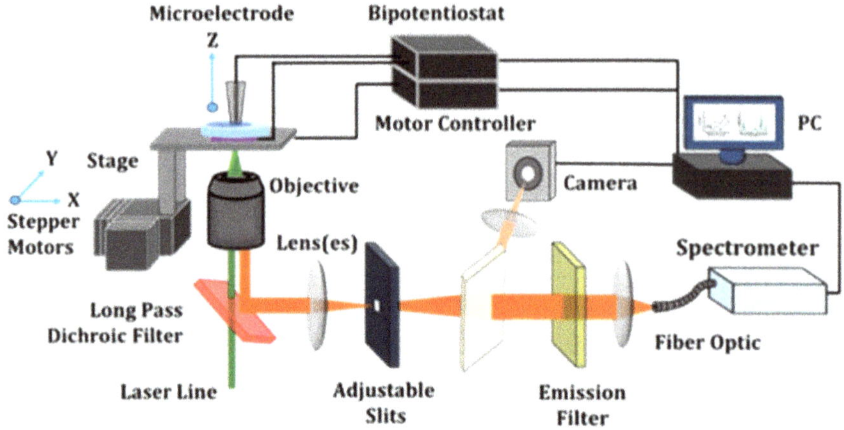

Figure 9.17. Schematic of in-house Raman SECM. Reproduced from [48]. Copyright (2018) American Chemical Society.

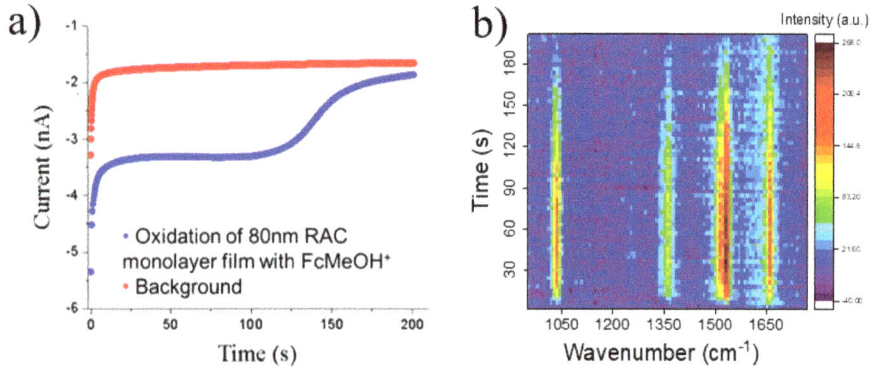

Figure 9.18. (Dis)charge of monolayer RAC films with SI-SECM with simultaneous co-localized Raman spectroscopic characterization. (a) SI-SECM response showing oxidation of RAC films with electrogenerated $FcMeOH^+$. Reproduced with permission from [134], copyright (2018) American Chemical Society. (b) Colocalized *in situ* Raman measurements illustrating multiple peaks being detected when V^{2+} is reduced to V^+ and oxidized back to V^{2+}. Reproduced from [172], copyright (2018) with permission from Elsevier.

locally titrate the film, as shown in figure 9.18(a). Quantifying titrated charge at the tip from SI-SECM allowed validation of the charge capacity of RACs measured by SECM tip contact measurements [47]. Simultaneously, Raman spectroscopy tracked peak intensities corresponding to the viologen (+1) redox state. Tracking peak intensities over time, as shown in figure 9.18(b), enabled characterization of charge–discharge rates as a function of RAC diameter. These measurements also indicated no charge trapping within the redox-active films after reduction/oxidation. Similar measurements during (dis)charge were also made for RAC particles of 830 nm diameter [134]. As with the advances using tip enhancement for local spectroscopic measurements, (e.g. tip-enhanced Raman spectroscopy) [32, 36], colocalizing an

SECM probe with spectroscopic methods opens great opportunity for analysis in battery systems with high spatial, chemical, and temporal information.

9.4.2.2 AFM-SECM

Combining AFM and SECM enables high resolution measurements of topography alongside electrochemical activity. Such instruments have been developed in-house [49, 67]. and are commercially available [50]. While a recent review covers multiple applications of AFM-SECM [190], this section's scope is restricted to a few examples on energy storage.

Using *in situ* AFM, Zampardi *et al* [67] monitored the formation of SEI on glassy carbon substrates. They found that the resulting SEI was 20–30 nm thick and started forming when potentials dropped below 0.6 V versus Li/Li^{+}. No SEI dissolution was observable when potentials were swept back to positive potentials. They subsequently used SECM to map SEI passivation. Successful operation of the instrument was verified by scratching an area with the AFM tip, exposing conductive glassy carbon. SECM imaging of this area revealed positive feedback, whereas all other regions were passivated, exhibiting negative feedback (figure 9.19). Another AFM-SECM study reported similar morphological-electrochemical reactivity interdependencies on Li–sulfur battery cathode particles [191].

9.4.2.3 SICM-SECM

The combination of SICM and SECM offers an alternative means to characterize sample topography and electrochemical reactivity within a single experiment. Morris *et al* [192] demonstrated the fabrication of a custom SICM nano-pipette with a conducting Au ring to enable current measurements. Figure 9.20 shows their electrode design and results from imaging topography and transport of electroactive

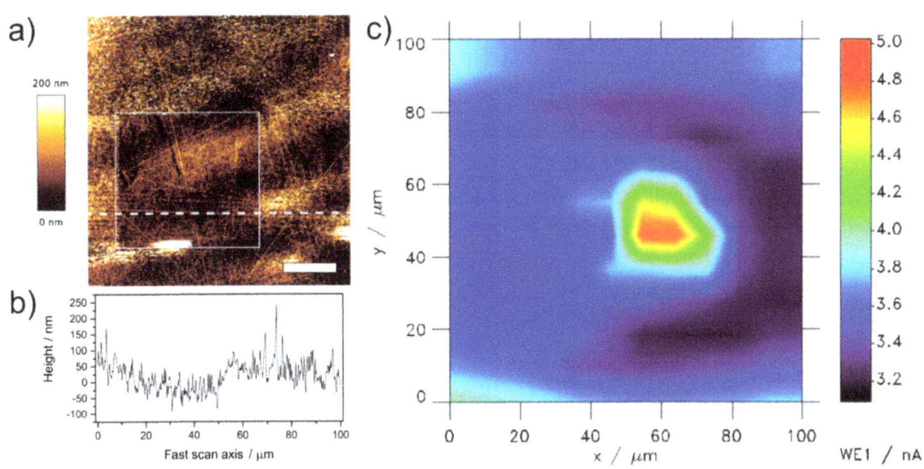

Figure 9.19. Monitoring the formation of SEI on glassy carbon substrates using *in situ* AFM and SECM. (a) AFM image showing scratched region (white rectangle) and (b) height profile (at white dotted line in (a)). Scale bar is 20 μm. (c) SECM feedback image illustrating positive feedback currents over a scratched region, and passivated, negative feedback in all other regions[67] John Wiley & Sons, copyright (2015).

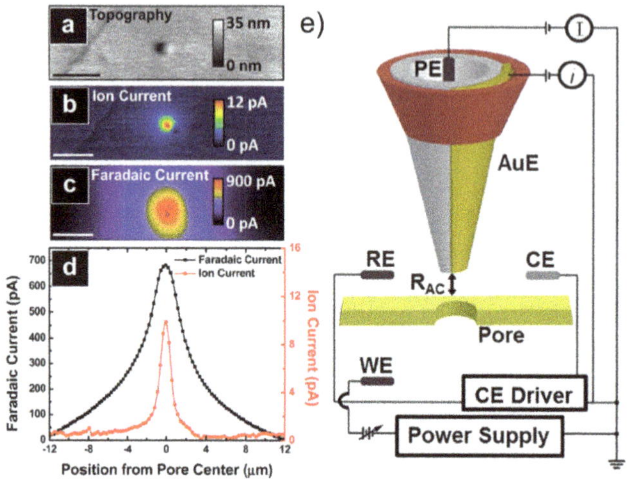

Figure 9.20. (a) 900 nm pore topography imaged with the nanopipet-electrode, (b) corresponding ionic current, (c) current from $[Fe(CN)_6]^{3-}$ reduction to $[Fe(CN)_6]^{4-}$ and (d) ionic and Faradaic currents across the pore from (b) and (c). (e) Schematic of the SICM-SECM tip for multi-modal characterization. Reproduced from [192] with permission from The Royal Society of Chemistry.

species from a pore in a polyimide membrane. While this measurement is not directly related to batteries, such approaches can be translated to characterize ion transport across membranes used in batteries [193–195] Combined SICM-SECM has otherwise been used in nanoparticle imaging [196] and biological [197] applications.

9.5 Outlook

The diverse nature of studies highlighted in the preceding sections illustrates the power of SECM for *in situ* investigation of battery interfaces. Different modes of this technique make it a versatile toolbox for probing electronic and ionic processes at various interfaces within batteries. However, SECM is far from being utilized to its full potential, evidenced by research groups around the world continually developing creative SECM studies in energy storage and expanding the applications of SECM in the process.

9.5.1 SECM technique development

One innovative approach is using AC-SECM to position the UME without the use of a redox-active mediator [156, 161]. Instead of mediator feedback, an impedance signal is monitored and used to position the tip close to a substrate [68]. AC-SECM has been utilized to visualize the surface topography of aqueous lithium-ion batteries [198]. To improve the resolution of SECM images, Stephens *et al* reported a finite element modelling based approach to generate super-resolution SECM images [199] by utilizing a point spread function (PSF) based deconvolution. Positioning the UME near the substrate is often difficult, especially over rough surfaces, due to tip crashing. To resolve this problem, Barforoush *et al* reported a novel fuzzy logic algorithm which automates this process and allows imaging at constant height [200]. The recent coupling of scanning probe techniques with a synchrotron beamline [201] exhibits promise for future powerful multi modal x-ray synchrotron/SECM characterization of battery materials.

9.5.2 Computational advancements

Coupling SECM with finite element computations enables extraction of more information than experimental data alone. To illustrate, researchers used COMSOL simulations to calculate parameters such as diffusion coefficient of species in porous electrodes [89] and reaction kinetics during SEI formation on graphite electrodes [77]. Such innovative computational approaches supplement SECM studies and will surely increase the capabilities of this technique, making it more popular in the coming years. In addition, open-source software development is another promising area for progress of SECM, with Python-based software being written to analyze raw SECM measurements [202].

9.5.3 Novel experiments

Although SECM is primarily limited to liquid systems, it can be used to detect evolving gases from various battery electrodes *in situ*. UMEs have already been used to detect dissolved gases such as O_2 and CO_2 in non-aqueous [203–205] and H_2 in aqueous [206] solutions. These gases are often by-products of decomposition processes during battery operation. [207, 208] SECM modes such as substrate generation/tip collection provide information about the operating parameters at which these reactions take place, possibly yielding critical information about safe operating limits for battery materials. Similarly, phase transformations in electrode materials leading to metal leaching could be detected using Hg UMEs [115]. Examples of such processes include metal dissolution from LIB cathodes [209–212].

Virtually no studies exist utilizing SECM on battery molecules/ion-transport across membranes. However, SECM has been utilized in characterizing transfer of chemical species such as hydroquinone [213] and $[Ru(NH_3)_6]^{3+/2+}$ ions [214] across Nafion membranes. In particular, Shi and Baker [214] used chemically degraded Nafion in their study, which shows that this approach can monitor degradation of membranes in battery materials. With increasing focus on membrane development, it is hoped that SECM investigations will provide deeper insight into the performance and durability of new membrane materials. Indeed, new methodologies based on impedance-based SECM approaches for detecting ion conductivity in solid electrolytes such as $Li_7La_3Zr_2O_{12}$ (LLZO) have been recently introduced [215], creating exciting opportunities to track ion transport and conductance in materials for solid-state batteries.

SECM holds much future promise, indicated by the recent advancements being made in both the theoretical and instrumental domain. We can expect its applicability to significantly increase in energy storage and beyond.

Acknowledgments

The authors gratefully acknowledge financial support from the National Science Foundation under grants CHE 1709391 and CHE 2004054. We also acknowledge financial support from the Joint Center for Energy Storage Research (JCESR), an Energy Innovation Hub funded by the U.S. Department of Energy, Office of Science, and Basic Energy Sciences. We acknowledge Professor Netzahualcóyotl Arroyo-Currás for providing content for use in this chapter.

References

[1] Whittingham M S 2012 History, evolution, and future status of energy storage *Proc. IEEE* **100** 1518–34

[2] Armand M and Tarascon J M 2008 Building better batteries *Nature* **451** 652–7

[3] Larcher D and Tarascon J M 2015 Towards greener and more sustainable batteries for electrical energy storage *Nat. Chem.* **7** 19–29

[4] Winter M, Barnett B and Xu K 2018 Before Li ion batteries *Chem. Rev.* **118** 11433–56

[5] Xu K 2014 Electrolytes and interphases in Li-ion batteries and beyond *Chem. Rev.* **114** 11503–618

[6] Robinson A L and Whittingham M S 2016 Pushing the frontiers of lithium-ion batteries raises safety questions *MRS Bull.* **41** 188–9

[7] Tarascon J M 2010 Key challenges in future Li-battery research. *Philos. Trans. A Math. Phys. Eng. Sci.* **368** 3227

[8] Trahey L, Brushett F R, Balsara N P, Ceder G, Cheng L and Chiang Y-M *et al* 2020 Energy storage emerging: a perspective from the Joint Center for Energy Storage Research *Proc. Natl. Acad. Sci.* **117** 12550

[9] Olivetti E A, Fu X, Ceder G and Gaustad G G 2017 Lithium-ion battery supply chain considerations: analysis of potential bottlenecks in critical metals *Joule* **1** 229–43

[10] Vikström H, Davidsson S and Höök M 2013 Lithium availability and future production outlooks *Appl. Energy* **110** 252–66

[11] Wu X, Leonard D P and Ji X 2017 Emerging non-aqueous potassium-ion batteries: challenges and opportunities *Chem. Mater.* **29** 5031–42

[12] Xue L, Li Y, Gao H, Zhou W, Kaveevivitchai W and Manthiram A *et al* 2017 Low-cost high-energy potassium cathode *JACS* **139** 2164–7

[13] Zhang L, Chen L, Zhou X and Liu Z 2015 Towards high-voltage aqueous metal-ion batteries beyond 1.5 V: The zinc/zinc hexacyanoferrate system *Adv. Energy Mater.* **5** 1400930

[14] Xu C, Li B, Du H and Kang F 2012 Energetic zinc ion chemistry: The rechargeable zinc ion battery *Angew. Chem.—Int. Ed.* **51** 933–5

[15] Blanc L E, Nazar L F and Kundu D 2020 Scientific challenges for the implementation of Zn-ion batteries *Joule* **4** 771–99

[16] Hwang J-Y, Myung S-T and Sun Y-K 2017 Sodium-ion batteries: present and future *Chem. Soc. Rev.* **46** 3529–614

[17] Kim H, Hong J, Park K-Y, Kim H, Kim S-W and Kang K 2014 Aqueous rechargeable Li and Na ion batteries *Chem. Rev.* **114** 11788–827

[18] Manthiram A, Fu Y, Chung S-H, Zu C and Su Y-S 2014 Rechargeable lithium–sulfur batteries *Chem. Rev.* **114** 11751–87

[19] Muldoon J, Bucur C B and Gregory T 2014 Quest for nonaqueous multivalent secondary batteries: magnesium and beyond *Chem. Rev.* **114** 11683–720
[20] Bruce D, Haresh K and Jean-Marie T 2011 Electrical energy storage for the grid: a battery of choices *Science* **334** 928
[21] Sui Y, Liu C, Masse R C, Neale Z G, Cao G and Atif M *et al* 2020 Dual-ion batteries: the emerging alternative rechargeable batteries *Energy Storage Mater.* **25** 1–32
[22] Yang Z, Zhang J, Kintner-Meyer M C W, Lu X, Choi D and Lemmon J P *et al* 2011 Electrochemical energy storage for green grid *Chem. Rev.* **111** 3577–613
[23] Ding Y, Zhang C, Zhang L, Zhou Y and Yu G 2019 Pathways to widespread applications: development of redox flow batteries based on new chemistries *Chem.* **5** 1964–87
[24] Luo J, Hu B, Hu M, Zhao Y and Liu T L 2019 Status and prospects of organic redox flow batteries toward sustainable energy storage *ACS Energy Lett.* **4** 2220–40
[25] Simon P, Gogotsi Y and Dunn B 2014 Where do batteries end and supercapacitors begin? *Science* **343** 1210
[26] Salanne M, Rotenberg B, Naoi K, Kaneko K, Taberna P L and Grey C P *et al* 2016 Efficient storage mechanisms for building better supercapacitors *Nat. Energy* **1** 16070
[27] Wang G, Zhang L and Zhang J 2012 A review of electrode materials for electrochemical supercapacitors *Chem. Soc. Rev.* **41** 797–828
[28] Ventosa E and Schuhmann W 2015 Scanning electrochemical microscopy of Li-ion batteries *Phys. Chem. Chem. Phys.* **17** 28441–50
[29] Bertoncello P 2010 Advances on scanning electrochemical microscopy (SECM) for energy *Energy Environ. Sci.* **3** 1620–33
[30] Bard A J, Fan F R F, Kwak J and Lev O 1989 Scanning electrochemical microscopy. introduction and principles *Anal. Chem.* **61** 132–8
[31] Schwager P, Bülter H, Plettenberg I and Wittstock G 2016 Review of local *in situ* probing techniques for the interfaces of lithium-ion and lithium–oxygen batteries *Energy Technol.* **4** 1472–85
[32] Hui J, Gossage Z T, Sarbapalli D, Hernández-Burgos K and Rodríguez-López J 2019 Advanced electrochemical analysis for energy storage interfaces *Anal. Chem.* **91** 60–83
[33] Polcari D, Dauphin-Ducharme P and Mauzeroll J 2016 Scanning electrochemical microscopy: a comprehensive review of experimental parameters from 1989 to 2015 *Chem. Rev.* **116** 13234–78
[34] Kumatani A and Matsue T 2020 Recent advances in scanning electrochemical microscopic analysis and visualization on lithium-ion battery electrodes *Curr. Opin. Electrochem.* **22** 228–33
[35] Yu X and Manthiram A 2018 Electrode–electrolyte interfaces in lithium-based batteries *Energy Environ. Sci.* **11** 527–43
[36] Tripathi A M, Su W-N and Hwang B J 2018 *In situ* analytical techniques for battery interface analysis *Chem. Soc. Rev.* **47** 736–851
[37] Laforge F O, Velmurugan J, Wang Y and Mirkin M V 2009 Nanoscale imaging of surface topography and reactivity with the scanning electrochemical microscope *Anal. Chem.* **81** 3143–50
[38] Mirkin M V, Sun T, Yu Y and Zhou M 2016 Electrochemistry at one nanoparticle *ACC. Chem. Res.* **49** 2328–35
[39] Slevin C J, Gray N J, Macpherson J V, Webb M A and Unwin P R 1999 Fabrication and characterisation of nanometre-sized platinum electrodes for voltammetric analysis and imaging *Electrochem. Commun.* **1** 282–8

[40] Shen M, Ishimatsu R, Kim J and Amemiya S 2012 Quantitative imaging of ion transport through single nanopores by high-resolution scanning electrochemical microscopy *JACS* **134** 9856–9

[41] Kai T, Zoski C G and Bard A J 2018 Scanning electrochemical microscopy at the nanometer level *Chem. Commun.* **54** 1934–47

[42] Bülter H, Peters F, Schwenzel J and Wittstock G 2014 Spatiotemporal changes of the solid electrolyte interphase in lithium-ion batteries detected by scanning electrochemical microscopy *Angew. Chem. Int. Ed.* **53** 10531–5

[43] Mirkin M V, Nogala W, Velmurugan J and Wang Y 2011 Scanning electrochemical microscopy in the 21st century. Update 1: five years after *Phys. Chem. Chem. Phys.* **13** 21196–212

[44] Sun P, Laforge F O and Mirkin M V 2007 Scanning electrochemical microscopy in the 21st century *Phys. Chem. Chem. Phys.* **9** 802–23

[45] Kempaiah R, Subramanian A and Vasudevamurthy G 2019 Scanning probe microscopy based characterization of battery materials, interfaces, and processes *Nano Energy.* **65** 103925

[46] Danis L, Gateman S M, Kuss C, Mauzeroll J and Schougaard S B 2017 Nanoscale measurements of lithium-ion-battery materials using scanning probe techniques *ChemElectroChem.* **4** 6–19

[47] Gossage Z T, Schorr N B, Hernández-Burgos K, Hui J, Simpson B H and Montoto E C *et al* 2017 Interrogating charge storage on redox active colloids via combined raman spectroscopy and scanning electrochemical microscopy *Langmuir* **33** 9455–63

[48] Schorr N B, Jiang A G and Rodríguez-López J 2018 Probing graphene interfacial reactivity via simultaneous and colocalized raman-scanning electrochemical microscopy imaging and interrogation *Anal. Chem.* **90** 7848–54

[49] Macpherson J V and Unwin P R 2000 Combined scanning electrochemical–atomic force microscopy *Anal. Chem.* **72** 276–85

[50] Huang Z, De Wolf P, Poddar R, Li C, Mark A and Nellist M R *et al* 2016 PeakForce Scanning electrochemical microscopy with nanoelectrode probes *Microsc. Today.* **24** 18

[51] Bard A J and Mirkin M V 2012 *Scanning Electrochemical Microscopy* 2nd edn (Boca Raton, FL: CRC Press)

[52] Amemiya S 2012 Scanning electrochemical microscopy *Characterization of Materials* ed E N Kaufmann (New York: Wiley) pp 1–13

[53] Wittstock G, Burchardt M, Pust S E, Shen Y and Zhao C 2007 Scanning electrochemical microscopy for direct imaging of reaction rates *Angew. Chem. Int. Ed.* **46** 1584

[54] Amemiya S, Bard A J, Fan F-R F, Mirkin M V and Unwin P R 2008 Scanning electrochemical microscopy *Annu. Rev. Anal. Chem.* **1** 95–131

[55] Michael C, Counihan D S and Rodriguez-Lopez J 2020 The chalkboard: picture your electrode: a primer on scanning electrochemical microscopy *Electrochem. Soc. Interface* **29** 30–2

[56] Bard A J and Faulkner L R 2001 *Electrochemical Methods: Fundamentals and Applications.* 2nd edn. (New York: John Wiley)

[57] Ching S, Dudek R and Tabet E 1994 Cyclic voltammetry with ultramicroelectrodes *J. Chem. Educ.* **71** 602–5

[58] Mauzeroll J, Hueske E A and Bard A J 2003 Scanning electrochemical microscopy. 48. Hg/Pt hemispherical ultramicroelectrodes: fabrication and characterization *Anal. Chem.* **75** 3880–9

[59] Walsh D A, Fernández J L, Mauzeroll J and Bard A J 2005 Scanning electrochemical microscopy. 55. Fabrication and characterization of micropipet probes *Anal. Chem.* **77** 5182–8

[60] Danis L, Polcari D, Kwan A, Gateman S M and Mauzeroll J 2015 Fabrication of carbon, gold, platinum, silver, and mercury ultramicroelectrodes with controlled geometry *Anal. Chem.* **87** 2565–9

[61] Lefrou C and Cornut R 2010 Analytical expressions for quantitative scanning electrochemical microscopy (SECM) *ChemPhysChem* **11** 547–56

[62] Cornut R, Poirier S and Mauzeroll J 2012 Forced convection during feedback approach curve measurements in scanning electrochemical microscopy: maximal displacement velocity with a microdisk *Anal. Chem.* **84** 3531–7

[63] Kiani A, Alpuche-Aviles M A, Eggers P K, Jones M, Gooding J J and Paddon-Row M N *et al* 2008 Scanning electrochemical microscopy. 59. Effect of defects and structure on electron transfer through self-assembled monolayers *Langmuir* **24** 2841–9

[64] Ritzert N L, Rodríguez-López J, Tan C and Abruña H D 2013 Kinetics of interfacial electron transfer at single-layer graphene electrodes in aqueous and nonaqueous solutions *Langmuir* **29** 1683–94

[65] Kathrin E and Wolfgang S 2008 Alternating current techniques in scanning electrochemical microscopy (AC-SECM) *Analyst* **133** 1486–97

[66] Sklyar O, Kueng A, Kranz C, Mizaikoff B, Lugstein A and Bertagnolli E *et al* 2005 Numerical simulation of scanning electrochemical microscopy experiments with frame-shaped integrated atomic force microscopy–SECM probes using the boundary element method *Anal. Chem.* **77** 764–71

[67] Zampardi G, Klink S, Kuznetsov V, Maljusch A, Schuhmann W and Ventosa E *et al* 2015 Combined AFM/SECM investigation of the solid electrolyte interphase in Li-ion batteries *ChemElectroChem.* **2** 1607–11

[68] Horrocks B R, Bard A J, Schmidtke D and Heller A 1993 Scanning electrochemical microscopy. 24. Enzyme ultramicroelectrodes for the measurement of hydrogen peroxide at surfaces *Anal. Chem.* **65** 3605–14

[69] Kwak J and Bard A J 1989 Scanning electrochemical microscopy. Theory of the feedback mode *Anal. Chem.* **61** 1221–7

[70] Claudio-Cintrón M A and Rodríguez-López J 2019 Scanning electrochemical microscopy with conducting polymer probes: validation and applications *Anal. Chim. Acta* **1069** 36–46

[71] Barton Z J and Rodríguez-López J 2017 Cyclic voltammetry probe approach curves with alkali amalgams at mercury sphere-cap scanning electrochemical microscopy probes *Anal. Chem.* **89** 2708–15

[72] Allen J B, Fu-Ren F F, David T P, Patrick R U, David O W and Feimeng Z 1991 Chemical imaging of surfaces with the scanning electrochemical microscope *Science* **254** 68

[73] Sánchez-Sánchez C M, Rodríguez-López J and Bard A J 2008 Scanning electrochemical microscopy. 60. Quantitative calibration of the SECM substrate generation/tip collection mode and its use for the study of the oxygen reduction mechanism *Anal. Chem.* **80** 3254–60

[74] Eckhard K, Chen X, Turcu F and Schuhmann W 2006 Redox competition mode of scanning electrochemical microscopy (RC-SECM) for visualisation of local catalytic activity *Phys. Chem. Chem. Phys.* **8** 5359–65

[75] Barton Z J and Rodríguez-López J 2017 Fabrication and demonstration of mercury disc-well probes for stripping-based cyclic voltammetry scanning electrochemical microscopy *Anal. Chem.* **89** 2716–23

[76] Barton Z J and Rodríguez-López J 2014 Lithium ion quantification using mercury amalgams as *in situ* electrochemical probes in nonaqueous media *Anal. Chem.* **86** 10660–7
[77] Gossage Z T, Hui J, Sarbapalli D and Rodríguez-López J 2020 Coordinated mapping of Li^+ flux and electron transfer reactivity during solid-electrolyte interphase formation at a graphene electrode *Analyst* **145** 2631–8
[78] Gossage Z T, Hui J, Zeng Y, Flores-Zuleta H and Rodríguez-López J 2019 Probing the reversibility and kinetics of Li^+ during SEI formation and (de)intercalation on edge plane graphite using ion-sensitive scanning electrochemical microscopy *Chem. Sci.* **10** 10749–54
[79] Rodríguez-López J, Alpuche-Avilés M A and Bard A J 2008 Interrogation of surfaces for the quantification of adsorbed species on electrodes: oxygen on gold and platinum in neutral media *JACS* **130** 16985–95
[80] Simpson B H and Rodríguez-López J 2015 Redox titrations via surface interrogation scanning electrochemical microscopy at an extended semiconducting surface for the quantification of photogenerated adsorbed intermediates *Electrochim. Acta* **179** 74–83
[81] Counihan M J, Setwipatanachai W and Rodríguez-López J 2019 Interrogating the surface intermediates and water oxidation products of boron-doped diamond electrodes with scanning electrochemical microscopy *ChemElectroChem.* **6** 3507–15
[82] Krumov M R, Simpson B H, Counihan M J and Rodríguez-López J 2018 *In situ* quantification of surface intermediates and correlation to discharge products on hematite photoanodes using a combined scanning electrochemical microscopy approach *Anal. Chem.* **90** 3050–7
[83] Newman J 1966 Resistance for flow of current to a disk *J. Electrochem. Soc.* **113** 501
[84] Scherson D A and Palencsár A 2006 Batteries and electrochemical capacitors *The Electrochem. Soc. Interface.* **15** 17–22
[85] Peled E 1979 The electrochemical behavior of alkali and alkaline earth metals in non-aqueous battery systems—the solid electrolyte interphase model *J. Electrochem. Soc.* **126** 2047–51
[86] Zampardi G and La Mantia F 2020 Solid–electrolyte interphase at positive electrodes in high-energy li-ion batteries: current understanding and analytical tools *Batteries Supercaps.* **3** 672–97
[87] Wang H, Rus E, Abruna H D, Sakuraba T, Kikuchi J and Kiya Y 2014 CO_2 and O_2 evolution at high voltage cathode materials of li-ion batteries: a differential electrochemical mass spectrometry study *Anal. Chem.* **86** 6197–201
[88] Teufl T, Müller P, Mendez M A, Strehle B and Gasteiger H A 2018 Oxygen release and surface degradation of Li- and Mn-rich layered oxides in variation of the Li_2MnO_3 content *J. Electrochem. Soc.* **165** A2718–31
[89] Hossain M S, Stephens L I, Hatami M, Ghavidel M, Chhin D and Dawkins J I G *et al* 2020 Effective mass transport properties in lithium battery electrodes *ACS Appl. Energy Mater.* **3** 440–6
[90] Bülter H, Peters F and Wittstock G 2016 Scanning electrochemical microscopy for the *in situ* characterization of solid–electrolyte interphases: highly oriented pyrolytic graphite versus graphite composite *Energy Technology.* **4** 1486–94
[91] Hui J, Burgess M, Zhang J and Rodríguez-López J 2016 Layer number dependence of Li^+ intercalation on few-layer graphene and electrochemical imaging of its solid–electrolyte interphase evolution *ACS Nano.* **10** 4248–57

[92] Zampardi G, Ventosa E, La Mantia F and Schuhmann W 2013 *In situ* visualization of Li-ion intercalation and formation of the solid electrolyte interphase on TiO_2 based paste electrodes using scanning electrochemical microscopy *Chem. Commun.* **49** 9347–9

[93] Yamada Y, Miyazaki K and Abe T 2010 Role of edge orientation in kinetics of electrochemical intercalation of lithium-ion at graphite *Langmuir* **26** 14990–4

[94] Krueger B, Balboa L, Dohmann J F, Winter M, Bieker P and Wittstock G 2020 Solid electrolyte interphase evolution on lithium metal electrodes followed by scanning electrochemical microscopy under realistic battery cycling current densities *ChemElectroChem.* **7** 3590–6

[95] Hatchard T D and Dahn J 2004 *In situ* XRD and electrochemical study of the reaction of lithium with amorphous silicon *J. Electrochem. Soc.* **151** A838–42

[96] Obrovac M N and Christensen L 2004 Structural changes in silicon anodes during lithium insertion/extraction *Electrochem. Solid State Lett.* **7** A93–6

[97] Su X, Wu Q, Li J, Xiao X, Lott A and Lu W *et al* 2014 Silicon-based nanomaterials for lithium-ion batteries: a review *Adv. Energy Mater.* **4** 1300882

[98] Tarnev T, Wilde P, Schuhmann W, Dopilka A, Chan C K and Ventosa E 2020 Surface properties of battery materials elucidated using scanning electrochemical microscopy: the case of type i silicon clathrate *ChemElectroChem.* **7** 665–71

[99] Scipioni R, Isheim D and Barnett S A 2020 Revealing the complex layered-mosaic structure of the cathode electrolyte interphase in Li-ion batteries *Appl. Mater. Today.* **20** 100748

[100] Whittingham M S 2004 Lithium batteries and cathode materials *Chem. Rev.* **104** 4271–302

[101] Liu S, Liu D, Wang S, Qian K, Kang F and Li B *et al* 2019 Understanding the cathode electrolyte interface formation in aqueous electrolyte by scanning electrochemical microscopy *J. Mater. Chem.* A **7** 12993–6

[102] Zampardi G, Trocoli R, Schuhmann W and La Mantia F 2017 Revealing the electronic character of the positive electrode/electrolyte interface in lithium-ion batteries *Phys. Chem. Chem. Phys.* **19** 28381–7

[103] Zhang H, Ning H, Busbee J, Shen Z, Kiggins C and Hua Y *et al* 2017 Electroplating lithium transition metal oxides *Sci. Adv.* **3** e1602427

[104] Kuwata N, Kumar R, Toribami K, Suzuki T, Hattori T and Kawamura J 2006 Thin film lithium ion batteries prepared only by pulsed laser deposition *Solid State Ionics.* **177** 2827–32

[105] Tao B, Yule L C, Daviddi E, Bentley C L and Unwin P R 2019 Correlative electrochemical microscopy of Li-ion (de)intercalation at a series of individual $LiMn_2O_4$ particles *Angew. Chem. Int. Ed.* **58** 4606–11

[106] Kumatani A, Takahashi Y, Miura C, Ida H, Inomata H and Shiku H *et al* 2019 Scanning electrochemical cell microscopy for visualization and local electrochemical activities of lithium-ion (de) intercalation process in lithium-ion batteries electrodes *Surf. Interface Anal.* **51** 27–30

[107] Ventosa E 2021 Why nanoelectrochemistry is necessary in battery research? *Curr. Opin. Electrochem.* **25** 100635

[108] Nepomnyashchii A B, Alpuche-Aviles M A, Pan S, Zhan D, Fan F-R F and Bard A J 2008 Cyclic voltammetry studies of Cd^{2+} and Zn^{2+} complexation with hydroxyl-terminated polyamidoamine generation 2 dendrimer at a mercury microelectrode *J. Electroanal. Chem.* **621** 286–96

[109] Fernández J L, Hurth C and Bard A J 2005 Scanning electrochemical microscopy #54. Application to the study of heterogeneous catalytic reactions hydrogen peroxide decomposition. *J. Phys. Chem.* B **109** 9532–9

[110] Selzer Y, Turyan I and Mandler D 1999 Studying heterogeneous catalysis by the Scanning Electrochemical Microscope (SECM): the reduction of protons by methyl viologen catalyzed by a platinum surface *J. Phys. Chem.* B **103** 1509–17

[111] Ciani I, Daniele S, Bragato C and Baldo M A 2003 Stability of mercury-coated platinum microelectrodes upon touching a solid surface in scanning electrochemical microscopy (SECM) experiments *Electrochem. Commun.* **5** 354–8

[112] Daniele S, Baldo M-A, Ugo P and Mazzocchin G-A 1989 Determination of heavy metals in real samples by anodic stripping voltammetry with mercury microelectrodes: Part 2. Application to rain and sea waters *Anal. Chim. Acta* **219** 19–26

[113] Hernández-Burgos K, Barton Z J and Rodríguez-López J 2017 Finding harmony between ions and electrons: new tools and concepts for emerging energy storage materials *Chem. Mater.* **29** 8918–31

[114] Barton Z J, Hui J, Schorr N B and Rodríguez-López J 2017 Detecting potassium ion gradients at a model graphitic interface *Electrochim. Acta* **241** 98–105

[115] Alpuche-Aviles M A, Baur J E and Wipf D O 2008 Imaging of metal ion dissolution and electrodeposition by anodic stripping voltammetry–scanning electrochemical microscopy *Anal. Chem.* **80** 3612–21

[116] Lhenry S, Leroux Y R and Hapiot P 2013 Chemically irreversible redox mediator for secm kinetics investigations: determination of the absolute tip–sample distance *Anal. Chem.* **85** 1840–5

[117] Zhang J, Shkrob I A, Assary R S, Tung S O, Silcox B and Curtiss L A *et al* 2017 Toward improved catholyte materials for redox flow batteries: what controls chemical stability of persistent radical cations? *J. Phys. Chem.* C **121** 23347–58

[118] Zhang J, Huang J, Robertson L A, Assary R S, Shkrob I A and Zhang L 2018 Elucidating factors controlling long-term stability of radical anions for negative charge storage in nonaqueous redox flow batteries *J. Phys. Chem.* C **122** 8116–27

[119] Zhang J, Huang J, Robertson L A, Shkrob I A and Zhang L 2018 Comparing calendar and cycle life stability of redox active organic molecules for nonaqueous redox flow batteries *J. Power Sources* **397** 214–22

[120] Watkins T S, Sarbapalli D, Counihan M J, Danis A S, Zhang J and Zhang L *et al* 2020 A combined SECM and electrochemical AFM approach to probe interfacial processes affecting molecular reactivity at redox flow battery electrodes *J. Mater. Chem.* A **8** 15734–45

[121] Li M, Rhodes Z, Cabrera-Pardo J R and Minteer S D 2020 Recent advancements in rational design of non-aqueous organic redox flow batteries *Sustain. Energy Fuels.* **4** 4370–89

[122] Rhodes Z, Cabrera-Pardo J R, Li M and Minteer S D 2020 Electrochemical advances in non-aqueous redox flow batteries *Isr. J. Chem.* **61** 101–12

[123] Sevov C S, Samaroo S K and Sanford M S 2017 Cyclopropenium salts as cyclable, high-potential catholytes in nonaqueous media *Adv. Energy Mater.* **7** 1602027

[124] Yan Y, Robinson S G, Sigman M S and Sanford M S 2019 Mechanism-based design of a high-potential catholyte enables a 3.2 V all-organic nonaqueous redox flow battery *JACS* **141** 15301–6

[125] Burgess M, Moore J S and Rodríguez-López J 2016 Redox active polymers as soluble nanomaterials for energy storage *ACC. Chem. Res.* **49** 2649–57

[126] Montoto E C, Nagarjuna G, Hui J, Burgess M, Hernández-Burgos K and Wei T S *et al* 2016 Redox active colloids as discrete energy storage carriers *JACS* **138** 13230–7

[127] Yuan M and Minteer S D 2019 Redox polymers in electrochemical systems: from methods of mediation to energy storage *Curr. Opin. Electrochem.* **15** 1–6

[128] Montoto E C, Nagarjuna G, Moore J S and Rodríguez-López J 2017 Redox active polymers for non-aqueous redox flow batteries: validation of the size-exclusion approach *J. Electrochem. Soc.* **164** A1688–94

[129] Currás N A 2015 Development of an alkaline redox flow battery: from fundamentals to benchtop prototype *PhD Thesis* The University of Texas at Austin

[130] Baran M J, Braten M N, Montoto E C, Gossage Z T, Ma L and Chénard E *et al* 2018 Designing redox-active oligomers for crossover-free, nonaqueous redox-flow batteries with high volumetric energy density *Chem. Mater.* **30** 3861–6

[131] Burgess M, Chénard E, Hernández-Burgos K, Nagarjuna G, Assary R S and Jingshu H *et al* 2016 Impact of backbone tether length and structure on the electrochemical performance of viologen redox active polymers *Chem. Mater.* **28** 7362–74

[132] Yan R, Ghilane J, Phuah K C, Pham Truong T N, Adams S and Randriamahazaka H *et al* 2018 Determining Li^{+}-coupled redox targeting reaction kinetics of battery materials with scanning electrochemical microscopy *J. Phys. Chem. Lett.* **9** 491–6

[133] Ventosa E, Zampardi G, Flox C, La Mantia F, Schuhmann W and Morante J 2015 Solid electrolyte interphase in semi-solid flow batteries: a wolf in sheep's clothing *Chem. Commun.* **51** 14973–6

[134] Gossage Z T, Hernández-Burgos K, Moore J S and Rodríguez-López J 2018 Impact of charge transport dynamics and conditioning on cycling efficiency within single redox active colloids *ChemElectroChem.* **5** 3006–13

[135] Burgess M, Hernández-Burgos K, Cheng K J, Moore J S and Rodríguez-López J 2016 Impact of electrolyte composition on the reactivity of a redox active polymer studied through surface interrogation and ion-sensitive scanning electrochemical microscopy *Analyst* **141** 3842–50

[136] Milshtein J D, Kaur A P, Casselman M D, Kowalski J A, Modekrutti S and Zhang P L *et al* 2016 High current density, long duration cycling of soluble organic active species for non-aqueous redox flow batteries *Energy Environ. Sci.* **9** 3531–43

[137] Kaur A P, Holubowitch N E, Ergun S, Elliott C F and Odom S A 2015 A highly soluble organic catholyte for non-aqueous redox flow batteries *Energy Technol.* **3** 476–80

[138] Yan R and Wang Q 2018 Redox-targeting-based flow batteries for large-scale energy storage *Adv. Mater.* **30** 1802406

[139] Xiong H, Guo J and Amemiya S 2007 Probing heterogeneous electron transfer at an unbiased conductor by scanning electrochemical microscopy in the feedback mode *Anal. Chem.* **79** 2735–44

[140] Xiong H, Kim J, Kim E and Amemiya S 2009 Scanning electrochemical microscopy of one-dimensional nanostructure: effects of nanostructure dimensions on the tip feedback current under unbiased conditions *J. Electroanal. Chem.* **629** 78–86

[141] Tan C, Rodríguez-López J, Parks J J, Ritzert N L, Ralph D C and Abruña H D 2012 Reactivity of monolayer chemical vapor deposited graphene imperfections studied using scanning electrochemical microscopy *ACS Nano.* **6** 3070–9

[142] Tsionsky M, Bard A J, Dini D and Decker F 1998 Polymer films on electrodes. 28. Scanning electrochemical microscopy study of electron transfer at poly(alkylterthiophene) films *Chem. Mater.* **10** 2120–6

[143] Schmehl R H and Murray R W 1983 The effect of redox site concentration on the rate of mediated oxidation of solution substrates by a redox copolymer film *J. Electroanal. Chem. Interfacial Electrochem.* **152** 97–109

[144] Bello L and Sing C E 2020 Mechanisms of diffusive charge transport in redox-active polymer solutions *Macromolecules* **53** 7658–71

[145] Li Y, Deng H, Dick J E and Bard A J 2015 Analyzing benzene and cyclohexane emulsion droplet collisions on ultramicroelectrodes *Anal. Chem.* **87** 11013–21

[146] Imanishi N, Bruce P and Luntz A C 2014 *The Lithium Air Battery: Fundamentals* (New York: Springer)

[147] Girishkumar G, McCloskey B, Luntz A C, Swanson S and Wilcke W 2010 Lithium-air battery: promise and challenges *J. Phys. Chem. Lett.* **1** 2193–203

[148] Grande L, Passerini S, Paillard E, Hassoun J, Park J B and Lee Y J *et al* 2015 The lithium/air battery: still an emerging system or a practical reality? *Adv. Mater.* **27** 784–800

[149] Shao Y, Park S, Xiao J, Zhang J G, Wang Y and Liu J 2012 Electrocatalysts for nonaqueous lithium-air batteries: status, challenges, and perspective *ACS Catal.* **2** 844–57

[150] Kraytsberg A and Ein-Eli Y 2011 Review on Li–air batteries—opportunities, limitations and perspective *J. Power Sources* **196** 886–93

[151] Christensen J, Albertus P, Lohmann T, Ahmed J, Kojic A and Sanchez-Carrera R S *et al* 2012 A critical review of Li/air batteries *J. Electrochem. Soc.* **159** R1–30

[152] Wang H, Liu Y, Li Y and Cui Y 2019 Lithium metal anode materials design: interphase and host *Electrochem. Energy Rev.* **2** 509–17

[153] Liu Y and Cui Y 2017 Lithium metal anodes: a recipe for protection *Joule.* **1** 649–50

[154] Lin D, Liu Y and Cui Y 2017 Reviving the lithium metal anode for high-energy batteries *Nat. Nanotechnol.* **12** 194–206

[155] Schwager P, Fenske D and Wittstock G 2015 Scanning electrochemical microscopy of oxygen permeation through air-electrodes in lithium-air batteries *J. Electroanal. Chem.* **740** 82–7

[156] Bülter H, Schwager P, Fenske D and Wittstock G 2016 Observation of dynamic interfacial layers in Li-ion and Li-O_2 batteries by scanning electrochemical microscopy *Electrochim. Acta* **199** 366–79

[157] Torres W R, Herrera S E, Tesio A Y, Pozo M d and Calvo E J 2015 Soluble TTF catalyst for the oxidation of cathode products in Li-oxygen battery: a chemical scavenger *Electrochim. Acta* **182** 1118–23

[158] Chen Y, Gao X, Johnson L R and Bruce P G 2018 Kinetics of lithium peroxide oxidation by redox mediators and consequences for the lithium-oxygen cell *Nat. Commun.* **9** 767

[159] Szot K, Nogala W, Niedziolka-Jönsson J, Jönsson-Niedziolka M, Marken F and Rogalski J *et al* 2009 Hydrophilic carbon nanoparticle-laccase thin film electrode for mediatorless dioxygen reduction: SECM activity mapping and application in zinc-dioxygen battery *Electrochim. Acta* **54** 4620–5

[160] Zhang Y, Wu X, Zeng X, Ding W, Fu Y and Shen W 2014 Carbon aerogel supported Pt-Zn catalyst and its oxygen reduction catalytic performance in magnesium-air batteries *J. Mater. Res.* **29** 2863–70

[161] Schwager P, Fenske D and Wittstock G 2015 Scanning electrochemical microscopy of oxygen permeation through air-electrodes in lithium–air batteries *J. Electroanal. Chem.* **740** 82–7

[162] Landa-Medrano I, Lozano I, Ortiz-Vitoriano N, Ruiz de Larramendi I and Rojo T 2019 Redox mediators: a shuttle to efficacy in metal–O_2 batteries *J. Mater. Chem.* A **7** 8746

[163] Bard A J 2010 Inner-sphere heterogeneous electrode reactions. electrocatalysis and photocatalysis: the challenge *JACS* **132** 7559–67

[164] Christensen J, Albertus P, Sanchez-Carrera R S, Lohmann T, Kozinsky B and Liedtke R *et al* 2012 A critical review of Li/air batteries *J. Electrochem. Soc.* **159** R1

[165] Choi N-S, Chen Z, Freunberger S A, Ji X, Sun Y-K and Amine K *et al* 2012 Challenges facing lithium batteries and electrical double-layer capacitors *Angew. Chem. Int. Ed.* **51** 9994

[166] Zhong-Li W, Dan X, Ji-Jing X and Xin-Bo Z 2014 Oxygen electrocatalysts in metal-air batteries: from aqueous to nonaqueous electrolytes *Chem. Soc. Rev.* **43** 7746

[167] Fernández J L and Bard A J 2003 Scanning Electrochemical microscopy. 47. Imaging electrocatalytic activity for oxygen reduction in an acidic medium by the tip generation –substrate collection mode *Anal. Chem.* **75** 2967–74

[168] Wain A J 2014 Scanning electrochemical microscopy for combinatorial screening applications: a mini-review *Electrochem. Commun.* **46** 9–12

[169] Jeon M K, Lee C H, Park G I and Kang K H 2012 Combinatorial search for oxygen reduction reaction electrocatalysts: a review *J. Power Sources* **216** 400–8

[170] Liang Y, Pfisterer J H K, McLaughlin D, Csoklich C, Seidl L and Bandarenka A S *et al* 2019 Electrochemical scanning probe microscopies in electrocatalysis *Small Methods.* **3** 1800387

[171] Barton Z J and Rodríguez-López J 2016 Emerging scanning probe approaches to the measurement of ionic reactivity at energy storage materials *Anal. Bioanal.Chem.* **408** 2707–15

[172] Schorr N B, Gossage Z T and Rodríguez-López J 2018 Prospects for single-site interrogation using *in situ* multimodal electrochemical scanning probe techniques *Curr. Opin. Electrochem.* **8** 89–95

[173] Chen C-C, Zhou Y and Baker L A 2012 Scanning ion conductance microscopy *Annu. Rev. Anal. Chem.* **5** 207–28

[174] Ebejer N, Güell A G, Lai S C S, McKelvey K, Snowden M E and Unwin P R 2013 Scanning electrochemical cell microscopy: a versatile technique for nanoscale electrochemistry and functional imaging *Annu. Rev. Anal. Chem.* **6** 329–51

[175] Williams C G, Edwards M A, Colley A L, Macpherson J V and Unwin P R 2009 Scanning micropipet contact method for high-resolution imaging of electrode surface redox activity *Anal. Chem.* **81** 2486–95

[176] Bentley C L, Edmondson J, Meloni G N, Perry D, Shkirskiy V and Unwin P R 2019 Nanoscale electrochemical mapping *Anal. Chem.* **91** 84–108

[177] Danis L, Gateman S M, Kuss C, Schougaard S B and Mauzeroll J 2017 Nanoscale measurements of lithium-ion-battery materials using scanning probe techniques *ChemElectroChem.* **4** 6–19

[178] Lipson A L, Ginder R S and Hersam M C 2011 Nanoscale *in situ* characterization of li-ion battery electrochemistry via scanning ion conductance microscopy *Adv. Mater.* **23** 5613–7

[179] Lipson A L, Puntambekar K, Comstock D J, Meng X, Geier M L and Elam J W *et al* 2014 Nanoscale investigation of solid electrolyte interphase inhibition on Li-ion battery MnO electrodes via atomic layer deposition of Al_2O_3 *Chem. Mater.* **26** 935–40

[180] Payne N A, Dawkins J I G, Schougaard S B and Mauzeroll J 2019 Effect of substrate permeability on scanning ion conductance microscopy: uncertainty in tip–substrate separation and determination of ionic conductivity *Anal. Chem.* **91** 15718–25

[181] Takahashi Y, Kumatani A, Munakata H, Inomata H, Ito K and Ino K *et al* 2014 Nanoscale visualization of redox activity at lithium-ion battery cathodes *Nat. Commun.* **5** 5450

[182] Snowden M E, Dayeh M, Payne N A, Gervais S, Mauzeroll J and Schougaard S B 2016 Measurement on isolated lithium iron phosphate particles reveals heterogeneity in material properties distribution *J. Power Sources* **325** 682–9

[183] Takahashi Y, Yamashita T, Takamatsu D, Kumatani A and Fukuma T 2020 Nanoscale kinetic imaging of lithium ion secondary battery materials using scanning electrochemical cell microscopy *Chem. Commun.* **56** 9324–7

[184] Inomata H, Takahashi Y, Takamatsu D, Kumatani A, Ida H and Shiku H *et al* 2019 Visualization of inhomogeneous current distribution on ZrO_2-coated $LiCoO_2$ thin-film electrodes using scanning electrochemical cell microscopy *Chem. Commun.* **55** 545–8

[185] Dayeh M, Ghavidel M R Z, Mauzeroll J and Schougaard S B 2019 Micropipette contact method to investigate high-energy cathode materials by using an ionic liquid *ChemElectroChem.* **6** 195–201

[186] Etienne M, Dossot M, Grausem J and Herzog G 2014 Combined Raman microspectrometer and shearforce regulated SECM for corrosion and self-healing analysis *Anal. Chem.* **86** 11203–10

[187] Steimecke M, Seiffarth G and Bron M 2017 *In Situ* characterization of Ni and Ni/Fe thin film electrodes for oxygen evolution in alkaline media by a Raman-coupled scanning electrochemical microscope setup *Anal. Chem.* **89** 10679–86

[188] Clausmeyer J, Grützke S, Kayran Y U, Schuhmann W and Nebel M 2018 Local surface modifications investigated by combining scanning electrochemical microscopy and surface-enhanced Raman scattering *ChemPlusChem.* **83** 414–7

[189] Bird C L and Kuhn A T 1981 Electrochemistry of the viologens *Chem. Soc. Rev.* **10** 49–82

[190] Shi X, Qing W, Marhaba T and Zhang W 2020 Atomic force microscopy—scanning electrochemical microscopy (AFM-SECM) for nanoscale topographical and electrochemical characterization: Principles, applications and perspectives *Electrochim. Acta* **332** 135472

[191] Mahankali K, Thangavel N K, Reddy and Arava L M 2019 *In Situ* electrochemical mapping of lithium-sulfur battery interfaces using AFM-SECM *Nano Lett.* **19** 5229–36

[192] Morris C A, Chen C-C and Baker L A 2012 Transport of redox probes through single pores measured by scanning electrochemical-scanning ion conductance microscopy (SECM-SICM) *Analyst* **137** 2933–8

[193] Baran M J, Braten M N, Sahu S, Baskin A, Meckler S M and Li L *et al* 2019 Design rules for membranes from polymers of intrinsic microporosity for crossover-free aqueous electrochemical devices *Joule.* **3** 2968–85

[194] Li C, Meckler S M, Smith Z P, Bachman J E, Maserati L and Long J R *et al* 2018 Engineered transport in microporous materials and membranes for clean energy technologies *Adv. Mater.* **30** 1704953

[195] Doris S E, Ward A L, Frischmann P D, Li L and Helms B A 2016 Understanding and controlling the chemical evolution and polysulfide-blocking ability of lithium-sulfur battery membranes cast from polymers of intrinsic microporosity *J. Mater. Chem.* A **4** 16946–52

[196] Panday N, Qian G, Wang X, Chang S, Pandey P and He J 2016 Simultaneous ionic current and potential detection of nanoparticles by a multifunctional nanopipette *ACS Nano.* **10** 11237–48

[197] Takahashi Y, Ida H, Matsumae Y, Komaki H, Zhou Y and Kumatani A *et al* 2017 3D electrochemical and ion current imaging using scanning electrochemical-scanning ion conductance microscopy *Phys. Chem. Chem. Phys.* **19** 26728–33

[198] Liu D, Zeng X, Liu S, Wang S, Kang F and Li B 2019 Application of alternating current scanning electrochemical microscopy in lithium-ion batteries: local visualization of the electrode surface *ChemElectroChem.* **6** 4854–8

[199] Stephens L I, Payne N A and Mauzeroll J 2020 Super-resolution scanning electrochemical microscopy *Anal. Chem.* **92** 3958–63

[200] Barforoush J M, McDonald T D, Desai T A, Widrig D, Bayer C and Brown M K *et al* 2016 Intelligent scanning electrochemical microscopy tip and substrate control utilizing fuzzy logic *Electrochim. Acta* **190** 713–9

[201] Vitorino M V, Fuchs Y, Dane T, Rodrigues M S, Rosenthal M and Panzarella A *et al* 2016 An *in situ* atomic force microscope for normal-incidence nanofocus x-ray experiments *J. Synchrotron Radiat.* **23** 1110–7

[202] Stephens L I and Mauzeroll J 2019 Flux: software for analysing SECM data *J. Electrochem. Soc.* **166** H861–5

[203] Wadhawan J D, Welford P J, McPeak H B, Hahn C E W and Compton R G 2003 The simultaneous voltammetric determination and detection of oxygen and carbon dioxide: a study of the kinetics of the reaction between superoxide and carbon dioxide in non-aqueous media using membrane-free gold disc microelectrodes *Sens. Actuators B: Chem.* **88** 40

[204] Dawson G A, Hauser P C, Kilmartin P A and Wright G A 2000 CO_2 gas sensing at microelectrodes in nonaqueous solvents *Electroanalysis* **12** 105

[205] Hahn C E W, McPeak H, Bond A M and Clark D 1995 The development of new microelectrode gas sensors: an odyssey. Part 1. O_2 and CO_2 reduction at unshielded gold microdisc electrodes *J. Electroanal. Chem.* **393** 61–8

[206] Jamali S S, Moulton S E, Wallace G G, Tallman D E, Forsyth M and Weber J 2015 Evaluating the corrosion behaviour of magnesium alloy in simulated biological fluid by using SECM to detect hydrogen evolution *Electrochim. Acta* **152** 294–301

[207] Holzapfel M, Würsig A, Scheifele W, Vetter J and Novák P 2007 Oxygen, hydrogen, ethylene and CO_2 development in lithium-ion batteries *J. Power Sources* **174** 1156–60

[208] Zeradjanin A R, Schilling T, Seisd S, Bron M and Schuhmann W 2011 Visualization of chlorine evolution at dimensionally stable anodes by means of scanning electrochemical microscopy *Anal. Chem.* **83** 7645–50

[209] Zhan C, Wu T, Lu J and Amine K 2018 Dissolution, migration, and deposition of transition metal ions in Li-ion batteries exemplified by Mn-based cathodes—a critical review *Energy Environ. Sci.* **11** 243–57

[210] Esbenshade J L and Gewirth A A 2014 Effect of Mn and Cu addition on lithiation and SEI formation on model anode electrodes *J. Electrochem. Soc.* **161** A513

[211] Yiling D, Long C and White R E 2013 Capacity fade model for spinel $LiMn_2O_4$ electrode *J. Electrochem. Soc.* **160** A182

[212] Thackeray M M, Kahaian A J, Kepler K D, Skinner E, Vaughey J T and Shao-Horn Y *et al* 1998 Structural fatigue in spinel electrodes in high voltage (4 V) $Li/Li_xMn_2O_4$ cells *Electrochem. Solid State Lett.* **1** 7–9

[213] Bath B D, Lee R D, White H S and Scott E R 1998 Imaging molecular transport in porous membranes. observation and analysis of electroosmotic flow in individual pores using the scanning electrochemical microscope *Anal. Chem.* **70** 1047–58

[214] Shi W and Baker L A 2015 Imaging heterogeneity and transport of degraded Nafion membranes *RSC Adv.* **5** 99284–90

[215] Catarelli S R, Lonsdale D, Cheng L, Syzdek J and Doeff M 2016 Intermittent contact alternating current scanning electrochemical microscopy: a method for mapping conductivities in solid Li ion conducting electrolyte samples *Front. Energy Res.* **4** 14

Chapter 10

Small-angle x-ray scattering for battery research

Yun Zhao, Kun Qian and Tao Li

The design of advanced batteries requires powerful characterization technology. Traditional characterization techniques, such as transmission electron microscope (TEM), scanning electron microscope (SEM), and atomic force microscope (AFM), can well obtain the morphology and structures of local areas of solid samples. Nevertheless, the demand to obtain the material's overall nanostructure information, especially the presence of phase-separated polymer membranes, changes in nanomaterials' structure, and sub-nano structures in liquid systems, are still challenging. Small-angle x-ray scattering (SAXS) uses the scattering signal generated in a small angle range close to the beam by x-ray after passing through the sample to obtain nanostructure information with different electron cloud densities. Therefore, SAXS has low requirements for test samples but can obtain abundant material structure information. In this chapter, we review the principles of the SAXS test and the corresponding testing sample requirements. The representative applications of SAXS in separator, liquid electrolyte, and redox flow battery and *in situ* or *operando* applications are then introduced and prospected.

10.1 Introduction

Due to the fossil energy crisis and environmental issues, battery technology, including lithium-ion batteries, fuel cells, and flow batteries, have emerged as one of the most active research areas in the past few decades. With the high energy storage efficiency (> 95%), batteries have become indispensable components in daily life, powered cell phones, laptops, and cameras. New areas such as large-scale energy storage and electric vehicles are also turning to the battery solution because of its high efficiency, considerable energy density, and safety. Many countries are promoting battery-powered cars to substitute traditional petrol cars, which will produce a vast revolution in automobile manufacturing, the transportation industry, and environmental protection.

doi:10.1088/978-0-7503-2682-7ch10

Battery research is a multidisciplinary field, including chemistry, electrochemistry, physics, materials, etc. The components in the working battery cover solids, liquids, and gases. Advanced characterization tools are critical to understanding the fundamental mechanisms of ions/electrons transportation, capacity origination, and cycle life. Small-angle x-ray scattering (SAXS) is of the capability to determine nanoparticle size distributions, resolve the size and shape of macromolecules, determine pore sizes, characteristic distances of partially ordered materials, probe the intermolecular spacing, cluster size, or inhomogeneity domains in 1–100 nm, which displays the representative structures of an entire sample. The SAXS methods can be used to characterize different components of a battery, such as electrode materials, separators, liquid electrolytes, polymer electrolytes, and solid-state electrolytes. It is an ideal complementary method for other characterization methods that provide local information, such as nuclear magnetic resonance (NMR), Raman, and Fourier transform infrared spectroscopy (FT-IR). This chapter summarizes the latest developments in battery-related SAXS characterization and introduces the latest understanding of battery characteristics.

10.2 The theory of small-angle x-ray scattering

Small-angle x-ray scattering (SAXS) means that when a beam of excellent x-ray passes through a material with uneven nano-sized electron density, the x-ray causes coherent scattering phenomenon in a small angular region of the original beam direction (scattering angle $2\theta \leqslant 5°$). The intensity of the coherent scattering phenomenon generally decreases with the increase of 2θ. The coherent scattering phenomenon is related to x-ray scattering. X-ray scattering is caused by electrons' transition in atoms between two energy levels with different energy levels. When an x-ray passes through the test sample, the x-ray interacts with the sample's electrons, resulting in x-ray scattering. X-ray scattering is divided into coherent scattering, incoherent scattering, and the fluorescent x-ray scattering generated by the photoelectric effect. Among them, incoherent scattering and fluorescent x-ray scattering will not interfere. Therefore, they can be ignored as background scattering in the scattering spectrum when the sample contains the differences in electron cloud density; the incident x-ray undergoes incoherent scattering. When there is a certain amount of periodically arranged atoms in the sample, the scattered waves will be superimposed, making the light intensity more strongly observed.

The SAXS experimental setup includes an x-ray light source, optical system, and detector (figure 10.1). The slow development of SAXS is mainly due to the limitation of the weak scattering signal. The collimation system of SAXS is relatively long, and the gratings and slits should be very thin to ensure that the focus of the light is narrow. Therefore, the light of an ordinary x-ray light source is weak, and the testing time for data collection is too long. Nowadays, the synchrotron radiation with high intensity and collimation used as the x-ray source dramatically improves the experimental efficiency, sensitivity, time, and spatial resolution and makes the weak scattering solution measurements *in situ* dynamic systems possible. Besides, after being equipped with corresponding facilities, various *in situ* experiments such

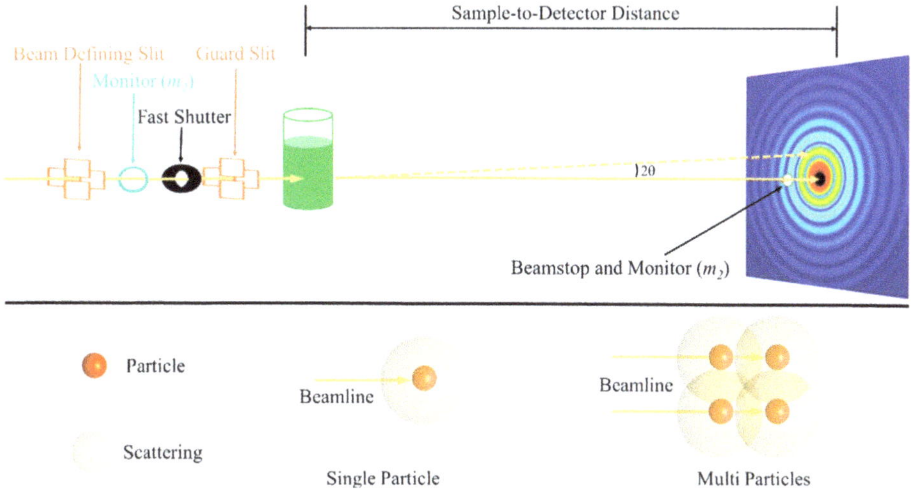

Figure 10.1. Schematic for the 12ID-B SAXS beamline's optical components at the Advanced Photon Source (APS) (top). Reprinted with permission from [1]. Copyright (2016) American Chemical Society. Schematic illustration of x-ray scattering in single-particle and multi particles (bottom).

as temperature change, stretching, and polymer self-assembly can be carried out. They can be combined with the transmission electron microscope (TEM), scanning electron microscope (SEM), and atomic force microscope (AFM), x-ray diffraction (XRD), infrared spectrometer, Raman spectrometer, and other characterization methods.

10.3 Potential applications and advantages of SAXS

10.3.1 Testing materials

The samples tested by SAXS should have an electron density difference, and no useful information can be obtained in a uniform system. The sample can be gas, liquid, solid, and their mixtures. Test objects often involved in lithium-ion batteries (LIBs) include nanomaterials, porous materials, and polymers. Nanomaterials refer to materials with at least one dimension in the nanometer size (0.1–100 nm) or are composed of them as basic units in a three-dimensional space, which is approximately equivalent to the scale of 10~1000 atoms closely arranged together. Nanomaterials are widely used as cathode and anode materials for high specific energy batteries. SAXS can study the necessary structure information of materials, composite materials during battery usage, and material structure changes. Porous materials with a network structure composed of interpenetrating or closed pores are often used as battery anode materials and sulfur cathode material hosts. Due to the electron density difference between the pores and the carbon material, and that these pores are mostly in the nanoscale range, SAXS is a useful technique to study the structure of such porous materials. The polymers studied by SAXS refer to block polymers that are used as the material for battery separators, polymer electrolytes, and other components. In practical applications, polymer film formation is often accompanied by phase separation, which are the main factors affecting its electrochemical performance.

The shape, size, porosity, pore distribution, volume fraction, specific surface parameters, and other microstructure information can be derived by calculating and analyzing the SAXS image or scattering curve. These microstructures can be pores, particles, defects, and irregular two-phase system structure in the material. Simultaneously, due to x-rays' penetrability, the SAXS signal is a statistical result of the scattering on the surface and inside the sample.

10.3.2 Advantages of SAXS comparing with other characterization techniques

For nanoscale materials, conventional characterization methods include TEM, SEM, and AFM, directly observing local material structure. At the same time, sample preparation is relatively complicated and requires a high experimental environment. It takes a long time to obtain available sample data or information. Moreover, it is often challenging to perform *in situ* dynamic experiments. Compared with traditional characterization methods, SAXS has the following advantages: (1) statistical information of a large area (several square millimeters) can be obtained, and the experimental data is more reliable; (2) the test is fast, and the test time is generally within one minute; (3) non-destructive analysis: samples can be tested multiple times, other tests can be performed after the test, and polymer materials can also be studied for a long time; (4) sample preparation is simple; (5) it has a wide range of applications, and the samples include liquids, powders, films, bulk materials, amorphous materials, and other materials; (6) it has absolute advantages over porous materials. SAXS can observe all the pore structures inside the material, while gas adsorption can only observe its open free surface porosity [2].

10.4 Data processing of SAXS

Through experiments, the scattering intensity's two-dimensional (2D) image is first obtained. A one-dimensional (1D) scattering curve can be further obtained by azimuthally integrating the 2D images. Background correction is very important to get the real scattering information from the samples. The background signal originates primarily from air, the beamline optics, including the x-ray windows, and the sample cell. This contribution can be removed by subtracting the scattering intensity of the empty container from that of the sample in the same container. For samples dispersed in solvents, the container with solvents should be measured and used as a background file. After subtracting the background, the sample's scattering information is presented; however, it is often impossible to directly obtain the structural parameters. This is because the scattering intensity is related to the convolution of distance inside particles and the distance between particles. Therefore, it is necessary to establish a model fitting. Usually, $P(q)$ form factor is used to describe the particles' shape information, and the $S(q)$ structure factor is derived from the relative position and interaction of the particles. The scattering intensity is obtained by Fourier transform, $I(q) = nP(q) \cdot S(q)$, where n is the number of particles.

10.5 Examples of SAXS applications in battery

Here we take LIBs as an example to describe the material characteristics of each battery component. The battery is composed of a cathode, anode, separator, and electrolyte. Lithium-ion battery (LIB) cathode materials consist of $LiCoO_2$ and $LiNi_{1/3}Mn_{1/3}Co_{1/3}O_2$ (working voltage ~3.7 V, > 150 mAh g^{-1}), spinel $LiMn_2O_4$ (3.7 V, 120 mAh g^{-1}), and olivine $LiFePO_4$ (3.2 V, 140 mAh g^{-1}). In the future, materials containing S (2.5 V, practical capacity ~1000 mAh g^{-1}) with high specific capacity may be used as cathode materials. The anode material currently being developed to replace the commercialized graphite anode is made of porous carbon materials and nanostructured high specific capacity anode materials (Si, Sn, P, metal oxides, etc). The separator is a highly porous membrane of polyethylene and/or polypropylene. The membrane is used to separate the cathode and anode electrodes and transport lithium ions. The electrolyte is a medium for ions transport and is generally composed of combustible carbonate molecules and lithium salts. Separators and electrolytes are developing in the direction of solid electrolytes to improve battery energy density and safety.

In the field of batteries, SAXS can be used to: (1) investigate the size and structure of materials, such as nanocomposite materials [3], layered cathode materials [4], metal lithium anode interface [5]; (2) observe the state of the liquid electrolyte, including information about the agglomeration between liquid molecules, and gaining a deeper understanding of the nature of the electrolyte; (3) monitor the morphology and state of polymer electrolytes, explain phenomena such as ions transport, and provide theoretical guidance for designing a new generation of electrolytes; (4) study the changes of porous or nano-electrode materials, explaining the failure mechanism of materials, and providing theoretical guidance for the design of new nanomaterials; (5) analyze the microstructure changes of various components of the battery *in situ*.

10.5.1 Porous carbon anode materials

The material and its structure determine the performance of the battery. Different post-treatment methods and different pores affect the material's electrochemical performance for synthesized carbon materials. For example, Dahn *et al* proved that the pyrolyzed coal has the optimal electrochemical performance when there are the most nanopores [6]. Therefore, the characterization of porous carbon materials is an essential part of fundamental theoretical research. SAXS has the advantages of atomic scale, mesoporous, and nanoscale detection and is gradually being valued.

SAXS is more convenient to characterize complex materials. The structural changes of carbon materials mixed or composited with other nanomaterials can be obtained through SAXS. For example, ordered mesoporous carbon with nitrogen doping to 3.6%, d-spacing gradually decreased from 9.7 to 9.3 nm. With the increase in iron oxide loading, the regularity of carbon materials is broken [7]. Complex carbon materials processed under different conditions are different such as the surface structure and porosity, and the low-frequency and high-frequency regions of SAXS can be used for characterization of macropores, nanopores, respectively [8].

The traditional soft carbon and microporous carbon model is often used to explain ions' carbon storage mechanism. Traditional characterization methods are incomplete or simply describe the morphology of carbon materials, often to assist the material in explaining its electrochemical properties. Casas-Cabanas *et al* found that soft carbon and microporous carbon have a micro-curled graphite layer inside, and the micropores are holes formed by winding the graphite layer [9].

10.5.2 Sulfur cathode materials

Sulfur is considered one of the ideal cathode materials for LIBs due to its high specific capacity. However, the polysulfide produced during the cycle is easily soluble in the electrolyte and shuttles between the cathode and anode electrodes and reacts with lithium metal anode to cause a severe capacity degradation. At present, the polysulfide shuttle issue is solved mainly through the combination of sulfur and porous carbon materials through physical confinement and chemical adsorption. Therefore, understanding the preparation mechanism of S/C materials is essential for further improvement and improvement of material properties.

Through SAXS characterization, the changes in the morphology, structure, and pores of S/C materials can be observed intuitively. Mascotte *et al* showed that S initially uniformly filled in the small pores of the carbon material, while macroporous deposition only occurs in high sulfur loading. Besides, the adsorption of polysulfides and the absorption of the electrolyte by carbon materials with different pores are significantly different during the cycling process. Therefore, combining the electrochemical performance and SAXS data can design a cathode material containing S more reasonably [10, 11]. This vapor deposition method is suitable for S and P and other electrode materials that can be vapor-deposited.

10.5.3 Separator

The separator is an essential component of the battery. The separator's primary function is to separate the cathode and anode electrodes and, simultaneously, as a medium for ion transport. Here, we will show the polymer membranes, including lithium-ion battery separators and proton exchange membranes. The phase separation of polymer membranes, including crystalline and amorphous regions, orientation, composite materials with inorganics, void changes, and aging studies, can all apply SAXS.

Under high-temperature conditions, the LIB separator's pore structure will change, resulting in a separator shut-down phenomenon, blocking lithium-ion transport, thereby improving battery safety. The separator shut-down research is significant for the design of the separator structure and improved battery safety. The conventional test is to process the separator at a high temperature and then test its electrochemical performance. The process is relatively cumbersome, and it can only test whether the hole is closed. SAXS can directly observe the changes of separator pores and the dissolution of internal crystals in real-time. The polypropylene separator was studied by ultra-SAXS and it was found that separator shut-down

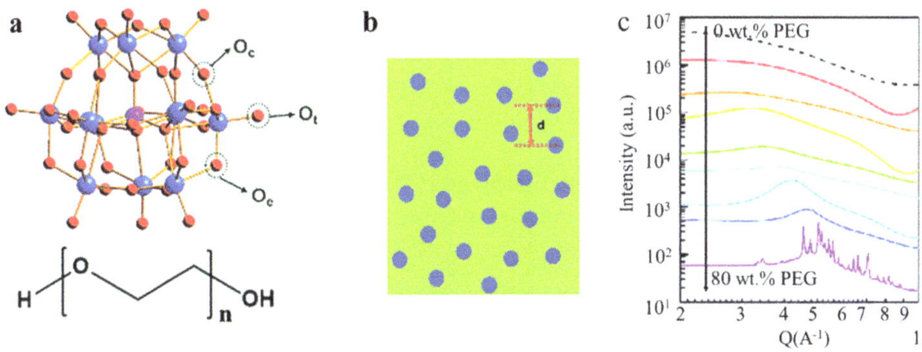

Figure 10.2. (a) Ball-stick model of polyoxometalates and molecular formula for polyethylene glycol (PEG400). (b) Schematic of the inter-cluster distance. (c) SAXS curves of the PEG400-polyoxometalates composites with different PEG400 concentrations. Reproduced from [15] with permission from the Royal Society of Chemistry.

temperature stably starts at 152 °C, close to the melting temperature of the internal crystals of the polypropylene separator [12].

The proton exchange membrane, as a critical component of all-vanadium redox flow batteries (VRFBs), is the most critical material for its commercial applications. The most basic of an ideal proton exchange membrane is to require high proton-conducting ability determined by the polymer state. Nafion membrane is one of the more successful commercial membranes, and its proton-conducting property depends on the membrane hydrophilic and hydrophobic domains, and it is difficult to observe the full view of these by conventional SEM. From the SAXS profile, different peaks corresponding to the hydrophilic structure and crystalline structure can be observed; the changes in the hydrophilic structure and crystalline structure, including the location and size of hydrophilic channels, can also be observed. The changes in the hydrophilic structure and crystalline structure and the influence of proton's conductivity can obtain the optimal parameters for the right polymer membrane. For example, the SAXS profile shows that the morphology changes with the tensile strength during the Nafion membrane's stretching process. It can be observed that the hydrophilic channels gradually become aligned with the increase in tensile strength, and they are along the tensile direction [13]. This is helpful for material development and mechanism explanation [14].

To develop new materials, especially copolymers or composite membranes, it is necessary to determine the hydrophilic structure and crystalline structure in different ratios between polymers or polymers, inferring and supporting why polymer membranes have a high conductivity of proton. In particular, Yin *et al* showed SAXS could obtain much structural information about polyethylene glycol (PEG400)-polyoxometalates composites (figure 10.2(a)). The contribution of intensity comes from the polyoxometalates cluster's contribution and the interaction between polyoxometalates clusters. The intensity of the PEG400-polyoxometalates sample (< 20 wt.%) is consistent with the theoretical calculation value of pure polyoxometalates (figure 10.2(b) and (c)). As the concentration of polyoxometalates is increased (20–70 wt.%), the interaction between polyoxometalates clusters is strengthened. As the concentration further increases, the average distance between the interactions between the corresponding polyoxometalates clusters gradually decreases. As the mass fraction of polyoxometalates in PEG400-polyoxometalate reaches 80%, a sharp scattering peak appears, indicating that some of the polyoxometalates are not entirely dissolved [15]. By testing the proton performance of the membrane, 70% polyoxometalates content shows optimal performance.

10.5.4 Electrolytes

10.5.4.1 Liquid electrolyte

The electrolyte's physical and chemical properties determine its electrochemical performance, such as cycling and rate performances. Although standard characterization techniques such as NMR, FTIR, and Raman can prove the chemical bonding between solvent molecules, they cannot prove the electrolyte system's agglomeration phenomenon. However, SAXS can be used to observe the manifestation of different electrolyte systems. In our group, we demonstrated with SAXS technique that the naked eye transparent mixture of electrolyte solvents with different chemical structures might form sub-nano clusters (figure 10.3(a)). Ethylene carbonate (EC) and propylene carbonate (PC) with the same cyclic structure can form a good solution system (figure 10.3(b)). However, after EC is miscible with linear ethyl methyl carbonate (EMC) and diethyl carbonate (DEC), the intensity of SAXS data in the q range from 0.02 to 0.5 Å^{-1} is enhanced, demonstrating that clusters may form (figure 10.3(c, d)). By data fitting, we found that cyclic EC molecules appear ~1 nm cluster in linear EMC or DEC solutions, while EC molecules with methyl groups (PC) will not form clusters, and this phenomenon is not affected by lithium salts except the cluster size [11]. Recently, our group and Cheng group found by the SAXS and molecular simulation that a high-concentration aqueous electrolyte is a heterogeneous system that consists of an ion region and water region [16]. The ion enriched region has excess anions, and the lithium ions diffuse in the region by hopping along with the anions. Lithium ions are relatively free in the water region.

New systems such as organic high-concentration electrolytes, water-based electrolytes, and local high-concentration electrolytes currently appearing are still in the initial stage of the application. Their basic scientific knowledge and electrolyte design concept are still vague, and it urgently needs advanced techniques to provide

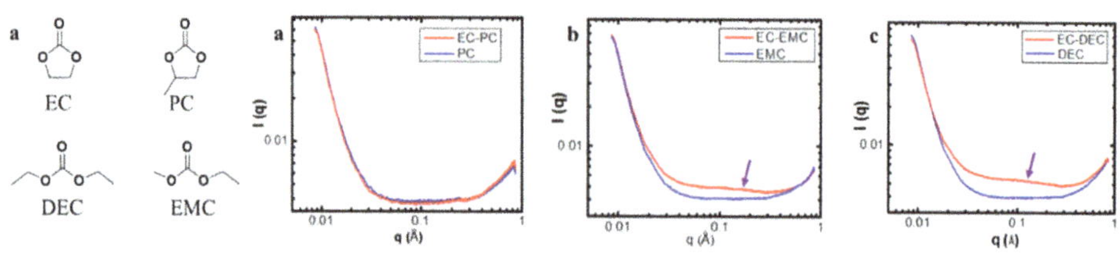

Figure 10.3. (a) Molecular chemical structure of electrolyte solvents. (b–c) SAXS data of different solvents and their mixture. Reproduced from [11]. Copyright IOP Publishing Ltd CC BY-NC-ND.

theoretical guidance. Therefore, we can predict that testing the nanometer manifestation of the electrolyte's inner molecules through SAXS will unveil the electrolyte's mysterious veil.

10.5.4.2 Solid polymer electrolyte

Solid polymer electrolytes (SPEs) are expected to replace current liquid electrolytes, enabling LIBs with excellent safety. The SPEs do not require an electrolyte but need to improve the polymer chain's movement ability to improve lithium-ion conductivity. Therefore, the critical method of polymer electrolytes is to reduce the crystal phase inside the membrane and increase the amorphous phases. For example, it can be observed from the SAXS profile that the crystal phase structure of polyethylene oxide (PEO) completely disappears at about 80 °C, and as the temperature continues to rise, the structures of the polymer do not change. This is consistent with the observed performance of lithium-ion conduction [17]. It is worth noting that it is a universal method to understand the structure of the crystalline phase in the polymer membrane and its composites, such as poly(vinylidene fluoride) (PVDF) [18] and PEO-based block polymers [19].

The understanding of the lithium-ion conductivity of the copolymer electrolyte is also beneficial. The SAXS study of PEO-b-polystyrene sulfonyl lithium (trifluoromethylsulfonyl) imide diblock copolymers found that in low-temperature conditions, an ordered lamellar phase is formed inside the membrane, and lithium ions are encapsulated in the PS phase, which limits lithium ions transport. As the temperature increases, when it is higher than the glass transition temperature, lithium ions are released and transport through the PEO chain, which significantly improves the ion conductivity. This kind of lithium-ion conducting mechanism is challenging to study by other techniques [20]. This discovery can provide new ideas for the future development of single-ion conductor electrolytes.

For inorganic composite polymer electrolytes, SAXS can confirm the size of inorganic particles, the dispersion state, and the distance between particles. For example, the combination of SAXS and TEM proves that the dispersion distance of nanoparticles in the SPEs is ~27 nm, and the dispersion is very uniform [21]. According to SAXS, it can be inferred that nanoparticles appear in the polymer as a three-dimensional structure, which is beneficial to destroy the crystallinity of the polymer, improve the ion conductivity and uniform lithium ion conductivity of the membrane, which is very important for safe Li metal-based batteries.

10.5.4.3 Redoxmer

Redox flow batteries (RFBs) are electric energy storage devices where chemical energy is provided by redoxmers dissolved in liquids that are pumped through the system on separate sides of a membrane, as shown in figure 10.4(a).[22] Their scalability, sustainability, and cost-efficiency are attracting increasing attention in large-scale energy storage. Unlike the traditional lithium-ion batteries that the electrolyte is inert to redox reactions, the electrochemical redox reactions are completed in the flow batteries' electrolytes. Therefore, the solubility, redox potential, chemical stability, and cost of redoxmers directly impact the energy

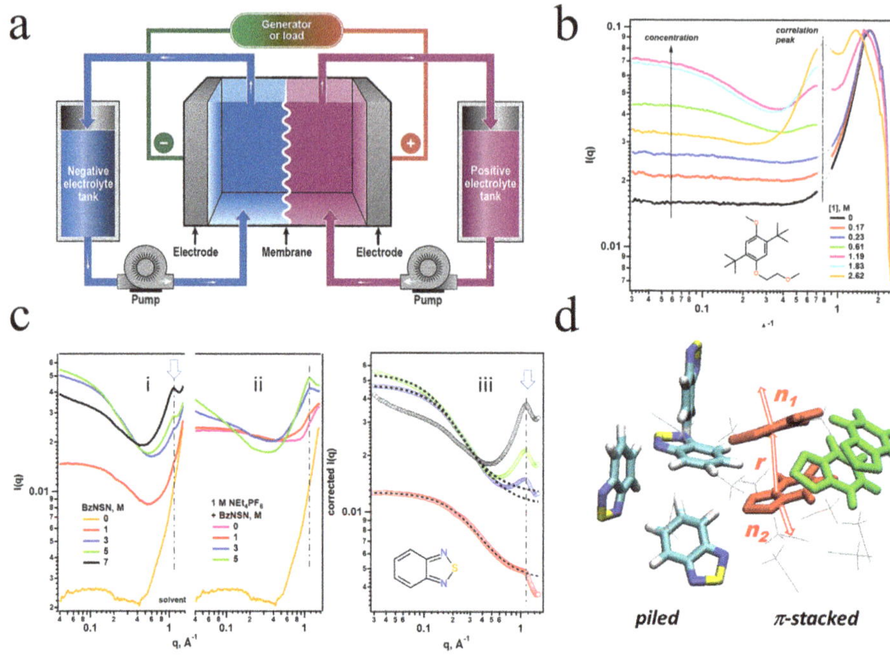

Figure 10.4. (a) Schematic view of a redox flow battery system. Reprinted with permission from [22]. Copyright (2017) American Chemical Society. (b) SAXS spectra for room temperature solutions of **1** in acetonitrile (0–2.6 M). The vertical arrow indicates the direction of the concentration increase. The dashed vertical line indicates the position of the solute correlation peak. Reprinted with permission from [23]. Copyright (2020) American Chemical Society. (c) (i, ii) The experimental background-corrected SAXS spectra for solutions indicated in the plots. (iii) The spectra from panel (i) after subtraction of the solvent contributions. (d) A cluster of BzNSN molecules in an MD snapshot for 3 M acetonitrile solution at 300 K. The solvent molecules are shown schematically with gray lines. Reprinted with permission from [24]. Copyright (2020) American Chemical Society.

density, cell voltage, cycle life, and cost of RFBs, respectively. SAXS method determines the intermolecular spacing, cluster size, or inhomogeneity domains in 1–100 nm, which provides a unique perspective on these amorphous type materials.

Our group and Shkrob *et al* [23] used SAXS to demonstrate the formation of self-assembled solute networks in crowded electrolyte solutions. Figure 10.4(b) displays the background-corrected $I(q)$ plots for 0–2.62 M solutions of **1** (molecular structure is in the illustration) in acetonitrile. The strong peak at 1.5 $Å^{-1}$ originates from the solvent structure. Apart from that, a solute correlation peak emerges at 0.6–0.7 $Å^{-1}$, which corresponds to a scattering length of 9.5 Å. Such peaks arise from spatial correlations between the positions of solute molecules or ions. Interestingly, for redoxmer using low molecular weight version catholyte molecules such that the arene core connected with two methyl groups in para or ortho position, there is no discernible correlation peak in the 0.2–1 $Å^{-1}$ region. This is because the arene ring in molecule **1** is substituted with two bulky tert-butyl groups, which add steric hindrance and preclude radical addition reactions. Such bulky groups strongly repulse the polar solvent, and the molecular aggregation can be strongly affected by their presence and position in the molecular structure. Another work reported by Tao Li and his co-workers [24] found that the packing motifs of a well-studied redoxmer, 2,1,3-benzothiadiazole (BzNSN), is competitive Pi-Stacking and H-bond Piling. This redoxmer is a small, symmetric, planar molecule with anomalously high solubility in electrolyte solutions. The molecular dynamics (MD) simulation found two packing motifs, face-to-face π-stacking, and random N–H bond piling, as shown in figure 10.4(d). In SAXS spectra of BzNSN dissolved in acetonitrile (panels (i) and (iii) in figure 10.4(c)), a correlation peak at 1.1 $Å^{-1}$ is observed at concentrations > 3 M. The correlation length $2\pi/q = 5.7$ Å corresponding to this peak position falls right in between the most probable c–c distances of 4.0 Å for π-stacking and 6.4 Å for N–H piling, i.e., the observation of this peak is the 'smoking gun' for the postulated competition between the two molecule packing motifs. Panel (ii) illustrates another qualitative prediction of the MD model: the correlation peak position does not depend on electrolyte presence even though the introduced salt strongly increases x-ray scattering (due to electron-rich atoms in the PF_6^- anion). This also suggests that the ordering responsible for the correlation peak at 1.1 $Å^{-1}$ stems exclusively through short-range interactions of BzNSN molecules.

10.5.5 *In situ* or operando SAXS

The test is divided into the *in situ* test and the *ex situ* test. The *ex situ* test is performed by disassembling the battery to characterize the material after the battery undergoes an electrochemical reaction. Therefore, the '*ex situ*' characterization is challenging to capture the transient state. The electrochemical process consists of a series of reactions, hardly exiting the equilibrium state, thus we cannot fully understand the battery changes. Besides, the test results may not respond to the real state. *Ex situ* tests require the battery to be disassembled, and the disassembled materials may react with the trace H_2O and O_2 in the glove box, causing irreversible damage. Each batch of batteries cannot be guaranteed to be the same. If continuous testing is required, numerous batteries should be disassembled, which is time-consuming. The '*in situ*' means that the whole characterization is real-time detecting of the same materials in the real operating environment (figure 10.5). This can be analyzed to maximize obtaining accurate data. The advantages are as follows: (1) detecting the same position of the battery and obtaining results with high comparability; (2) taking less time and obtaining plentiful information; (3) simple operation.

SAXS can well observe the changes in the pore structure. The intensive synchronous light can assist in *operando* characterization of the mesopores and microstructures of carbon materials. Csencsits *et al* collected data every 5 min through SAXS and found that the porous carbon material structure did not change during the charging and discharging process, which indicates that the electrolyte will not enter the porous material [26]. Porous carbon materials often show higher specific capacity in the process of lithium and sodium storage. The reason has not been clearly explained. Dahn *et al* combined charge–discharge curves and *in situ*

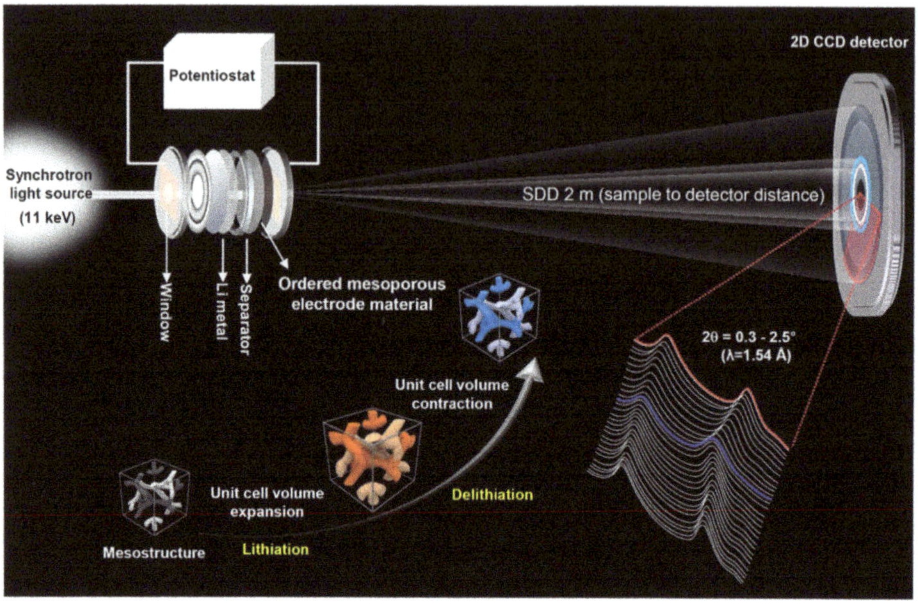

Figure 10.5. Schematic illustration for in operando SAXS with the *in situ* coin cell to monitor pore dynamics of ordered mesoporous metal oxide electrodes during cycling. Reprinted with permission from [25]. Copyright (2015) American Chemical Society.

SAXS data to explore the micropore changes of carbon materials. The micropores of the porous carbon material have not changed when sodium ions are inserted.

When the discharge voltage is close to 0V, sodium ions begin to deposit in the micropores to form metallic sodium, which blocks the micropores. As the metal sodium is released during charging, pores begin to appear gradually, proving that the carbon material's sodium metal deposition is reversible [27]. Also, Dahn *et al* proved that the pores of hard carbon materials remain unchanged before and after (de)lithiation [28].

Currently, battery anode materials are mainly carbon materials. New conversion metal oxides and alloyed silicon anode materials are gradually being developed to increase anode materials' specific capacity. However, these materials' huge lithiation volume change has become a stumbling block for their commercial applications. SAXS is sensitive to pore structure and material size changes, thereby providing some theoretical guidance for applying these materials in batteries. Using *in operando* SAXS, Kim *et al* demonstrated the pore structures of meso-TiO_2 during intercalation reactions do not change during the charge and discharge process, indicating that the volume was almost unchanged (figure 10.6(a)). The discharge process of meso-Co_3O_4 is accompanied by transforming the Co_3O_4 structure to metallic Co and Li_2O (figure 10.6(b)). As the amount of lithium intercalation increases, the pore structure gradually decreases. During the charging process, the metal Co is oxidized to CoO, which increases the pore structure of the material slightly. The pore peaks of SAXS gradually disappear during the charging process of

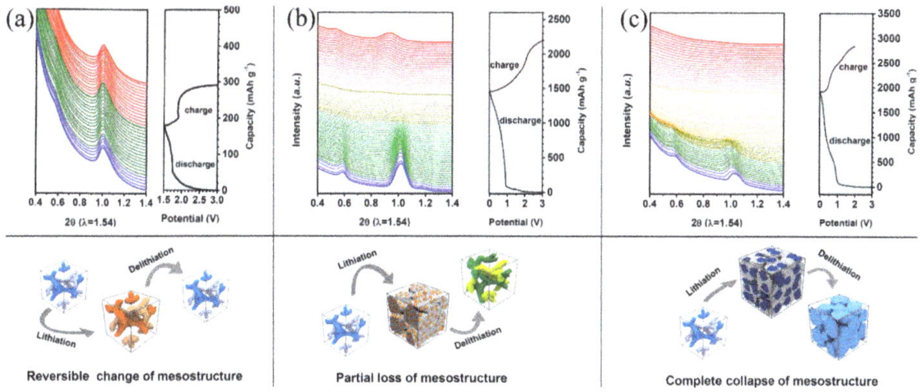

Figure 10.6. (a), (b) and (c) *In operando* SAXS spectra, the corresponding voltage profiles and volume changes models of meso-TiO_2, meso-Co_3O_4, and meso-SnO_2, respectively. Reprinted with permission from [25]. Copyright (2015) American Chemical Society.

meso-SnO_2, and these peaks cannot be reproduced after discharge (figure 10.6(c)). Therefore, the material structure of meso-SnO_2 has completely changed after the first charge and discharge [25]. Using a similar strategy, Amine and coworkers demonstrated that the volume change of sodium ions intercalated into porous cobalt oxide is much smaller than that of lithium-ion intercalation, thus speculating that sodium ions may be intercalated into cobalt oxide in a smaller amount or that it is an intercalation reaction instead of the conversion reaction. It is further proved that the volume change is not the main reason for the capacity decay of the metal oxide as the anode electrode in the sodium ion battery [29]. Sartori *et al* found amorphous red phosphorus in a carbon matrix remained intact during the lithiation process, but the porous structure is produced during the delithiation process [30]. These findings can provide useful theoretical guidance for the design of large-volume materials.

SAXS can observe the manifestation of silicon in silicon/carbon materials. In 2019, Lyonnard *et al* found that the volume expansion of silicon materials does not have a linear relationship with the amount of lithium inserted in the anode, and the porosity of carbon materials can stabilize silicon materials [31]. They further demonstrated that pre-lithiation silicon/carbon materials could significantly improve the materials' electrochemical stability [32]. It is shown that 66% prelithiation of the anode electrode material prolongs the cell to 1760 cycles with a capacity retention of 89%. With the help of SAXS, they found that the average volume change of the material after prelithiation becomes small, and the lithium ions will be evenly distributed in the anode electrode during the cycling process.

During battery usage, side reactions always occur, leading to decreased battery capacity, gas production and heat generation, and the increase of battery internal resistance, which has a significant impact on the battery's service life and safety. Hence, an in-depth understanding of battery internal material changes is essential for improving battery performance. For example, the polymer electrolyte's lithium-ion conductivity depends on the polymer's morphology and crystallinity, and different preparation methods have different effects on the long-term battery cycle.

Through *in situ* SAXS research, it is found that the mismatch between the polymer electrolyte and the electrode material interface is the main reason for the long-term battery capacity degradation [33].

10.6 Conclusion

This chapter discussed SAXS principles, test applications and advantages, data processing, and application examples. The characterization application of SAXS in battery material characterizations, electrolytes, separators, and polymer electrolytes and *in situ* and *operando* SAXS research is emphasized. Although SAXS has relatively few applications in the battery field, its unique advantages make it more critical in batteries based on nanomaterials. The statistical characterization of electrolytes, separators, polymer electrolytes, and porous materials also plays a vital role in the advanced materials with volume expansion or structural evolution. Therefore, more information will be mined by SAXS to guide higher performance battery research in the future. Combining the reported literature and our own experience, we provide the following directions for the research of SAXS in the field of LIBs for readers' reference: (1) investigating the electrolyte system, including the agglomeration of solvent molecules, the interaction between high-concentration electrolyte molecules, and the electrolyte system that may contain nanoclusters; (2) understanding of the microstructure changes of multi-level structure materials during the battery process; (3) observing the failure mechanism of battery materials, for example, to observe the changes in the interface between the cathode and anode electrodes; (4) *operando* studying the shuttle effects of sulfur batteries or organic batteries.

References

[1] Li T, Senesi A J and Lee B 2016 Small angle x-ray scattering for nanoparticle research *Chem. Rev.* **116** 11128–80

[2] Dou X, Hasa I, Saurel D, Vaalma C, Wu L, Buchholz D, Bresser D, Komaba S and Passerini S 2019 Hard carbons for sodium-ion batteries: Structure, analysis, sustainability, and electrochemistry *Mater. Today* **23** 87–104

[3] Zhu B, Guo G, Wu G, Zhang Y, Dong A, Hu J and Yang D 2019 Preparation of dual layers N-doped Carbon@Mesoporous Carbon@Fe_3O_4 nanoparticle superlattice and its application in lithium ion battery *J. Alloys Compd.* **775** 776–83

[4] Wu N, Wu H, Liu H and Zhang Y 2016 Solvothermal coating $LiNi_{0.8}Co_{0.15}Al_{0.05}O_2$ microspheres with nanoscale Li_2TiO_3 shell for long lifespan Li-ion battery cathode materials *J. Alloys Compd.* **665** 48–56

[5] Stone G M, Mullin S A, Teran A A, Hallinan D T Jr, Minor A M, Hexemer A and Balsara N P 2012 Resolution of the modulus versus adhesion dilemma in solid polymer electrolytes for rechargeable lithium metal batteries *J. Electrochem. Soc.* **159** A222–7

[6] Zheng T, Xing W and Dahn J R 1996 Carbons prepared from coals for anodes of lithium-ion cells *Carbon* **34** 1501–7

[7] Qiang Z, Chen Y-M, Gurkan B, Guo Y, Cakmak M, Cavicchi K A, Zhu Y and Vogt B D 2017 Cooperatively assembled, nitrogen-doped, ordered mesoporous carbon/iron oxide nanocomposites for low-cost, long cycle life sodium-ion batteries *Carbon* **116** 286–93

[8] Xie F *et al* 2019 Unveiling the role of hydrothermal carbon dots as anodes in sodium-ion batteries with ultrahigh initial coulombic efficiency *J. Mater. Chem.* A **7** 27567–75
[9] Saurel D, Segalini J, Jauregui M, Pendashteh A, Daffos B, Simon P and Casas-Cabanas M 2019 A SAXS outlook on disordered carbonaceous materials for electrochemical energy storage *Energy Storage Mater.* **21** 162–73
[10] Petzold A, Juhl A, Scholz J, Ufer B, Goerigk G, Froba M, Ballauff M and Mascotto S 2016 Distribution of sulfur in carbon/sulfur nanocomposites analyzed by small-angle x-ray scattering *Langmuir* **32** 2780–6
[11] Feng Z, Sarnello E, Li T and Cheng L 2019 Communication—microscopic view of the ethylene carbonate based lithium-ion battery electrolyte by x-ray scattering *J. Electrochem. Soc.* **166** A47–9
[12] Xiong B, Chen R, Zeng F, Kang J and Men Y 2018 Thermal shrinkage and microscopic shutdown mechanism of polypropylene separator for lithium-ion battery: In-situ ultra-small angle x-ray scattering study *J. Membr. Sci.* **545** 213–20
[13] So S, Cha M S, Jo S W, Kim T H, Lee J Y and Hong Y T 2018 Hydrophilic channel alignment of perfluoronated sulfonic-acid ionomers for vanadium redox flow batteries *ACS Appl. Mater. Interfaces* **10** 19689–96
[14] Kim H G, Kim R, Kim S, Choi C, Kim B, Guim H and Kim H-T 2018 Propylene carbonate-derived size modulation of water cluster in pore-filled Nafion/polypropylene composite membrane for the use in vanadium redox flow batteries *J. Ind. Eng. Chem.* **60** 401–6
[15] Zheng Z, Zhou Q, Li M and Yin P 2019 Poly(ethylene glycol) nanocomposites of sub-nanometer metal oxide clusters for dynamic semi-solid proton conductive electrolytes *Chem. Sci.* **10** 7333–9
[16] Yu Z, Curtiss L A, Winans R E, Zhang Y, Li T and Cheng L 2020 Asymmetric composition of ionic aggregates and the origin of high correlated transference number in water-in-salt electrolytes *J. Phys. Chem. Lett.* **11** 1276–81
[17] Sandi G, Humberto J, Kizilel R, Seifert S and Carrado K A 2003 *In situ* SAXS studies of the structural changes of polymer nanocomposites used in battery applications *Chem. Mater.* **15** 838–43
[18] Abbrent S, Plestil J, Hlavata D, Lindgren J, Tegenfeldt J and Wendsjö Å 2001 Crystallinity and morphology of PVdF–HFP-based gel electrolytes *Polymer* **42** 1407–16
[19] Porcarelli L, Aboudzadeh M A, Rubatat L, Nair J R, Shaplov A S, Gerbaldi C and Mecerreyes D 2017 Single-ion triblock copolymer electrolytes based on poly(ethylene oxide) and methacrylic sulfonamide blocks for lithium metal batteries *J. Power Sources* **364** 191–9
[20] Inceoglu S, Rojas A A, Devaux D, Chen 1 0 C, Stone G M and Balsara N P 2014 Morphology–conductivity relationship of single-ion-conducting block copolymer electrolytes for lithium batteries *ACS Macro Letters* **3** 510–14
[21] Choudhury S 2019 *Rational Design of Nanostructured Polymer Electrolytes and Solid–Liquid Interphases for Lithium Batteries* (Cham: Springer)
[22] Wei X *et al* 2017 Materials and systems for organic redox flow batteries: status and challenges *ACS Energy Lett.* **2** 2187–204
[23] Shkrob I A *et al* 2020 Self-assembled solute networks in crowded electrolyte solutions and nanoconfinement of charged redoxmer molecules *J. Phys. Chem.* B **124** 10226–36
[24] Zhao Y *et al* 2020 Competitive Pi-stacking and H-bond piling increase solubility of heterocyclic redoxmers *J. Phys. Chem.* B **124** 10409–18

[25] Park G O *et al* 2015 In operando monitoring of the pore dynamics in ordered mesoporous electrode materials by small angle x-ray scattering *ACS Nano* **9** 5470–7
[26] Sandi G, Carrado K A, Winans R E, Johnson C S and Csencsits R 1999 Carbons for lithium battery applications prepared using sepiolite as inorganic template *J. Electrochem. Soc.* **146** 3644–8
[27] Stevens D A and Dahn J R 2000 An *in situ* small-angle x-ray scattering study of sodium insertion into a nanoporous carbon anode material within an operating electrochemical cell *J. Electrochem. Soc.* **147** 4428–31
[28] Buiel E, George A E and Dahn J R 1998 On the reduction of lithium insertion capacity in hard-carbon anode materials with increasing heat-treatment temperature *J. Electrochem. Soc.* **145** 2252–7
[29] Xu G L *et al* 2017 Insights into the distinct lithiation/sodiation of porous cobalt oxide by in operando synchrotron x-ray techniques and *ab initio* molecular dynamics simulations *Nano Lett.* **17** 953–62
[30] Berhaut C L, Dominguez D Z, Kumar P, Jouneau P H, Porcher W, Aradilla D, Tardif S, Pouget S and Lyonnard S 2019 Multiscale multiphase lithiation and delithiation mechanisms in a composite electrode unraveled by simultaneous operando small-angle and wide-angle x-ray scattering *ACS Nano* **13** 11538–51
[31] Povia M, Sottmann J, Portale G, Knudsen K D, Margadonna S and Sartori S 2018 Operando SAXS/WAXS on the a-P/C as the anode for Na-ion batteries *J. Phys. Chem.* C **122** 5917–23
[32] Berhaut C L *et al* 2020 Prelithiation of silicon/graphite composite anodes: benefits and mechanisms for long-lasting Li-Ion batteries *Energy Storage Mater.* **29** 190–7
[33] Möhl G E, Metwalli E and Müller-Buschbaum P 2018 In operando small-angle x-ray scattering investigation of nanostructured polymer electrolyte for lithium-ion batteries *ACS Energy Lett.* **3** 1525–30

www.ingramcontent.com/pod-product-compliance
Ingram Content Group UK Ltd.
Pitfield, Milton Keynes, MK11 3LW, UK
UKHW051903100926
8556IPUK00005B/750

9 780750 326803